Broadening the Genetic Base of Grain Legumes

Mohar Singh • Ishwari Singh Bisht
Manoranjan Dutta
Editors

Broadening the Genetic Base of Grain Legumes

Springer

Editors
Mohar Singh
Ishwari Singh Bisht
Manoranjan Dutta
Germplasm Evaluation Division
National Bureau of Plant Genetic Resources
New Delhi, India

ISBN 978-81-322-2022-0 ISBN 978-81-322-2023-7 (eBook)
DOI 10.1007/978-81-322-2023-7
Springer New Delhi Heidelberg New York Dordrecht London

Library of Congress Control Number: 2014950143

Printed on acid-free paper

Springer is part of Springer Science+Business Media (www.springer.com)

Preface

Most of the grain legume crop species worldwide have an intrinsically narrow genetic base that has been exploited to the maximum level of productivity. Further breakthroughs in increasing yield and improving stability in future crop cultivars will require the identification and incorporation of new sources of variation into the cultivated background. Knowledge, access, and use of diversity available in cultivated and wild relatives are essential for widening the genetic base of commercial legume crop species. In view of this need, an effort has been made to bring together the rather scattered research work done in this important area in the form of an edited collection, a compilation that should be of benefit to legume researchers across the world. The book comprises 10 chapters contributed by researchers from reputed organizations of the world. The introductory chapter presents key issues linked to domestication bottlenecks and current trends on wide hybridization. The subsequent chapters (Chaps. 2, 3, 4, 5, 6, 7, 8, 9, 10) deal with various aspects related to broadening the genetic base of grain legumes. Each chapter provides detailed information on the following aspects: crop genepool, evolutionary relationships and systematic' assessment of gene flow for crop improvement, gene flow constraints, level of diversity in crop germplasm, production related problems, traits of importance for base broadening, wide hybridization, barriers to wide hybridization, methods of wide hybridization, and genome mapping and genomics status of each grain legume. The editors are extremely thankful to all the authors for their significant contributions to this book. The entire process of preparing the manuscript was marked by cordial collegiality. We are also grateful to Prof. Kailash Chander Bansal, The Director, National Bureau of Plant Genetic Resources, New Delhi, India, for providing guidance and a supportive atmosphere for the development of this edited collection. Thanks are also due to Ms. Megha Bakshi working as project assistant with the first editor for her assistance during the preparation of this multi-authored edited book in terms of compilation, edited chapter processing, and typographical scientific work. We are also indebted to Springer India for excellent professional support in the completion of this project. Despite several rounds of proofreading and our best efforts, the book may still have some scientific, technical, and printing errors. We will appreciate it if these omissions are brought to our notice, so that they may be rectified in future editions. Finally, we hope the book will be used and enjoyed by legume researchers and other readers across the world.

New Delhi, India Editors

Contents

About the Editors

Mohar Singh is presently working as a Senior Scientist (Plant Breeding) at the National Bureau of Plant Genetic Resources Pusa, New Delhi, India. Dr. Singh obtained his Doctoral degree in Plant Breeding from the Himachal Pradesh Agricultural University, Palampur, India, in 2001. He has been working on genetic and genomic resources of pulses for the last 12 years. Presently, he is working on pre-breeding and genetic enhancement for breaking yield barriers in lentil and chickpea. He has identified several useful gene sources for various traits of interest in wild *Cicer* and *Lens* species, and some of them have been introgressed into the cultivated background for widening the gene pool. Dr. Singh has published more than 50 research papers in the journals of national and international repute like *Plant Breeding*, *Crop Science*, *PLoS ONE*, *DNA Research*, *Journal of Agricultural Science*, *Journal of Genetics and Breeding*, *Journal of Genetics, Genetic Resources and Crop Evolution*, *Plant Genetic Resources: Characterization and Utilization*, and *Advances in Horticultural Science* and also authored two text books and edited one book published by Elsevier Insights.

Ishwari Singh Bisht is working as a Principal Scientist at the National Bureau of Plant Genetic Resources (ICAR), Pusa Campus, New Delhi. As Professor of Plant Genetic Resources (PGR) discipline of Postgraduate School, Indian Agricultural Research Institute (a deemed university), he also coordinates the postgraduate teaching and research of the discipline.

Dr. Bisht has been actively associated with various agro-biodiversity management issues being addressed at the national level including crop genetic resources exploration, collection, characterization, conservation, germplasm enhancement, etc. He has been working in the field of plant genetic resources management for over 30 years, and his academic achievements in this area have been widely recognized nationally and internationally.

Dr. Bisht's research mainly focuses on a thorough understanding of the factors that influence the level of crop genetic diversity maintained on-farm and addressing the complex range of factors shaping the conservation or erosion of genetic diversity in farmers' fields. These range from farmers' decision-making to local environmental change to interactions between and within crop populations. His research also explores modalities on mainstreaming biodiversity in production landscapes and linking biodiversity conservation to livelihood security of farmer households in smallholder subsistence farming. Dr. Bisht has also significantly contributed in

PGR education addressing country's requirement for trained manpower to carry out various biodiversity management tasks.

Dr. Bisht has published 84 research papers in peer-reviewed reputed journals. He is also the co-editor of a book titled *Genetic and Genomic Resources of Grain Legume Improvement* published by Elsevier in 2013.

Manoranjan Dutta is a well-known researcher and academician involved in teaching and research in the areas of crop improvement and genetic resource management for more than 30 years. He had a 2 years' postdoctoral research stint at the International Crops Research Institute for the Semi-Arid Tropics, India, before joining the Govind Ballabh Pant University of Agriculture and Technology where he held several important positions including Professor and Head, Department of Crop Improvement; Director, Communication and Publication among others. He was the Head, Division of Germplasm Evaluation at the National Bureau of Plant Genetic Resources, ICAR from where he retired recently. Currently, he is a consultant in the Ministry of Agriculture, Govt. of India.

During his long research career, he has developed 15 crop varieties, registered several genetic stocks, and published more than 60 research papers in reputed international and national journals, 4 books, and 17 book chapters.

1 Introduction

Mohar Singh, Ishwari Singh Bisht, and Manoranjan Dutta

Over the last about 60 years, the global scientific community has become increasingly aware of the implications of genetic erosion in terms of its impact on environmental and agricultural sustainability (Ford-Lloyd et al. 2008). Reductions in both the number of species and the level of intraspecific variation have resulted in crops becoming more vulnerable to unpredictable weather patterns, epidemics of pests and diseases and fluctuations in global markets. All these in combination directly affect the food availability for human consumption. The ability to respond constructively to these situations requires continuing access to a broad range of novel forms of genetic variation.

Plant breeding is one way to confront the challenge of bridging the ever-widening gap between the demand and supply of food. Despite the importance, however, plant breeding has its own negative side effects (Keneni et al. 2012). The replacement of landraces with a few genetically uniform varieties depletes genetic diversity and provides ideal conditions for diseases and insect pests, resulting in genetic vulnerability. The increasingly growing human population and the consequent rise in demand for more food, on the one hand, and the success of such efforts like the "green revolution" through adoption of genetically uniform varieties in many parts of the world, on the other, are the main contributing factors towards the narrow genetic base of present-day crop cultivars. It is, therefore, important to understand the phenomena and adopt appropriate strategies to minimize the risks from genetic vulnerability. Under marginal conditions, where resource-poor farmers predominate, the current plant breeding strategies and variety release, registration and certification procedures leading to genetic uniformity must be reconsidered, and some level of genetic diversity should be deliberately maintained in variety development programmes. Genetic diversity can be introduced at different levels and in different ways which may include intra-varietal, inter-varietal, inter-parental and interspecific diversities. Breeding for specific adaptation instead of wide adaptation, systematic spatial and temporal gene deployment, use of interspecific varietal mixtures and integration of horizontal and vertical resistances have been suggested as possible solutions (Keneni et al. 2012).

Crop improvement through plant breeding, like crop evolution in general, occurs through selection operating on genetic variability. Directed and intense selection by plant breeders or by farmers can be intense and has resulted in major crop improvements. However, continued success in plant breeding can only be realized in so far as new variability is available for selection to operate. Genetic diversity is therefore absolutely essential for sustained crop improvement. On the other hand, there is a perception and concern that genetic diversity is limited, both within

M. Singh (✉) • I.S. Bisht • M. Dutta
Germplasm Evaluation Division, National Bureau of Plant Genetic Resources,
Pusa, New Delhi 110012, India
e-mail: moharsingh@nbpgr.ernet.in; bishtis@nbpgr.ernet.in; mdutta@nbpgr.ernet.in

M. Singh et al. (eds.), *Broadening the Genetic Base of Grain Legumes*,
DOI 10.1007/978-81-322-2023-7_1, © Springer India 2014

production systems and in breeding programmes, and that there is a need for continuing and concerted efforts to broaden the genetic base of crops. In a wide sense, broadening the genetic base of crops can be considered at three levels: (a) increasing the extent of useful diversity available to breeders, i.e., broadening the genetic base of breeders' material through genetic enhancement or pre-breeding; (b) increasing the range of useful diversity available to farmers as planting material; and (c) increasing the diversity of crops and varieties grown in the field. All three of these areas have been critically reviewed and the main concepts related to broadening the genetic base of crops, including the greater use of participatory plant breeding and *on-farm* conservation of genetic resources, have been discussed and deliberated in greater detail by Cooper et al. (2001).

Genetic base as revealed by the pedigree records of released varieties appears to be narrow in major pulse crops, because of the frequent use of the same parents and their derivatives in breeding programmes. Pedigree analysis of 86 chickpea cultivars, for example, that have been released through hybridization in India traced back to only 95 ancestors (Kumar et al. 2004). This appears to be an insignificant part of chickpea germplasm accessions that are conserved in various gene banks across the world. Knowledge, access and use of diversity available in the cultivated and wild relatives are essential for broadening the genetic base of cultivars to sustain crop improvement. An overview of the existing level of diversity and the genetic base of major grain legumes is presented in this chapter based on the contributions made by respective crop experts collated in this book.

1.1 Common Bean

The origin, domestication and organization of genetic diversity in *Phaseolus*; major production constraints and traits deficient in the common bean cultivars; sources of useful germplasm, and progress achieved in broadening the genetic base of cultivars need to be better understood. *Phaseolus* that originated in the Americas has been critically reviewed by Singh (2001). Only five species, *P. acutifolius* A. Gray, *P. coccineus* L., *P. lunatus* L., *P. polyanthus* G. and *P. vulgaris* L., have been domesticated. The interspecies diversity in common bean is organized in primary, secondary, tertiary and quaternary gene pools. Unique among crop plants, common bean consists of two geographically distinct evolutionary lineages that predate domestication and trace back to a common, still extant ancestor. Common bean, cultivated across the world, is comprised of two major gene pools, the Andean originating from the Andes Mountains of South America and the Mesoamerican from Mexico and Central America along with well-established races. Gene flow between domesticated and wild beans led to substantial introgression of alleles from the domesticated gene pool into the wild gene pool and vice versa. Cultivars are further divided into races, each with their distinguishing characteristics. Also, cultivars of dry seed and snap bean exist. Abiotic and biotic stresses limit common bean production in most parts of the world. The genetic base of cultivars within market classes is narrow, and the level of resistance to common bacterial blight [caused by *Xanthomonas campestris* pv. *phaseoli* (Smith) Dye] and white mould [caused by *Sclerotinia sclerotiorum* (Lib) de Bary] is inadequate although high levels of resistance to these and other desirable traits exist in the relatives and gene pools of *P. vulgaris*. Early maturity, adaptation to higher latitude, upright plant type, high pod quality and seed yield and resistance to *bean common mosaic virus* (a potyvirus) and/or rust [caused by *Uromyces appendiculatus* (Pers.) Unger] have been bred into modern cultivars. However, most of the genetic variability available in the common bean races, gene pools and wild relatives remains to be utilized yet. Resistance to various diseases has been transferred from *P. coccineus*, *P. acutifolius* and *P. costaricensis* to *Phaseolus vulgaris*. In addition, *P. acutifolius* and *P. parvifolius* have been explored to transfer high-iron-content trait to *P. vulgaris*. In order to maximize and sustain bean production, high-yielding, high-quality cultivars that are less dependent on water, fertilizer, pesticides and manual labour should be developed. This warrants sustained, comprehensive and integrated genetic improvement in which favourable alleles from cultivated and wild relatives are accumulated in superior cultivars. A three-tiered breeding approach involving (1) gene introgression from alien germplasm, (2) pyramiding favourable alleles from different sources and (3) simultaneous improvement of multiple traits for common bean cultivars would be the most appropriate strategy to meet these needs in comming future.

1.2 Chickpea

Efforts to improve the yield and quality of cultivated chickpea (*Cicer arietinum* L.) are constrained by a low level of intraspecific genetic diversity. Increased genetic diversity can, however, be achieved through hybridization of the cultivated species with the "wild" relatives from within the 43 species of the *Cicer* genus. To date, the 8 species sharing an annual growth habit and common chromosome number with *C. arietinum* have been the primary focus of screening and introgression efforts (Siddique et al. 2000). Screening of these species has uncovered novel morphological characteristics and resistance to a number of abiotic and biotic stresses that are of potential value to chickpea improvement programmes. Detailed analysis of protein and DNA, karyotyping and crossability studies have begun to elucidate the relationships between the annual *Cicer* species. In comparison, perennial species have received little attention due to difficulties in collection, propagation and evaluation. A recent review by Croser et al. (2003) discusses the progress towards an understanding of genetic relationships between the *Cicer* species and the introgression of genes from wild *Cicer* species into the cultivated background. Considerable diversity among members of the primary gene pool of the chickpea, including *C. arietinum*, *C. reticulatum* and *C. echinospermum*, has been recorded. Better understanding of the structure, diversity and relationships within and among the members of this gene pool will contribute to more efficient identification, conservation and utilization of chickpea germplasm for allele mining; association genetics; mapping and cloning gene(s); and applied breeding to widen the genetic base of cultivated species, for the development of elite lines with superior yield and improved adaptation to diverse environments. Limiting factors to crop production, possible solutions and ways to overcome them, importance of wild relatives and barriers to alien gene introgression and strategies to overcome them and traits for base broadening have been discussed. It has been clearly demonstrated that resistance to major biotic and abiotic stresses can be successfully introgressed from the primary gene pool comprising progenitor species (Ahmad and Slinkard 1992). However, many desirable traits including high degree of resistance to multiple stresses are present in the species belonging to secondary and tertiary gene pools, that can also be introgressed by using special techniques to overcome pre- and post-fertilization barriers. Besides resistance to various biotic and abiotic stresses, the yield QTLs have also been introgressed from wild *Cicer* species to the cultivated varieties. Status and importance of molecular markers, genome mapping and genomic tools for chickpea improvement are duly elaborated. By using major genes for various biotic and abiotic stresses, the transfer of agronomically important traits into elite cultivars has been made easy and practical through marker-assisted selection and marker-assisted backcross breeding. Usefulness of molecular markers such as SSR and SNP for the construction of high-density genetic maps of chickpea and for identification of genes/QTLs for stress resistance, quality and yield-contributing traits has also been discussed.

1.3 Faba Bean

Broadening of the genetic base and systematic exploitation of heterosis in faba bean (*Vicia faba* L.) requires reliable information on the genetic diversity in germplasm. Three divergent groups of faba bean inbred lines were examined by molecular means, viz., European small-seeded lines, European large-seeded lines and Mediterranean lines (Link et al. 1995). The results were in harmony with published archaeobotanical findings.

Faba bean germplasm of over 38,000 accessions of landraces and varieties is conserved worldwide in at least 43 national gene banks as well as at ICARDA centre (Duc et al. 2010). The largest genetic diversity study of faba bean landraces was of 802 landraces and varieties from China, North Africa, Europe and Asia, as revealed by ISSR markers (Wang et al. 2012). The Chinese landraces were widely separated from a combined grouping of African, European and Asian (outside of China) germplasm, which were closely similar though still distinct. Further, to attain breakthrough in enhancing yield and improving stability in commercial cultivars, new sources of variation need to be incorporated from wild crop relatives for genetic base widening.

1.4 Cowpea

Cowpea (*Vigna unguiculata* L. Walp.) is a widely adapted, stress-tolerant grain legume, vegetable and fodder crop grown over seven million ha in warm to hot regions of Africa, Asia and the Americas. Major breeding achievements, current objectives and future opportunities for cowpea improvement have been reviewed by Ehlers and Hall (1997). Early maturing cultivars have been developed with regionally acceptable grain quality and resistance to some important diseases and pests including bacterial blight (*Xanthomonas campestris*), cowpea aphid-borne mosaic virus (CABMV), cowpea aphid (*Aphis craccivora*), cowpea curculio (*Chalcodermus aeneus*), root-knot nematodes (*Meloidogyne incognita* and *M. javanica*), cowpea weevil (*Callosobruchus maculatus*) and the parasitic weeds *Striga gesnerioides* and *Alectra vogelii*. Earliness is an important character in Africa and other regions because early cultivars can escape drought and some insect infestations, can provide the first food grain and marketable product and can be grown in a diverse array of cropping systems. New early maturing cultivars with indeterminate growth habits have been very effective in the extremely dry and hot environment of the Sahel. Heat-tolerant breeding lines have been developed that have markedly higher pod set than most cultivars under high-night-temperature conditions. Development of cultivars with multiple resistance to biotic and abiotic stresses is an important current breeding objective. Earliness, delayed leaf senescence and indeterminate growth habit are characteristics, which are being combined to improve drought adaptation. In the future, high levels of resistance to important insect pests such as flower thrips (*Megalurothrips sjostedti*), maruca pod borer (*Maruca testulalis*), lygus (*Lygus hesperus*) and pod bugs (*Clavigralla tomentosicollis*) need to be identified. Introducing genes from wild cowpeas or related *Vigna* species, through genetic engineering, may be necessary to breed cultivars with high level of resistance to several of the major insect pests. A wide range of biotic (virus, bacteria, fungi, insects, nematodes and plants) and abiotic (like low phosphorus availability, soil acidity or salinity, drought and high temperature at night) factors are limiting cowpea production in different parts of the world. To overcome these constraints, diverse programmes were implemented for base broadening using interspecific hybridization between cowpea and other members of its genus with limited success because of pre- and post-zygotic barriers associated in the crop. These failures led the investigators to implement protocols to introduce foreign genes into cowpea. Currently, several genes of interest such as *herbicide imazapyr*, *α-amylase inhibitor 1* (against bruchids), *Cry1Ab* and *Cry1Ac* (against *Maruca*) have been introduced successfully into commercially important cultivars, and the genes are transmitted in Mendelian fashion. In addition, significant genomic resources and a consensus genetic map where agronomic, growth habit, disease, pest resistance and other trait loci have been placed and are ready for use in breeding programmes.

1.5 Lentil

Cultivation of lentil (*Lens culinaris* ssp. *culinaris*) in South Asia dates back to around 2000 BC. Because of its nutritive value, soil ameliorative properties and suitability to the existing cropping systems, it attains the status of main pulse crop in the region. Today, about half of the world's lentil is grown in South Asia for human food, animal feed and diversification and intensification of cropping systems. However, the indigenous lentils are of a specific ecotype (*pilosae*), which lack marked variability for morphological, phenological and yield related traits and resistance to key stresses. The narrow genetic base, which limited the breeding progress in the region for a long time has been widened through the introduction of exotic germplasm and their use in hybridization programmes. The progenies developed through diverse crosses have led to the selection of transgressive segregants with enormous variability for crop duration, seed traits, plant height, growth habit, biomass production, resistance to rust, *Stemphylium* blight, drought tolerance, etc. The promising progenies have led to the development of important lentil varieties in Afghanistan, Bangladesh, India, Nepal and Pakistan. Some of these cultivars with diverse genetic background are adapted to new niches emerging in the context of climate change, consumers' preference and market opportunities. Adoption and impact of these varieties have enhanced farmer's income and national production and are contributing to food and nutritional security to the people of South Asia (Sarker et al. 2010). Wild and cultivated accessions need to be further evaluated with regard to major biotic and abiotic stresses, placing emphasis on the development of more accurate screening methods. In addition, further inheritance studies of resistance/tolerance to biotic and abiotic stresses are required for a better understanding of the respective genetic systems controlling these responses. This knowledge will be useful to design appropriate breeding programmes based on regional and local requirements. There is a need to include reliable morphological and molecular markers to develop a comprehensive consensus genetic linkage map in lentil, allowing for molecular tagging of resistance genes against biotic and abiotic stresses in order to exploit them in breeding with increased selection efficiency. New molecular tools will increase the speed and precision of introgression of newly identified alleles from both the adapted and wild *Lens* species and subspecies into advanced breeding populations. The understanding of allele values from the adapted source and the wild gene pool will definitely increase both the efficiency and worth of germplasm utilization in lentil breeding programmes.

1.6 Pigeon Pea

Pigeon pea (*Cajanus cajan* L. Millspaugh), with ample genetic and genomic information becoming available in recent times, is now considered a trendy and pace-setter crop. The results of Diversity Arrays Technology (DArT) revealed most of the diversity was among the wild relatives of pigeon pea or between the wild and cultivated pigeon pea. The genetic relationships among the accessions are consistent with the available information and systematic classification of species and the cultivated *C. cajan.* The narrow genetic base of cultivated pigeon pea is likely to represent a serious impediment to breeding progress in pigeon pea. Further, it is now possible to cross wild relatives not only from the *Cajanus* group placed in secondary and tertiary gene pool, but also the related genera placed in quaternary gene pool. The most significant achievement is the development of unique cytoplasmic male sterility systems. This is no small achievement for a legume, which is an important crop of Asia and Africa and plays a major role in the diet of majority of the people of this region. Recently, scientists from the International Crops Research Institute for the Semi-Arid Tropics, Hyderabad, have developed interspecific crosses using *C. acutifolius* and generated recombinants for high seed weight. A good success has also been achieved while attempting interspecific crosses with *C. albicans* (Subbarao et al. 1990), *C. platycarpus* (Mallikarjuna and Moss 1995), *C. reticulatus* var. *grandifolius* (Reddy et al. 2001) and *C. acutifolius* (Mallikarjuna and Saxena 2002).

1.7 Groundnut

Groundnut (*Arachis hypogea* L.) germplasm consists of the cultivated allotetraploid species *Arachis hypogaea* L. and a large number of wild species, which are nearly all diploids. A very low level of genetic variability in American and other cultivars have been reported. However, in unadapted germplasm lines from various centres of origin in South America, Africa and China, considerable morphological and physiological variability has been reported to exist. Wild species of section *Arachis* were included in the evaluations since they show a high degree of variation when assayed by molecular analysis. In all cases, a very low level of variability was found in cultivated groundnut, while abundant variability was present among wild diploid species showcasing their significance to groundnut breeding (Halward et al. 1992). Cultivated groundnut contains a fraction of the genetic diversity present in their closely related wild relatives, which is not more than 13 %, due to domestication bottleneck. Closely related ones are placed in section *Arachis*, which have not been extensively utilized until now due to ploidy differences between the cultivated and wild relatives. In order to overcome *Arachis* species utilization bottleneck, a large number of tetraploid synthetics were developed at Legume Cell Biology Unit, Grain Legumes Program, ICRISAT, India. Evaluation of synthetics for some of the production constraints showed that these were good sources of multiple disease and pest resistances. Some of the synthetics were utilized by developing ABQTL mapping populations, which were further screened for important biotic and abiotic constraints. Phenotyping experiments showed ABQTL progeny lines with traits of interest were necessary for the improvement of groundnut.

1.8 *Vigna* species

The Asian *Vigna* species are very sensitive to photoperiod and temperatures, and these two variables have a very high bearing on the plant type and its adaptability in all these crops. Tickoo et al. (1994) elaborated the aspect in greater details in context of green gram. Jain (1975) has argued that grain legumes, as a group, are still undergoing domestication. Not long back, in the cultivation history of these crops and even today in most areas of the growing countries, these crops are being grown under conditions, not very much different than the ones in which their wild relatives have been grown. Under conditions of low-input management, the evolution has been for the survival of the crop species itself rather than for grain yield which is important from breeders' point of view. The silver lining has been the evolution of symbiotic relationship of these crops and the nitrogen fixing *Rhizobia* and the subsequent high protein content of their seeds. However, it will always be debatable, whether the evolution of symbiosis in grain legumes is a curse or a blessing to them. Other characters like indeterminate growth habit, photo- and thermo-insensitivity, low harvest index, shattering of ripe pods, seed hardiness, zero seed dormancy, etc., have all evolved more to natural than human selection (Tickoo et al. 1994). Further, information on intraspecific diversity, particularly in *mungo-radiata* complex, is lacking. This information is essential for effective use of wild species germplasm in crop improvement programmes. A wild mung bean accession, *Vigna radiata* var. *sublobata*, was reported to be highly resistant to bruchid, *Callosobruchus chinensis* (L.), a serious pest of grain legumes during storage at the AVRDC (Talekar 1994). Mung bean yellow mosaic virus (MYMV) has been a major problem in green gram. The wild species *Vigna radiata* var. *sublobata* is an important source to incorporate resistance to MYMV into cultivated varieties (Singh 1994). In addition to the landraces and cultivars, the wild species therefore needs to be collected, characterized and conserved carefully for use in crop improvement programmes.

In common with most grain legume crop species, the wild related *Vigna* species do not form a particularly extensive or accessible genetic resource. More variability in wild conspecific forms is to be exploited in addition to the fullest range of landraces and cultivars (Singh et al. 2006; Bisht and Singh 2013). Greater exploitation of the conspecific wild species with valuable characters needs to be undertaken to make extended cultivation economically attractive (Smartt 1990). Some populations of *V. mungo* var. *sylvestris*, *V. radiata* var. *sublobata* and *V. radiata* var. *setulosa* with valuable characters like more number of pod-bearing clusters and pods per cluster have great agronomic potential for use in crop improvement programmes beside in the resistance/tolerance to biotic stresses (Bisht et al. 2005). Sources of resistance available in *Vigna radiata var. sublobata* (Singh 1994) need to be exploited more vigorously with the use of precise molecular tools.

1.9 Horse Gram

Horse gram (*Macrotyloma uniflorum*) is a dual-purpose pulse and fodder crop native to the Southeast Asia and tropical Africa. India is the only country cultivating horse gram on a large scale, where it is being used for human food purpose. About 1,161 germplasm accessions have been systematically characterized and evaluated for various stress parameters and important agromorphological traits in different research institutions of India. Disease and pest resistance and improving grain quality and protein percentage would be desirable traits for broadening the genetic base through interspecific hybridization.

There is increasing evidence that the formal breeding sector has not well served the needs of farmers in developing countries. Modern plant breeding leads inevitably to an extremely dangerous reduction of diversity in major world food crops. Indigenous locally adapted races of cultivated plants usually known as primitive cultivars are the product of agro-ecotypic differentiation during thousands of years of conscious and unconscious selection by man in the diverse conditions of cultivation in many habitats. Genetically, they are marked by high level of adaptation to climate and disease and high variability, and these factors render primitive cultivars essential in plant breeding. Unfortunately, under advanced farming systems, the essentials of *on-farm* evolution, that is, the generation of variation and its subsequent natural and farmer selection, are no longer possible. This leads to an extremely dangerous reduction of diversity in major world crops, as noted by Simmonds (1979). Successful crop breeding tends to narrow the genetic base of a crop in direct proportion to its success; in all advanced, technology-based agriculture, few excellent varieties, themselves often interrelated, tend to cover large areas of land to the exclusion of all else (Simmonds 1979). In addition to the "bottleneck" effect during plant breeding, the diversity on which many important varieties were based was very limited, so there is also a "founder" effect. As an example, most varieties of the US soybean crop can be traced to 50 original introductions, only 10 of which contributed to 80 % of northern, and 7–80% of southern cultivars. Most introductions originated from Northeast China (Delannay et al. 1983). There is a particular danger with varieties based on limited sources of cytoplasmic male sterility—found, for example, in many food crops (Chang 1985). It has been a common consensus now that a reasonable level of genetic diversity should be deliberately maintained in variety development programmes. Farmers should be involved and get encouraged to maintain, improve and enhance the locally adapted and diverse genetic materials with appropriate technical and policy supports. In the marginal areas where problems of diseases, insects and environmental fluctuations are more and hence risks are great, we should reconsider the current breeding strategies leading to genetic uniformity. Since risk aversion strategies condition farmers' responses to new options, technologies and policies should conform to and reinforce these strategies (Tilahun 1995).

Crop production characterized by genetic diversity is normally stable as compared to those characterized by genetic uniformity (Simmonds 1979; de Boef et al. 1996) for a number of reasons. The problem with narrow genetic base is particularly aggravated by the widespread use of one or a few genetically uniform varieties over a large hectarage (spatial dominance), which is considered the other prerequisite for the widespread of disease and insect pests as it provides ideal condition and creates vulnerability (Simmonds 1979; Sharma 2001; Rubenstein et al. 2005; Smolders 2006). The continuous production of a single variety of the same crop year after year will facilitate disease epidemics particularly when host plant resistance is defeated by the counteractive mutation of disease pests and pathogens (Marshall 1977; Wolfe and Barrett 1977; Agrios 1978;

Simmonds 1979; de Boef et al. 1996; Rubenstein et al. 2005; Smolders 2006).

Conventional breeding will continue to have a valuable role in providing resistance to biotic and abiotic stresses. Nevertheless, it appears that in some systems there will continue to be barriers to achieving resistance using this approach (Edwards and Singh 2006). Molecular tools give us an opportunity to develop genotypes that carry resistance genes, and these tools have been utilized in DNA fingerprinting for identification of cultivars, marker-assisted selection and, to a limited level, genetic modification in breeding for resistance (Acosta-Gallegos et al. 2008). Problems associated with the need for long backcrossing cycles, gene pyramiding and the difficulties of crossing heterozygous clonally propagated crops with the conventional breeding method have already been resolved with the use of modern biotechnological tools to a great extent (Higgins et al. 1998; Witcombe and Hash 2000).

The habit of pushing varieties for release under marginal situations to a state of extreme uniformity by modern plant breeding has been criticized by several researchers (Marshall 1977; Wolfe and Barrett 1977; Simmonds 1979; de Boef et al. 1996). Under such situations, it is rather believed that uniformity is not biologically necessary or even desired, but diversity can, at least sometimes, enhance performance and stability (Simmonds 1979). Traits of interest to the resource-poor farmers in the marginal areas include yield stability; resistance to diseases, insects and abiotic calamities; and low dependence on external inputs (de Boef et al. 1996). Farmers achieve these by deliberately creating genetic diversity at intra-varietal and/or interspecific levels (Weltzien and Fiscbeck 1990; Broerse and Visser 1996; de Boef et al. 1996). Breeding activities to address this group of farmers should, therefore, build on farmers' practices to complement them and not to substitute their practices (Bunders et al. 1996). The physical environment and price ratios between external inputs and farm outputs do not allow the use of large quantities of purchased inputs, especially agrochemicals in large parts of the tropics and subtropics (de Boef et al. 1996) and which are also not ecofriendly.

Several co-workers have suggested that emphasis should be given to the maintenance of diversity in some planned fashion (Marshall 1977; Wolfe and Barrett 1977; Simmonds 1979; de Boef et al. 1996) for sustainability. The dangers of genetic uniformity can also largely be avoided if plant breeders use different sources of genes (inter-parental diversity) in their breeding material, and it would certainly be dangerous to rely too much on any one individual source of resistance (Russell 1978). There are also speculations that crosses between parents with high inter-parental divergence would not only be more responsive to improvement since they are likely to produce higher heterosis and desirable genetic recombination and segregation in their progenies (Wallace and Yan 1998; Chahal and Gosal 2002) but also help to develop varieties with broad genetic base (Russell 1978; Chandel and Joshi 1983) that are not liable to genetic vulnerability (Chandel and Joshi 1983). There is a possibility that, like many other traits including grain yield, heterosis may be manifested through diseases and insect pest resistance (Singh 2002) and climatic resilience.

Farmers, particularly in the tropics and subtropics, have been employing interspecific varietal mixtures not only as a yield maximization and diversification but also as risk aversion strategies since time immemorial (Weltzien and Fiscbeck 1990; Broerse and Visser 1996; de Boef et al. 1996). In marginal areas where resource-poor farmers dominate and environments are highly variable, the initiation and enhancement of local breeding should be thought as one of the best strategies to overcome problems associated with genetic vulnerability of modern crop cultivars (Smolders 2006). Farmers have been breeders since time immemorial, and the key feature in local crop development has been its maintenance of genetic diversity, both between and within species (de Boef et al. 1996). Local breeding involves the maintenance of local varieties (conservation through utilization), their enhancement (through selection and enrichment with exotic materials) and the seed system (production, selection, treatment, storage and exchange).

To conclude that, narrow genetic base of crops and less use of germplasm are the prime factors limiting grain legume production and productivity worldwide. Therefore, exploitation of diverse sources of variability is required through pre-breeding for the genetic enhancement of grain legumes. It also further necessitates the identification and use of wild germplasm resources to develop new high-yielding cultivars with a broad genetic base.

References

Acosta-Gallegos AJ, Kelly JD, Gepts P (2008) Prebreeding in common bean and use of genetic diversity from wild germplasm. Crop Sci 47:44–59

Agrios GN (1978) Plant pathology, 2nd edn. Academic, London

Ahmad F, Slinkard AE (1992) Genetic relationships in the genus *Cicer* L. as revealed by polyacrylamide gel electrophoresis of seed storage proteins. Theor Appl Genet 84:688–692

Bisht IS, Singh M (2013) Asian *vigna*. In: Singh M, Upadhyaya HD, Bisht IS (eds) Genetic and genomic resources of grain legume improvement. Elsevier Science, London, pp 237–267

Bisht IS, Bhat KV, Lakhanpaul S, Latha M, Jayan PK, Biswas BK, Singh AK (2005) Diversity and genetic resources of wild *Vigna* species in India. Genet Resour Crop Evol 52:53–68

Broerse JEW, Visser B (1996) Assessing the potential. In: Bunders J, Haverkort B, Hiemstra W (eds) Biotechnology: building on farmers' knowledge. Macmillan, London/Basingstoke, pp 131–180

Bunders J, Loeber A, Broers JEW, Havertkort B (1996) An integrated approach to biotechnology development. In: Bunders J, Haverkort B, Hiemstra W (eds) Biotechnology; building on farmers' knowledge. Macmillan, London/Basingstoke, pp 201–227

Chahal GS, Gosal SS (2002) Principles and procedures of plant breeding. Narosa Publishing House, New Delhi, p 603

Chandel KPS, Joshi BS (1983) Multivariate analysis in green-seeded pea. Indian J Agric Sci 53(4):198–200

Chang TT (1985) Principles of genetic conservation. Iowa State J Res 59:325–348

Cooper HD, Spillane C, Hodgkin T (2001) Broadening the genetic base of crop production. CABI, Oxford, p 452

Croser JS, Ahmad F, Clarke HJ, Siddique KHM (2003) Utilization of wild *Cicer* in chickpea improvement – progress, constraints, and prospects. Aust J Agric Res 54:429–444

de Boef WST, Berg T, Haverkort B (1996) Crop genetic resources. In: Bunders J, Haverkort B, Hiemstra W (eds) Biotechnology; building on farmers' knowledge. Macmillan, London/Basingstoke, pp 103–128

Delannay X, Rodgers DM, Palmer RG (1983) Relative genetic contributions among ancestral lines to North American soybean cultivars. Crop Sci 23:944–949

Duc G, Bao SY, Baum M, Redden B, Sadiki M, Suso MJ, Vishniakova M, Zong XX (2010) Diversity maintenance and use of *Vicia faba* L. genetic resources. Field Crop Res 115:270–278

Edwards O, Singh KB (2006) Resistance to insect pests: what do legumes have to offer? Euphytica 147:273–285

Ehlers JD, Hall AE (1997) Cowpea (*Vigna unguiculata* L. Walp.). Field Crop Res 53:187–204

Ford-Lloyd BV, Brar D, Khush GS, Jackson MT, Virk PS (2008) Genetic erosion over time of rice landrace agrobiodiversity. Plant Genet Resour Charact Util 7: 163–168

Halward TM, Stalker HT, LaRau EA, Kochert GD (1992) Genetic Variation detectable with molecular markers among unadapted germplasm resources of cultivated peanut and related wild species. Genome 34:1013–1020

Higgins VJ, Huogen Lu, Xing Ti, Gelli A, Blumwald E (1998) The gene-for-gene concept and beyond: interactions and signals. Can J Plant Pathol 20:150–157

Jain HK (1975) Breeding for yield and other attributes in grain legumes. Indian J Genet Plant Breed 35: 169–187

Keneni G, Bekele E, Imtiaz M, Dagne K (2012) Genetic vulnerability of modern crop cultivars: causes, mechanism and remedies. Int J Plant Res 2(3):69–79

Kumar S, Gupta S, Chandra S, Singh BB (2004) How wide is the genetic base of pulse crops. In: Ali M, Singh BB, Kumar S, Dhar V (eds) Pulses in new perspective. Indian Society of Pulse Research and Development, IIPR, Kanpur, pp 211–221

Link W, Dixkens C, Singh M, Schwall M, Melchinger AE (1995) Genetic diversity in European and Mediterranean faba bean germplasm revealed by RAPD markers. Theor Appl Genet 90:27–32

Mallikarjuna N, Moss JP (1995) Production of hybrids between *Cajanus platycarpus* and *Cajanus cajan*. Euphytica 83(1):43–46

Mallikarjuna N, Saxena KB (2002) Production of hybrids between *Cajanus acutifolius* and *C. cajan*. Euphytica 124(1):107–110

Marshall DR (1977) The advantages and hazards of genetic homogeneity. In: Day PR (ed) The genetic basis of epidemics in agriculture. The New York Academy of Sciences, New York, pp 1–20

Reddy LJ, Kameswara Rao N, Saxena KB (2001) Production and characterization of hybrids between *Cajanus cajan* x *C. reticulatus* var. *grandifolius*. Euphytica 121:93–98

Rubenstein DK, Heisey P, Shoemaker R, Sullivan J, Frisvold G (2005) Crop genetic resources: an economic appraisal. Economic Information Bulletin No. 2. United States Department of Agriculture (USDA), Washington, DC. www.ers.usda.gov

Russell GE (1978) Plant breeding for pest and diseases resistance. Butterworths, London

Sarker A, Agrawal S, Shrestha R, Kumar J, Uddin MJ (2010) Broadening the genetic base of lentil in South Asia. In: ASA, CSSA, and SSSA international annual meetings. https://a-c-s.confex.com/crops/2010am/webprogram/Paper61096.html

Sharma JR (2001) Principles and practice of plant breeding. Tata McGraw-Hill, New Delhi

Siddique KHM, Brinsmead RB, Knight R, Knights EJ, Paull JG, Rose IA (2000) Adaptation of chickpea and faba bean to Australia. In: Knights R (ed) Linking research and marketing opportunities for pulses in the 21st century. Kluwer Academic, Dordrecht, pp 289–303

Simmonds NW (1979) Principles of crop improvement. Longman, London/New York

Singh DP (1994) Breeding for resistance to diseases in mungbean: problems and prospects. In: Asthana AN, Kim DH (eds) Recent advances in pulses research. Indian Society of Pulses Research (IIPR), Kanpur, pp 152–164

Singh SP (2001) Broadening the genetic base of common bean cultivars. Crop Sci 41:1659–1675

Singh BD (2002) Plant breeding: principles and methods. Kalyani Publishers, New Delhi

Singh M, Bisht IS, Sardana S, Gautam NK, Husain Z, Gupta S, Singh BB, Dwivedi NK (2006) Asiatic *vigna*. In: Dhillon BS, Saxena S, Agrawal A, Tyagi RK (eds) Plant genetic resources: food grain crops. Narosa Publishing House, New Delhi, pp 275–301

Smartt J (1990) The old world pulses: vigna species. In: Grain legume: evolution and genetic resources. Cambridge University Press, Cambridge

Smolders H (2006) Enhancing farmers' role in crop development: framework information for participatory plant breeding in farmer field schools. Centre for Genetic Resources, Wageningen University and Research Centre, Wageningen

Subbarao GV, Johansen C, Kumar Rao JVDK, Jana MK (1990) Salinity tolerance in F1 hybrids of pigeonpea and a tolerant wild relative. Crop Sci 30(4):785–788

Talekar NS (1994) Sources of resistance to major insect pests of mungbean in Asia. In: Asthana AN, Kim DH (eds) Recent advances in mungbean research. Indian Society of Pulses Research (IIPR), Kanpur, pp 40–49

Tickoo JL, Gajraj Mahto R, Manji C (1994) Plant type in mungbean (*Vigna radiata* L. Wilczek). In: Asthana AN, Kim DH (eds) Recent advances in mungbean research. Indian Society of Pulses Research (IIPR), Kanpur, pp 197–213

Tilahun M (1995) Indigenous risk management strategies of small farms in the Central Rift Valley of Ethiopia. Food security, nutrition and poverty alleviation in Ethiopia: problems and prospects. In: Demeke M, Wolday A, Simeon E, Zegeye T (eds) Proceedings of the inaugural and first annual conference of the Agricultural Economics Society of Ethiopia, Addis Ababa, 8–9 June 1995, pp 85–108

Wallace DH, Yan W (1998) Plant breeding and whole-system crop physiology. University Press, Cambridge

Wang H, Xuxiao Zong X, Jianping Guan J, Yang T, Sun X, Yu Y, Redden R (2012) Genetic diversity and relationship of global faba bean (I L.) germplasm revealed by ISSR markers. Theor Appl Genet 124:789–797

Weltzien E, Fiscbeck G (1990) Performance and variability of local barley landraces in Near-Eastern environments. Plant Breed 104:58–67

Witcombe JR, Hash CT (2000) Resistance gene deployment strategies in cereal hybrids using marker-assisted selection: gene pyramiding, three-way hybrids, and synthetic parent populations. Euphytica 112:175–186

Wolfe MS, Barrett JA (1977) Population genetics of powdery mildew epidemics. In: Day PR (ed) The genetic basis of epidemics in agriculture. The New York Academy of Sciences, New York, pp 151–163

Common Bean

2

Anju Pathania, Surinder Kumar Sharma, and Prem Nath Sharma

Abstract

The chapter on the common bean reviews the origin and domestication, gene pool organization and their evolutionary relationship, genetic diversity and gene flow assessment, production constraints, crop production-limiting factors, and crop improvement strategies. While the common bean originated in the Americas, it is now widely grown in all continents of the world. Several bean germplasm banks have been established and contain diverse genetic resources comprising five domesticated and wild *Phaseolus* species, as well as an incipient stock collection. Unique among crop plants, the common bean consists of two geographically distinct evolutionary lineages that predate domestication and trace back to a common, still extant ancestor. The common bean cultivated across the world comprises of two major gene pools: the Andean originating from the Andes mountains of South America and the Mesoamerican from Mexico and Central America along with well-established races. Gene flow between domesticated and wild beans led to substantial introgression of alleles from the domesticated gene pool into the wild gene pool and vice versa. Like other crops, the common bean also suffers from various biotic and abiotic stresses; however, these constraints vary with the agroecological regions experiencing tropical to temperate environments. The important biotic and abiotic constraints limiting bean production are bean anthracnose, angular leaf spot, bean common mosaic and necrosis virus, bean golden mosaic virus, bacterial blights, drought, and phosphorus deficiency. The common bean improvement program in Europe and the USA is mainly focused on biotic and abiotic factors, mainly diseases, drought, and biofortification involving intra- and interspecific hybridization programs.

A. Pathania (✉) • S.K. Sharma
Mountain Agriculture Research and Extension Centre Sangla, Kinnaur, Himachal Pradesh, India
e-mail: anju_pathania@rediffmail.com; surender.sharma01@gmail.com

P.N. Sharma
Department of Plant Pathology, CSK, Himachal Pradesh Agricutlural University, Palampur 176 062, Himachal Pradesh, India
e-mail: pns1960@gmail.com

M. Singh et al. (eds.), *Broadening the Genetic Base of Grain Legumes*,
DOI 10.1007/978-81-322-2023-7_2, © Springer India 2014

Various breeding methods, namely, recurrent backcrossing, congruity backcrossing, inbred backcross line and gamete selection using multiple-parent crosses, and recurrent selection, have been used to transfer the trait of interest from related species. Resistance to various diseases has been transferred from *P. coccineus*, *P. acutifolius*, *P. costaricensis* to *Phaseolus vulgaris*. In addition, *P. acutifolius* and *P. parvifolius* have been explored to transfer high iron content to *P. vulgaris*. The crop improvement program over the last two decades involves the use of a marker-assisted selection strategy as a number of useful genes vis-à-vis molecular markers linked to them have been identified. In addition, bean transformation protocols have also been developed to facilitate introgression of alien genes. The sequencing of whole common bean genome is going to open an era of functional genomics to understand, identify, and overcome the constraints experienced by researchers across the world.

2.1 Introduction

The common bean (*Phaseolus vulgaris* L.) is one of the most important legume crops grown in all continents of the world except Antarctica, because of its high protein, fiber, and complex carbohydrate content. Beans (*Phaseolus* spp.) are extremely diverse crops in terms of cultivation methods, in diverse environments and elevations ranging from sea level to 3,000 m height, morphological variability, and utilities (dry as pulse and green as vegetable). The tremendous variability for plant types and growth behavior makes beans part of the most diverse production systems of the world (Broughton et al. 2003). They are cultivated in monoculture, in associations, or in rotations. Dry beans are consumed both as pulse and vegetable (both leaves and pods). Their genetic resources exist as a complex array of major and minor gene pools, races, and intermediate types, with occasional introgression between wild and domesticated types. Beans are thus a crop that is adapted to many niches, both in terms of agronomic and consumer preference. While its production is mainly centered on small holdings, beans are grown in monoculture bush beans to a complex association of indeterminate or climbing beans with maize, sugarcane, and other cereals or fruit crops. Short bush growth habits offer minimal competition and permit interplanting with other species, for example, in reforestation projects or among fruit trees or coffee plantations during the early years until the main crop can be exploited. At the other extreme, aggressive climbers are found at higher altitudes on subsistence farms where a few plants are maintained as a sort of insurance and are continually harvested for about 6 months. Over the past 20 years, beans have also been increasingly cultivated on a commercial scale and are now offered in national, regional, and international markets. The common bean regarded as a "grain of hope" is an important component of subsistence agriculture grown worldwide over an area of about 28.78 million hectares with an annual production of 23.14 million tonnes (FAO 2013), and feeds about 300 million people in the tropics and 100 million people in Africa alone. Besides it is emerging as an important income generation especially in Central America where beans have top ranking as income generators among field crops. In terms of global pulse production, the common bean alone with 23 million tonnes accounts for about half of the total pulse production (FAO 2012). However, this number is an overestimate as FAO does not report data for *Phaseolus* and non-*Phaseolus* species separately. Many comprehensive reviews on the different aspects of the common bean like origin and evolution (Gepts 1998), genetic diversity (Singh et al. 1991a, b, c; Blair et al. 2012a, b), population structure and evolutionary dynamics (Zizumbo-Villarreal et al.

2005), biotic and abiotic stresses (Miklas et al. 2006b; Singh and Schwartz 2010), breeding for insect pests (Singh and Schwartz 2011), genomics (Gepts et al. 2008), genetic transformation (Kwapata et al. 2012), etc., have been published from time to time. In this chapter, an attempt has been made to compile the information on various aspects of the common bean from various sources to provide an overview about this important legume to the readers. The different sections consist of common bean origin, gene pool, evolutionary relationship and systematics, gene flow, gene flow constraints, diversity in germplasm, production-limiting factors, important traits for broadening the genetic base, conventional and contemporary approaches for gene transfer, and the impact of molecular advances for the improvement of production and productivity of this important legume crop.

2.2 Crop Gene Pool, Evolutionary Relationship, and Systematics

The existence of archaeological, botanical, and historical evidences indicated that the common bean is a New World crop that originated ~7,000 years back in a long arc from the present-day northern Mexico (Chihuahua) through Central America and the Andes mountains to northwest Argentina (San Luis) (Broughton et al. 2003). Wild beans first described in Argentina (Burkart 1941; Burkart and Brucher 1953) and in Guatemala (MacBryde 1947) are grown in a wide arc stretching from north Mexico (approx. 30°N) to northwestern Argentina (approx. 35°S) at an altitude ranging from 500 to 2,000 m, experiencing annual rainfall of 500–1,800 mm. The two taxonomic subdivisions of *Phaseolus* have been described as *P. vulgaris* var. *aborigineus* and *P. vulgaris* var. *mexicanus* (Delgado-Salinas 1985) based on morphological and molecular differentiation. There is no accurate record of the transition of wild beans to cultivated beans due to gaps in the archaeological record, but this event is estimated to occur somewhere between 7,000 and 5,000 years ago (Broughton et al. 2003).

The individuals of this important legume are now well characterized on the basis of morphological and molecular traits into two gene pools: the Andean originating from the Andes mountains of South America (southern Peru, Bolivia, and northern Argentina) and Mesoamerican from Central America and Mexico with at least seven races (Gepts et al. 1986; Koenig and Gepts 1989a, b; Singh et al. 1991a, b, c; Becerra Velásquez and Gepts 1994; Tohme et al. 1996; Beebe et al. 2000, 2001; Díaz and Blair 2006; Blair et al. 2007; Kwak and Gepts 2009). Important parameters separating individuals into different gene pools include seed size, phaseolin (seed storage protein) patterns, plant morphology, isozymes, and RFLP, RAPD, AFLP, and microsatellite markers (Evan 1976; Gepts et al. 1986; Koenig and Gepts 1989a, b; Singh et al. 1991a; Khairallah et al. 1992; Becerra Velásquez and Gepts 1994; Blair et al. 2006a, 2007, 2010, 2012a, b). The morphological and agroecological adaptation traits have been further used to distinguish the races among these gene pools (Singh et al. 1991a, b; Gepts 1988). Race structure studies of Singh et al. (1991a, b, c) demonstrated that members of each race have distinctive and specific physiological, agronomic, biochemical, and molecular characteristics and diverge from other races in the allelic frequencies at specific loci. Domestication is an outcome of the selection process that leads to increased adaptation of plants to cultivation and utilization by humans. During domestication, plants and animals are subjected to natural as well as human selection. The traits that distinguished the domesticated forms from wild forms are collectively called as domestication syndrome (Hammer 1984). These traits include lack of seed dispersal and dormancy, compact plant architecture, higher yield, a synchronicity and early flowering, photoperiod sensitivity, harvest index and seed pigmentation (Gepts and Debouck 1991; Koinange et al. 1996). The domesticated common bean originated from wild gene pools through independent domestication events (Gepts and Debouck 1991; Kwak and Gepts 2009; Kwak et al. 2009) at two centers of domestications, i.e., Mesoamerica (western Mexico) and southern Andes of South America.

Two primary centers, located in Mesoamerica and in the southern Andes, gave rise to two major cultivar groups. The Mesoamerican cultivars consisted of small-seeded beans with a predominantly "S" phaseolin type, whereas Andean beans included large-seeded "T" phaseolin-type cultivars (Gepts et al. 1986; Gepts and Bliss 1988; Koenig and Gepts 1989a, b; Sonnante et al. 1994; Chacón et al. 2005). However, another minor domestication center located in Colombia has been shown to possess small-seeded "B" phaseolin-type seed protein cultivars (Gepts and Debouck 1991; Chacon et al. 1996; Tohme et al. 1996).

The major common bean gene pools have been further divided into races based on plant morphology, adaptation range, and agronomic traits. The Mesoamerican gene pool includes the races Durango (prostrate bush types with medium-sized seed from the dry highlands Mexico), Jalisco (climbing beans from the moist highlands of central Mexico), and Mesoamerica (small-seeded types, mostly bush habits, from the lowlands Central America and Mexico; Singh et al. 1991a), whereas the Andean gene pool includes the races Peru (predominantly highland climbing beans), Nueva Granada (mostly bush beans with mid-altitude adaptation), and Chile (prostrate bush or weak climbers, with temperate adaptation to higher altitudes). However, Beebe et al. (2000) suggested the existence of a fourth race—Guatemala in the Meso gene pool (mostly climbing beans from Guatemala and southern Mexico)—as well as some systematic variation within races. Chacón et al. (2005) found that the races Durango and Jalisco shared a common chloroplast DNA pattern, while the races Guatemala and Mesoamerica each presented a distinctive pattern, emphasizing their evolutionary uniqueness. Díaz and Blair (2006) found a dichotomous structure in the Middle American gene pool, with the grouping of the Durango and Jalisco races apart from the race Mesoamerica, and novel diversity in some climbing bean accessions potentially from the race Guatemala. The differentiation of three races in the Andean gene pool (Singh et al. 1991a, b, c) by restriction fragment length polymorphism (RFLP) or random amplified polymorphic DNA (RAPD) markers is not as clear as in the Middle American gene pool (Becerra Velásquez and Gepts 1994; Beebe et al. 2001). These races display a common chloroplast DNA composition, suggesting domestication of a single population (Chacón et al. 2005). However, Andean races are clearly distinguished by microsatellite alleles (Blair et al. 2007) and have different growth habit prevalence and adaptation ranges. Significant introgression into the Andean gene pool from Middle American types that have filtered into the northern Andes since pre-Colombian times has been reported (Islam et al. 2004; Blair et al. 2007).

After domestication, common beans were introduced into other regions of the world (Evan 1976) with initial dissemination mainly in the Americas and Europe. Mesoamerican cultivars became predominant in the lowlands of South America and southwestern USA, whereas Andean cultivars became predominant in Africa, Europe, and northeastern USA. The cultivars of Mesoamerican origin may have been disseminated along a route from Mexico or Guatemala to Colombia and Venezuela, and eventually into Brazil (South America). The introduction route of African common bean cultivars could have been directly from the Andes or indirectly through the Iberian Peninsula (Purseglove 1976) or Western Europe during the colonial period. In pre-Columbian times, a limited bean germplasm exchange took place between Mesoamerica and South America, but much more extensive seed movement occurred after the 1500s. The common bean was introduced into the Iberian Peninsula (Spain and Portugal) from Central America around 1506 (Ortwin-Sauer 1966) and from the southern Andes after 1532 by sailors and traders who brought with them the nicely colored and easily transportable seeds as a curiosity (Debouck and Smartt 1995). Some studies have provided evidences for the existence, in the Iberian Peninsula, of varieties of the Andean and Mesoamerican genetic pools that could have been disseminated through other countries, along the European Mediterranean basin, such as Greece, Cyprus, and Italy, as indicated by the phaseolin types found in such areas (Lioi 1989).

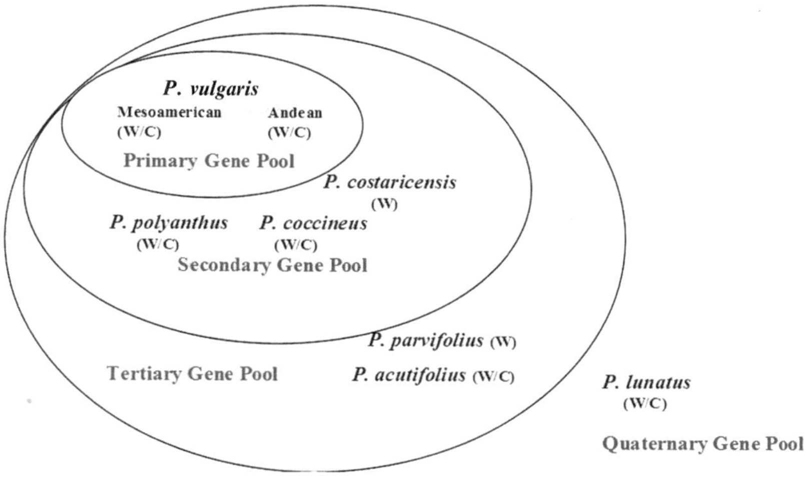

Fig. 2.1 Interspecies diversity-based common bean gene pool

The primary center of diversity for the genus *Phaseolus* lies in central and south Mexico, bordering Guatemala of the Mesoamerica that harbors 70–80 wild species and the ancestors of five domesticated species (Freytag and Debouck 2002). The presence of secondary centers of diversification in Spain, Turkey, and Ethiopia (Harlan 1971; Santalla et al. 2002; Ocampo et al. 2005) is controversial (Kami et al. 1995). The addition to the secondary centers of common bean diversity in Asia has been made in recent analysis of Chinese and Indian beans (Zhang et al. 2008; Sharma et al. 2013).

The genus *Phaseolus* belongs to the tribe *Phaseoleae* (family *Leguminosae*, subfamily *Papilionoideae*), which includes two other genera of domesticated species *Glycine* (soybean) and *Vigna* (cowpea). *Phaseolus* is a diverse genus with at least 50 species (Verdecourt 1970; Marechal et al. 1978; Lackey 1983). Out of these, only five species (*Phaseolus vulgaris* (common bean), *Phaseolus lunatus* (lima bean), *Phaseolus acutifolius* A. Gray (tepary bean), *Phaseolus coccineus* L. (runner bean), and *Phaseolus dumosus* (=*polyanthus*) L. (yearlong bean)), are cultivated. Each of these species has a distinct wild ancestor, distinct geographical origin, life history, and reproductive system (Marechal et al. 1978; Delgado-Salinas 1985; Debouck 1991; Smartt 1990). The phylogenetic relationship among domesticated *Phaseolus* spp. by using morphological, biochemical, and molecular tools (Marechal et al. 1978; Debouck 1991; Sullivan and Freytag 1986; Jaaska 1996; Pueyo and Delgado-Salinas 1997; Delgado-Salinas et al. 1993; Schmit et al. 1993; Liaca et al. 1994; Hamann et al. 1995; Vekemans et al. 1998) and combined analysis by Delgado-Salinas et al. (1999) suggested that the *Phaseolus* is monophyletic. The interspecies diversity in relation to the common bean is organized in primary, secondary, tertiary, and quaternary gene pools (Fig. 2.1).

2.2.1 Primary Gene Pool

The primary gene pool comprises both *Phaseolus vulgaris* cultivars and wild populations. Wild *Phaseolus vulgaris* forms are immediate ancestors of common bean cultivars (Berglund-Brucher and Brucher 1976; Brucher 1988), and they are found from north Mexico to northwest Argentina (Toro et al. 1990; Singh 2001). Within domesticated and wild populations, two major gene pools can be distinguished: an Andean and a Mesoamerican gene pool (Gepts and Bliss 1985; Koenig and Gepts 1989a; Kairaillah et al. 1990). Only few domesticated and wild forms of *Phaseolus vulgaris* are sexually compatible and in some Andean and Mesoamerican parents F_1 hybrid weakness is found. These hybridization barriers are due to the presence of incompatibility genes in some Mesoamerican (Dl1) and Andean (Dl2) wild and domesticated populations (van Reheenen et al. 1979; Shii et al. 1980; Singh and Gutierrez 1984; Vieira et al. 1989). These genes lead to inviable F_1 hybrids (Shii et al. 1980; Gepts and Bliss 1985) and hybrid weakness and phenotypic abnormalities in later generations (Koinange and Gepts 1992; Debouck et al. 1993; Freyre et al. 1996; Singh and Molina 1996). Gene exchange and introgression in certain wild and domesticated populations of different gene pools is therefore limited (Kornegay et al. 1992; Mumba and Galwey 1998, 1999).

2.2.2 Secondary Gene Pool

The secondary gene pool of the common bean comprises *P. coccineus*, *P. dumosus* (=*P. polyanthus* Greenman), and *P. costaricensis* Freytag and Debouck. These three species cross easily with each other and with domesticated common beans under natural and controlled conditions mainly when the common bean is the female parent (Baggett 1956; Camarena and Baudoin 1987; Singh et al. 1997). When the common bean is the pollen donor parent, the cross is difficult because progenies tend to revert to the female parent (Smartt 1970; Ockendon et al. 1982; Hucl and Scoles 1985; Debouck 1999). Some crosses result partially sterile progeny (Ibrahim and Coyne 1975; Manshardt and Basset 1984). *Phaseolus coccineus* is infertile with common beans (Smartt 1979, 1990; Manshardt and Bassett 1984; Hoover et al. 1985; Debouck and Smartt 1995).

2.2.2.1 *Phaseolus coccineus*

Introgression is possible in both directions (Smartt 1970; Ibrahim and Coyne 1975; Ockendon et al. 1982). However, the level of cross compatibility between two species varies and seems to be influenced by parental genotypes (Bemis and Kedar 1961; Smartt and Haq 1972; Zagorcheva and Poriazov 1983). Natural hybrids between *P. coccineus* and *P. dumosus* are abundant in areas of sympatry in Colombia and Ecuador and not in Guatemala (Freytag and Debouck 2002).

2.2.3 Tertiary Gene Pool

The tertiary gene pool includes *P acutifolius* and *P. parvifolius* Greene (Debouck 1999). Crosses between these two species with the common bean are possible through embryo rescue. Viable interspecific F_1 progeny is sterile (Menjia-Jimenez et al. 1994; Singh et al. 1998). Hybrid fertility can be restored by chromosome doubling or by one or more generations of backcrossing with recurrent common bean parents (Prendota et al. 1982; Pratt et al. 1985; Thomas and Waines 1984). In exceptional cases, fertile hybrids are obtained (Honma 1956). The level of cross compatibility between *P. vulgaris* and *P. acutifolius* is strongly influenced by the genotype, due to the action of two complimentary dominant genes that are present in some populations (Parker and Michaels 1986). Choosing compatible common bean genotypes as female parents facilitates interspecific hybridization and increases hybrid fitness (Federici and Waines 1988). Singh (1991) suggested that the F_1 hybrid incompatibility between *P. vulgaris*×*P. acutifolius*, *P. vulgaris*×*P. coccineus*, and other interspecific crosses might be due to the presence of *Dl-1* and *Dl-2* genes.

2.2.4 Quaternary Gene Pool

Common beans have a so-called quaternary gene pool which includes *Phaseolus lunatus*, *P. carteri* Freytag and Debouck, *P. filiformis* Benth, and *P. angustissimus* A. Grey. Crosses of the common bean have been attempted between P. *filiformis* (Benth), *P. angustissimus* (A. Grey), and *P. lunatus*, but none of these reports documented that viable and fertile hybrid progeny could be obtained. Thus, for the purpose of gene introgression, these species should be included in the quaternary gene pool of common beans (Fig. 2.1).

Cultivated forms of the common bean are herbaceous annuals with determinate or indeterminate growth habits. On germination, the plant is predominantly tap-rooted and bears adventitious roots. Papilionaceous flowers are borne in axillary and terminal racemes and may bear one to many flowers.

Table 2.1 Growth habit classes of the common bean

Growth habit	Morphological features
Type I	1. Determinate growth habit
	2. Terminal bud reproductive
	3. Stems and branches erect or prostrate
	4. Terminal guide absent
	5. Pods distributed along the length of the stem
Type II	1. Indeterminate growth habit
	2. Terminal bud vegetative
	3. Stems and branches erect
	4. Terminal guide absent or medium
	5. Pods distributed along the length of the stem
Type III	1. Indeterminate growth habit
	2. Terminal bud vegetative
	3. Stems and branches prostrate with little or no climbing ability
	4. Terminal guide small or long pods distributed mainly in the basal portion
Type IV	1. Indeterminate growth habit
	2. Terminal bud vegetative
	3. Stems and branches twining with strong climbing ability
	4. Terminal guide long or very long
	5. Pods distributed along the length of the stem or mainly in the upper portion

The flowers are zygomorphic with a bi-petalled keel and two lateral wing petals and are large, genetically independent seed color, but the association between particular flowers and seed colors is common. The flowers may be white, pink, purple, or red. The flower contains ten stamens and a single multi-ovuled ovary, is predominantly self-fertilized, and develops into a straight or slightly curved pod. Seeds may be round, elliptical, somewhat flattened, or rounded elongate in shape, and a rich assortment of coat colors and patterns exists. Seed size ranges from 50 mg seed in wild accessions collected in Mexico, to more than 2,000 mg seed in some large-seeded Colombian varieties (detailed description by Debouck 1991).

Singh (1982) classified the common bean growth habits into four (Table 2.1). Determinate plants of the common bean (growth habit I) may have 3–7 trifoliate leaves on the main stem before the terminal double raceme (as found in 'bush' or 'dwarf' cultivars selected for earliness in Europe and the USA) or may have many nodes with 7–15 (Middle American) or 15–25 (Andean) trifoliate leaves on the main stem (Debouck 1991). Three indeterminate growth habits are distinguished in the common bean (Schoonhoven and Pastor-Corrales 1987):

Growth habit II: upright habit, with an erect stem and branches, and often without a guide

Growth habit III: bush habit with weak and prostrate stem and numerous branches; having a short or long guide and with variable ability to climb

Growth habit IV: climbing if supported on a suitable tutor, with a weak, long and twisted stem and reduced branching

Germination in *P. vulgaris* is epigeal and requires 5–7 days at a soil temperature of 16 °F. Time to flowering varies with cultivar, temperature, and photoperiod. Flowering is usually initiated 28–42 days after planting, but significantly delayed among climbing varieties grown at higher elevation. Flowering in cultivars of growth habit I is concentrated over a very short period of time (usually 5–6 days). Indeterminate cultivars produce additional nodes after initial flowering, leading to extendable flowering for 15–30 days. As many as two-thirds of all the flowers produced may abscise, and under temperature or water stress, young fruits and/or developing seeds may also abort. Abscission is greatest among flowers formed on the upper nodes and branches and within a raceme is greatest among the later flowers to form (CIAT 1975). Seed filling periods may vary between 23 and 50 days in determinate cultivars and indeterminate climbing varieties. Physiological maturity (the stage beyond which no further increase in seed dry matter takes place) may occur in 60–65 days after planting early maturing varieties grown under a short growing season or may extend to 200 days in climbing varieties grown in cooler upland elevations (Graham and Ranalli 1997).

2.3 Assessment of Gene Flow for Crop Improvement

Wild and cultivated forms of common beans are often found together throughout the areas of its distribution (Zizumbo-Villareal et al. 2005). Gene flow between domesticated and wild beans has been shown to result in substantial introgression of alleles from the domesticated gene pool to the wild gene pool (Freyre et al. 1996; Beebe et al. 1997). Although the common bean is a predominantly self-pollinated species, still outcrossing of 2–3 % has been observed (Ibarra-Pérez et al. 1997; Ferreira et al. 2000). Ellstrand et al. (1999) suggested that in highly self-pollinated species, gene flow is likely to limit the independent evolution of wild and domesticated populations. Ferreira et al. (2007) assessed the gene escape from transgenic bean plants into non-transgenic and wild populations of beans and found no natural outcrossing when cultivars are planted beyond 3.25 m. Over time, the level of introgression has been apparently such that genetic diversity from the domesticated gene pool has been displacing genetic diversity in the wild gene pool, with the exception of the genome regions surrounding domestication genes. In these regions, selection against deleterious domesticated alleles prevent or delay introgression of domesticated alleles that lead to the maintenance of the original wild alleles (Papa et al. 2005). Genetic compatibility between wild and domesticated populations leads to wild–weedy–domesticated hybrid complexes in sites with sympatric distribution by introgression of genes from wild populations to domesticated ones or vice versa. The resultant of this introgression usually show morphological traits reminiscent of one or the other parent, such as larger seeds than the wild parent or seed color or color patterns similar to those observed in wild beans, thereby constitute a valuable source of genes to the farmer or plant breeder (Debouck and Smartt 1995; Beebe et al. 1997). However, at the same time, they also impose a risk for the massive introduction of genes from the domesticated to wild populations (Gepts et al. 1999; Papa and Gepts 2003).

Studies on gene dynamics of wild–weedy–domesticated complexes within the Mesoamerican domestication are important, since 4 of the 5 domesticated species and 45 of the 50 species of the genus *Phaseolus* dominate this area, and their natural reproductive relationships are still unknown. Genetic bridges may exist between these species, as reported by Escalante et al. (1994) regarding gene flow between wild populations of *P. vulgaris* and *P. coccineus* L. Recent studies on wild populations of Mesoamerica, in areas where traditional agriculture continues, indicate that opportunities for wild–domesticated hybridization are common in spite of a predominantly autogamous reproductive system. They suggest that gene flow is asymmetric, at least three times greater from the domesticated to the wild (Papa and Gepts 2003; Chacón et al. 2005; González et al. 2005; Payro de la Cruz et al. 2005; Zizumbo-Villareal et al. 2005). This asymmetric gene flow may be due to the differences in population size (domesticated > wild), dominance of wild phenotypes, and/or human selection against hybrids in domesticated populations (Papa and Gepts 2003, 2004; Papa et al. 2005). Thus, a displacement of genetic diversity can take place in wild populations due to the gene flow from domesticated populations (Papa and Gepts 2003). The primitive cultivars could have spread from an area of domestication to other regions where they crossed with local wild populations, giving rise to present-day landraces (Gentry 1969; Beebe et al. 2000). For example, in Mexico, Vanderborght reported up to a 50 % introgression rate in wild populations. However, Lentini et al. (2006) observed a predominant gene flow from wild to domesticated type in Costa Rica, attributing it to the longer flowering period of the wild type and floral characteristics rendering them prone to outcrossing. This hypothesis is further supported by studies on chloroplast gene flow estimated by allozyme and microsatellite marker analysis (Singh et al. 1991b, c; Gonzalez-Torres 2004). In lima bean, the magnitude of both recent and long-term intraspecific gene flow between domesticated and wild populations is low (Hardy et al. 1997; Maquet et al. 2001; Ouedraogo and Baudoin 2002; Martinez-Castillo et al. 2007).

2.4 Level of Diversity in Crop Germplasm

Wild accessions of crops and landraces are valuable genetic resources for plant breeding and conservation of alleles and gene combinations in plants. Information on the organization of genetic diversity also helps us to prioritize the significant segments of genetic diversity for conservation. The common bean is in a unique situation among various crops as its germplasm consists of two major evolutionary lineages tracing back to a common ancestor, which still exists and has been identified. Cultivated common bean is a morphologically diverse crop (Hedrick 1931), and wide variations exist for growth habit, pigmentation, pod, seed, phenology, and other characters (Leakey 1988; Singh 1989). Singh et al. (1991a) suggested the existence of subgroups within each of the major Andean and Mesoamerican groups, with distinctive morphology, adaptation, and disease resistance. The domestication process has led to a reduction in the genetic diversity within each of the bean gene pools (Sonnante et al. 1994). Several evolutionary forces played a role in shaping the *P. vulgaris* genetic diversity, which include the introgressive hybridization between *P. vulgaris* and *P. coccineus* in Mesoamerica, gene flow and selection between wild and domesticated populations, evolution of disease resistance, and introduction of beans into the Old World. Blair et al. (2012a, b) suggested that geographical isolation, founder effects, or natural selection could have created the semi-discrete populations of wild beans and that multiple domestications and introgression were involved in creating the diversity in cultivated beans. In spite of the ample germplasm resources available in national and international gene banks, only a small proportion of the genetic variability has been utilized by bean breeders for the identification and development of high-yielding bean in national and international bean breeding programs (Miklas 2000; Singh 2001).

2.4.1 Wild Gene Pool

Genetic drift and selection for domestication traits along with local adaptation are the major reasons for the reduction in genetic diversity that has accompanied domestication in crops (Gepts 2004). Based on this theoretical argument, it is therefore likely that the genetic diversity of the domesticated gene pool is reduced compared to that of its wild, ancestral gene pool. This has been verified numerous times at the molecular level (Gepts 2004). Blair et al. (2012a, b) reported high diversity among wild beans as compared to cultivated ones. The diversity in wild accessions of the species can be divided into various subpopulations from specific geographical regions for the species (Chacon et al. 2007; Miklas and Singh 2007). The number of subpopulations has been a matter of discussion since the division of wild *P. vulgaris* is not as simple as for the domesticated beans which are easily separated into Andean and Mesoamerican gene pools. In addition, the morphological and molecular differences among groups of wild accessions are not as clear as among races in the cultivated types and rely on differences in seed size, flower coloration, bracteole size, seed protein (phaseolin) type, and in large part on molecular marker evaluations (Chacon et al. 2007; Kwak and Gepts 2009; Rossi et al. 2009; Gepts et al. 1986; Koenig and Gepts 1989b; Gepts and Debouck 1991).

Based on DNA fingerprinting with amplified fragment length polymorphism (AFLP) markers, wild common bean accessions have been divided into four groups or gene pools (Tohme et al. 1996). These include Mesoamerican, Andean, Colombian, and Ecuadorian northern Peruvian gene pools. Some studies with Andean wild and cultivated common beans with the same marker system found no grouping of wild accessions within the Andean gene pool (Beebe et al. 2001). However, northern Argentinean and southern wild Bolivian accessions have been suggested to be most similar to cultivated Andean beans. Rossi et al. (2009) also found geographical separation of wild bean populations with AFLP markers, suggesting that Colombian wild beans were closely related to Mesoamerican wild beans which could be separated into accessions from Mexico and Central America. They also suggested a reduction in diversity in the Andean gene pool. Kwak et al. (2009) found that wild beans from Mexico varied in their simple sequence repeat (SSR) fingerprint and that domestication of the cultivated Mesoamerican gene pool was likely to have occurred in the Lerma valley. Introgression of wild-derived genes from other subgroups of wild beans has been postulated to explain current SSR-based race structure in cultivated common beans (Díaz and Blair 2006; Blair et al. 2007, 2009a). Finally, based on sequence information for five gene fragments, Bitocchi et al. (2012) proposed a Mesoamerican origin rather than a South American origin for wild populations of *P. vulgaris* based on the detection of a strong bottleneck in the actual Andean gene pool of wild beans.

2.4.2 Domesticated Gene Pool

The interspecies diversity in relation to the common bean is organized in primary, secondary, tertiary, and quaternary gene pools. Intraspecies diversity in the common bean is separated into two major gene pools (Andean and Middle American). Cultivars are further divided into races, each with their distinguishing characteristics. Based on the distribution of most *Phaseolus* species in Mexico and Central America, it was thought that the Mesoamerican gene pool of the common bean was also the original gene pool of that species and that the Andean gene pool was a derived gene pool. If the Andean gene pool had been derived from the Mesoamerican gene pool, one would have expected that the Andean gene pool only contained a subset of the diversity of the Mesoamerican gene pool (except for variation having appeared after the separation of the two gene pools). However, data suggest that both gene pools are derived gene pools. Thus, both gene pools should be conserved because they may likely contain different genes of agronomic interests. Evidence from disease resistance analyses shows that the two gene pools contain distinct resistance specificities (Kelly and Vallejo 2004; Geffroy et al. 1999), illustrating the need for conservation of the two gene pools. Nutritional quality differed between the gene pools with respect to seed iron and zinc concentration, while genotypes from the intermediate group were notably high in both minerals (Blair et al. 2010). The genetic diversity in common bean gene pools elucidated by phaseolin types, allozyme alleles and molecular markers revealed that Mesoamerican gene pool is more diverse than Andean gene pool (Gepts et al. 1986; Koenig and Gepts 1989a; Becerra-Velásquez and Gepts 1994; Koenig et al. 1990; Blair et al. 2006a, 2009a; Chacon et al. 2007; Bitocchi et al. 2012).

2.4.3 Landraces

Landraces are distinct but variable populations, which usually have a local name, lack "formal" crop improvement, are characterized by a specific adaptation to the environmental conditions of the area of cultivation (tolerant to the biotic and abiotic stresses of that area), and are closely associated with the uses, knowledge, habits, dialects, and celebrations of the people who developed and continue to grow it (Negri et al. 2009; Polegri and Negri 2010). Although landraces are an important component of agrobiodiversity, most of them are at risk of extinction because they are grown by old farmers and need to be preserved in situ and ex situ (Negri et al. 2009). Molecular markers have been widely used in the common bean to assess the genetic diversity and genetic structure of several important collections and to develop maps (Baudoin et al. 1998; Yu et al. 2000; Blair et al. 2003, 2006a, 2009a, 2012a, b; Hornakova et al. 2003; Sharma et al. 2006; Zhang et al. 2008; Kwak and Gepts 2009; McClean et al. 2012; Raggi et al. 2013).

Both molecular data and screening of germplasm showed that landraces harboring significant levels of diversity are not found in the advanced breeding gene pools. Thus, recently bean breeders have actively sought to broaden the domesticated gene pool by crossing with landraces of the two gene pools in the centers of origin. Landraces can provide sources of genetic diversity for disease and pest resistance and tolerance to abiotic stresses (Beebe et al. 1997; Singh et al. 1991c). A further justification for the conservation of landraces is supported by the fact that local, indigenous, or traditional agriculture is gradually being replaced with industrial agriculture. A concomitant trend is the replacement of native crop diversity with genetic monoculture.

2.4.4 Commercial Cultivars

Most of the present-day bean cultivars have resulted from crosses between closely related cultivars belonging to either of the gene pools. Pedigree and RFLP data revealed that commercial classes of common bean possess low levels of genetic diversity, presumably because of little or no introduction of genetic diversity from outside these classes. Limited diversity in commercial cultivars has serious consequences for bean breeding and emphasizes broadening of the cultivated gene pool through effective breeding programs. Many workers have reported low genetic diversity in elite and commercial common bean cultivars as compared to landraces and wild accessions (Cardoso 2009; Cabral et al. 2011; Blair et al. 2012a, b).

2.5 Production Related Problems

Constraints to bean productivity vary with region and various other concerns like marginal landholdings of farmers in Asia, Africa, and Latin America and the large scale producers commonly found in the USA and Europe. For the former, the greatest need is to stabilize yield under marginal conditions and using limited technical inputs, whereas the latter is mainly concerned with yield improvement per se. In addition, the common bean suffers both from biotic and abiotic production constraints (Schwartz and Pastor-Corraies 1989; Graham and Ranalli 1997; Singh and Schwartz 2010, 2011).

2.5.1 Bean Diseases

The bean crop may be attacked by a wide range of insect pests, diseases, and nematodes. Among diseases, the major constraints in bean production are bean anthracnose (*Colletotrichum lindemuthianum*), angular leaf spot (*Phaeoisariopsis griseola*), halo blight (*Pseudomonas phaseolicola*), rust (*Uromyces phaseoli*), and *bean common mosaic virus* (Miklas et al. 2006b). However, seed-borne diseases are the major issue with bean cultivation. The extent of damage generally depends on the type of cultivar, prevailing environments, and above all the production area. For example, none of the above-listed diseases are a problem in dry areas except viruses and dry root rots. The warm and humid environment of the tropics and subtropics favors pathogen development, while 2–3 crop planting cycles per year^{-1} in some regions provides a continuity of inoculum. Further constraints include the small land availability that limits the possibilities for crop rotation and the scarcity of disease-free seed. Beebe and Pastor-Corrales (1991) suggest that more pathogens and more virulent isolates of these pathogens are associated with bean production in Latin America and Africa than are found in the bean-producing regions of the USA and Europe. Good seed quality, uniform emergence, early seedling vigor, and freedom from disease are prerequisites for high yield. Seed contaminants also pose a problem with seed germination and may reduce the germination by 40 % (Ellis et al. 1976; Tseng et al. 1995a, b). Seed coat characteristics also influence resistance to mechanical damage during harvest (Dickson and Boettger 1976, 1977).

The coevolution of hosts and pathogens has led to development of numerous physiologic races in most of the important pathogens listed above across the world, and the pathotypes originating on a particular gene pool primarily infect cultivars of the same gene pool, but with passage of time, complex virulences have also evolved that attack both the gene pools (Mahuku and Riascos 2004). A similar coevolution of pathogen virulence with common bean gene pools has been observed for the angular leaf spot (Guzmán et al. 1995; Pastor-Corrales and Jara 1995), anthracnose (Balardin and Kelly 1998; Islam et al. 2002; Sharma et al. 1999, 2007; Rodriguez-Guerra et al. 2003; Mahuku and Riascos 2004), common bacterial blight (Mkandawire et al. 2004), and rust pathogens (Sandlin et al. 1999). The coevolution of pathogen virulence within gene pools affects resistance-gene deployment strategies. A number of resistance genes have been identified against almost all the biotic stresses in bean gene pools and efforts made to introgress them in commercial classes. For example, more than 13 R-genes have been identified to condition resistance against bean anthracnose, of which 12 major genes *Co-2* to *Co-13* are of Middle American origin and *Co-1* is the only locus from the Andean gene pool. However, only one recessive gene *co-8* is known to impart anthracnose resistance (Kelly and Vallejo 2004; Campa et al. 2009). Resistance genes of Middle American origin are very effective when transferred to beans of Andean background and deployed in regions where Andean isolates prevail (East Africa, Colombia, Ecuador). Similarly, genes of Andean origin are very effective when transferred to beans of Middle American background and deployed in regions where isolates of Middle American origin prevail (Central America, Mexico, and the USA).

The introgression of resistance genes from both gene pools is a recognized strategy for developing improved and durable bean varieties. The VAX lines with combined resistance from *P. vulgaris* and *P. acutifolius* possess the highest level of CBB resistance developed to date (Singh et al. 2001a). AND 277 and the derived line CAL 143 represent an important breakthrough as the first Andean beans with useful levels of resistance to angular leaf spot (Aggarwal et al. 2004). The most recent breeding advance has been interspecific crosses that have yielded adapted lines (Beaver et al. 2005) with novel resistance to BGYMV derived from *P. coccineus*, a well-known but underutilized source of resistance from the secondary gene pool (Beebe and Pastor-Corrales 1991; Bianchini et al. 1994). The widespread nature and importance of *F. solani* as the predominant root rot pathogen in the common bean emphasizes the need for effective control through the development of resistant cultivars (Schneider et al. 2001; Chowdhury et al. 2002; Navarro et al. 2003). However, complex inheritance combined with genetic incompatibility between gene pools have limited attempts to incorporate *Fusarium* root rot resistance into large-seeded Andean bean cultivars, despite the existence of extensive information on sources of resistance in the Middle American gene pool (Beebe and Bliss 1981; Wallace and Wilkinson 1975). The highly variable nature of the rust pathogen, causal organism *Uromyces appendiculatus*, and the rapid breakdown of major gene resistance present in bean cultivars have challenged bean breeders working to develop durable resistance to bean rust. Pyramiding different resistance genes and mechanisms (specific, adult plant, slow rusting, reduced pustule size, and pubescence) should prolong the life of a bean cultivar by creating a more durable resistance complex (Mmbaga et al. 1996).

2.5.2 Bean Insects

The major bean insect pests including leafhoppers, thrips, weevils, whiteflies, bean flies or bean stem maggots (*Ophiomyia phaseoli* and *O. spencerella*), chrysomelid beetles (*Ootheca bennigseni* and *O. mutabilis*), pod borers (*Maruca vitrata* and *Helicoverpa armigera*), and mites (*Tetranychus cinnabarinus* Boisd., *Polyphagotarsonemus latus* Banks) attack crops at various stages (Schwartz et al. 1999); however, aphids (*Aphis fabae*, *A. craccivora*) are sometimes a problem on beans during dry spells, especially in the early stages of crop growth. The important insect pests in stored beans are the bean bruchids (*Acanthoscelides obtectus* and *Zabrotes subfasciatus*) (Schwartz and Pastor-Corraies 1989; Allen et al. 1998). Among nematodes, the root-knot nematodes (*Meloidogyne incognita* and *M. javanica*) can be a problem in sandy loam soils (Ijani et al. 2000). Insect pests attack all parts of the bean plant from the roots and lower stem through to the pods and seeds. Rotation with other susceptible crops such as tomato, pepper, and potato aggravates problems for the succeeding common bean crop (Schwartz and Pastor-Corraies 1989; Abawi and Widmer 2000; Schwartz et al. 2005).

In addition to causing direct damage to foliar and underground plant parts, insects also act as vectors of numerous common bean viruses (Schwartz et al. 2005). For example, *bean common mosaic*, *bean common mosaic necrosis*, and *bean yellow mosaic viruses* are worldwide aphid-vectored potyviruses. *Bean golden mosaic*, *beet curly top*, and *bean golden yellow mosaic viruses* occurring in Argentina, Bolivia, and Brazil found in tropical and subtropical Central America, coastal Mexico, the Caribbean, and southeastern USA are caused by whitefly-transmitted Gemini viruses. Singh and Schwartz (2010) recently reviewed advances in genetics and breeding for resistance to many of these viral and other diseases. Average yield losses caused by major insect pests and nematodes in the American continent may range from 35 to 100 % depending upon the occurrence and severity of the individual and collective insect pests and nematodes occurring in the same field, production systems and management practices used, environmental conditions, common bean cultivars grown, and crop stages when affected. The efforts towards insect-pest resistance in the common bean are still meager as compared to diseases; thus, more efforts are required for resistance breeding to insect pests and nematodes (Singh and Schwartz 2011). There are no defined methods of germplasm screening, data about genetic nature of resistance, and breeding reports against major common bean insect pests.

2.5.3 Weeds

Beans, like other annual crops, are highly susceptible to early weed competition, but their yield is also sensitive to late-emerging weeds which are favored by crop foliage loss during the reproductive stage. The critical period of weed competition is between 10 and 30–40 days after crop emergence. However, the importance and the role played by weeds vary with the production system. The weed flora prevalent in the common bean include broad-leafed species like *Amaranthus* spp., *Baltimora recta* L., *Bidens pilosa* L., *Melampodium divaricatum* DC., *Tridax procubens* L., *Chamaesyce hirta* (L.) Millsp., *Euphorbia heterophylla* L., *Mimosa pudica* L., *Portulaca oleracea* L., *Parthenium hysterophorus* L., *Solanum nigrum* L., *Commelina benghalensis,* and *Ageratum conozoides* (important grassweeds and sedges include *Cenchrus* spp., *Digitaria* spp., *Eleusine indica* (L.) Gaertn., *Echinochloa colona* (L.) Link, *Setaria* spp., *Ixophorus unisetus* (Presl) Schlecht., *Sorghum halepense* (L.) Pers., *Cynodon dactylon* (L.) Pers., *Cyperus esculentus* L., and *Cyperus rotundus* L.). Weeds can be managed by adopting suitable cultural practices like intercropping and chemical measures. Beans are commonly intercropped with maize, and interrow cultivation is useful for weed control during the first month after crop emergence and can be combined with the application of selective herbicide spray. Several herbicides can be used selectively in beans; however, the feasibility of their use depends on the farm economy. Preemergence herbicides like pendimethalin (1.0–1.5 kg ai/ha), metolachlor (92.0–2.5 kg ai/ha), and metobromuron (0.75–1.0 kg ai/ha) and postemergence herbicides like fomesafen (0.25–0.38 kg ai/ha), diclofop-methyl (1.1–1.44 kg ai/ha), and quizalofop-ethyl (0.10–0.15 kg ai/ha) can be used for annual grassweeds. The composition of weed population and most dominant weeds in bean fields vary from region to region and depend on many factors like growing environment and agronomic management. In many regions of Latin America and Africa, beans are introduced with limited tillage or pesticide application into cereal residues or slashed/burned fallow areas. While the limited inputs used will often constrain yields, Tapia and Comacho (1988) and Thurston (1992) found less damage from insect pests and pathogens in such systems, and better control of soil erosion. Hand-weeding is often used in subsistence agriculture and is critical during the period until flowering, as weeds can compete for light, nutrients, and water. Liebman et al. (1993) compared dry bean seed yield in a no-tillage system without fertilizers and herbicide, and attributed lower yields without tillage to limited N availability and competition from weeds. In further experiments with a no-tillage-rye mulch system, weed production was greater than for a conventionally tilled planting, and yield without N fertilization was much reduced (Liebman et al. 1995). N fertilization improved crop yield in both conventional and no-till systems, but the effect was much greater for the no-till system under normal rainfall conditions.

2.5.4 Abiotic Stresses

Among the most important abiotic constraints limiting bean production are drought and phosphorus and nitrogen deficiency due to poor nitrogen fixation (Rao 2001). Aluminum and manganese toxicity are also frequent problems (Thung 1990; Thung and Rao 1999; Wortmann et al. 1998). Water stress is a frequent problem in beans after rainy season (i.e. during October–November) in those regions where moderate water stress occurs frequently. High temperature (>30 °C day or 20 °C night) during anthesis or seed setting at lower elevations (<650 m) or at higher altitude during summer, especially when relative humidity is low, severely reduces bean production. Low temperature (<10 °C) and frost during the beginning and end of the growing season at higher elevations (>2,000 m amsl) significantly lower the yield of beans.

Abiotic stress resistance is by nature more complex physiologically, is typically subject to large environmental effects, and has been less well studied than biotic stress resistance in the common bean (Rao 2001). Therefore, compared to pest and disease resistance, much less is known about genetics of resistance to abiotic constraints or physiological stress. Plant response to one stress may be conditioned by the presence or absence of other stresses. For example, inadequate soil fertility or compacted soil structure may reduce root growth and thus limit the potential of the plant to express drought resistance. Abiotic stress resistance is typically governed by polygenic inheritance and may be conditioned by multiple interacting mechanisms, thus complicating the study of this stress. Yet abiotic stress tolerance may be the key to improve the common bean yields both in stressed and unstressed environments if experience in other crops is an indicator (Beebe et al. 2004).

2.5.5 Drought

Drought is a worldwide constraint in common bean production (Fairbairn 1993; Wortmann et al. 1998) and is endemic to a number of American continents in northeastern Brazil (1.5 million ha) and in the central and northern highlands of Mexico (1.5 million ha). In Central America, moderate drought towards the end of the cropping season (November and December) is a routine. In tropical and subtropical Latin America, commercial bean-growing environments experience intermittent drought which may result in complete crop failures (Acosta-Gallegos et al. 1999; Schneider et al. 1997). However, in dryland farming systems in intermountain regions of the USA, characterized by inadequate or no summer rainfall, terminal drought is more likely to affect bean production. Moderate to high drought stress reduces the biomass, number of seeds and pods, days to maturity, harvest index, seed yield, and seed weight in common bean (Acosta-Gallegos and Adams 1991; Ramirez-Vallejo and Kelly 1998). Foster et al. (1995) recorded a 41 % reduction in bean yield under moderate drought stress without altering nitrogen (N) partitioning, whereas severe drought stress reduced the yield by 92 % along with N harvest index and N- and water-use efficiency in the common bean. Root rots caused by *Macrophomina phaseolina* (Tassi) Goid., *Fusarium solani* f. sp. *phaseoli* (Burk.) Snyder & Hansen, and other fungi may aggravate drought stress. Similarly, DS bean crops may become prone to damage by leafhoppers (*Empoasca kraemeri* Ross & Moore) in the tropics and subtropics.

The effect of drought can vary when it occurs during the different stages of plant development. In general, drought mainly affects bean seed yield when it occurs during the reproductive phase of plant development. Morphological and phenological traits such as plant type, root systems, and early flowering play a major role in the adaptation of plants to specific drought conditions. For example, the bean cultivar 'Pinto Villa' has broad adaptation and yield stability in the semiarid highlands of Mexico (intermittent drought) partially due to phenotypic plasticity and the ability to continue seed filling at low night temperatures (Acosta-Gallegos et al. 1995). Beaver and Rosas (1998) found that the selection for early flowering, a greater rate of partitioning, and a shorter reproductive period permitted the selection of small red bean breeding lines having 1 week earlier maturity without sacrificing yield potential. Lines with earlier maturity would be less vulnerable to terminal drought; however, an association between early maturity and lower yields exists in some of the plant species. Rooting pattern, especially root length in lower soil strata, is an important drought-resistance mechanism (Sponchiado et al. 1989). More recently, the capacity to partition a greater proportion of carbohydrate to the seed under stress has emerged as an important trait (Rao 2001). The potential to select drought tolerance with QTL analysis and MAS was investigated by Schneider et al. (1997) in seven environments in Michigan and highland

Table 2.2 Sources of resistance to drought in different seed classes

Name	Seed color/type	Reference
Negro Vizcaya, L88-63, B98311	9/black	Acosta-Gallejos et al. (2004) and Frahm et al. (2004)
SEA 5, SEA 10, SEA 13	2/cream	Singh et al. (2001b)
RAB 651, RAB 655	6/small red	CIAT (2002)
Matterhorn	1/great northern	Kelly et al. (1999)
San Cristobal 83, ICA Palmar	6M/red mottled 5K/light red kidney	White and Singh (1991) and Rosales-Serna et al. (2004)

Mexico. The selection based on MAS was effective under severe drought in Michigan but not for moderate drought in Mexico. The genotype–environment interaction apparently affected the expression of QTL, and furthermore, genome coverage was incomplete and some unidentified QTL might have determined yield in the Mexican environments. Additional preliminary drought QTLs have been identified in bean accession BAT 477 under nonirrigated conditions at CIAT (Blair et al. 2002). Dehydration-responsive element-binding protein (DREB) genes have been identified in beans, but their significance to drought tolerance remains to be demonstrated (Galindo et al. 2003). However, if these are proven to have drought-resistance functions in the common bean, then gene-based MAS might be feasible (Ishitani et al. 2004). Sources of resistance against drought have been identified in different seed classes of common bean (Table 2.2).

Local adaptation is another important component of drought resistance, as evidenced by a common set of genotypes evaluated in several countries in the 1980s (White 1987). Conventional genetic studies of White et al. (1994) suggested that drought-resistant genotypes available at that time and selected respectively in the Mexican highlands and in Colombia did not adapt in the other environment, and that the value of drought-resistance sources as parents was closely associated with the yield in the given environment. If the component of local adaptation is greater than that of drought tolerance per se, the expression of genes or QTL associated with draft could be limited due to the inhibitory effects of local adaptation. On the other hand, the stability of drought tolerance could be achieved by enhancing levels of resistance and accumulation of "drought-specific" genes in elite lines and might become as important as local adaptation.

2.5.6 Phosphorus Deficiency

Genetic variability in the common bean has been widely documented as the capacity to produce grain in low soil phosphorus availability conditions (Rao 2001; Beebe et al. 1997). P availability is greater in the topsoil compared to subsoil; the selection for root trait QTL markers associated with adventitious rooting and topsoil foraging may enhance P acquisition. For example, Liao et al. (2004) showed that under phosphorus stress, the basal roots of bean accession G 19833 could reorient to explore more shallow soil strata where phosphorus is concentrated. Phosphorus absorption in the common bean is a complex trait with multiple mechanisms (Yan et al. 1995a, b, 1996). Higher levels of leaf phosphates might be expected to improve phosphorus use efficiency by facilitating better remobilization of phosphorus within the plant, but a major gene difference in this trait did not alter phosphorus use efficiency in the common bean (Yan et al. 2001). QTL analysis has been very useful in determining which root traits contribute to phosphorus uptake. Beebe et al. (2006) identified 26 QTLs associated with basal root development and greater P acquisition efficiency under P stress in the common bean.

2.5.7 Symbiotic Nitrogen Fixation

Dry beans have the ability to form an association with *Rhizobium* spp. which provides nitrogen fixed from the atmosphere to the bean plant. Dry beans are generally considered poor nitrogen fixers and nitrogen is applied to achieve good yield. There exists variability among genotypes to fix nitrogen. Bliss (1993) and Snoeck et al. (2003)

extensively reviewed the genetics and breeding for symbiotic nitrogen fixation. The potential to improve SNF has been amply demonstrated in the common bean (Bliss 1993; Barron et al. 1999). However, SNF improvement typically does not form a part of routine cultivar improvement programs, and incorporating selection criteria for SNF such as nodule mass, nitrogenase activity, and xylem ureide content into breeding schemes while attending to other breeding objectives remains a challenge. If MAS for SNF could be integrated into breeding programs that are already practicing MAS for other traits, this would avoid the necessity of additional phenotypic selection methodologies purely for nitrogen or nodule determination. As many as six chromosomal regions for the degree of nodulation were identified on an RFLP map based on a cross of BAT 93 and Jalo EEP558, which are Mesoamerican and Andean genotypes, respectively (Tsai et al. 1998). In tropical soils, the low availability of soil phosphorus often limits SNF, and genetic variability for the ability to fix N at low P has been reported (Vadez et al. 1999). In subsequent trials, BAT 477 displayed this trait and also exhibits superior nitrogen fixation under optimal conditions (Kipe-Nolt et al. 1993) and under drought conditions (Castellanos et al. 1996). If MAS can become the primary selection criterion of SNF and can be coordinated with the selection for other traits, MAS could stimulate a more directed selection for SNF and thus permit capitalizing on the vast body of knowledge that exists about genetic variability in SNF in the common bean.

2.6 Traits of Importance for Base Broadening

Knowledge, access, and the use of diversity available in cultivated and wild relatives of a given crop are essential for broadening the genetic base of cultivars to sustain improvement, particularly those which are not present in commercial classes. For example, the genetic base of bean cultivars within market classes is narrow and inadequate for resistance to common bacterial blight (caused by *Xanthomonas campestris* pv. *phaseoli* (Smith) Dye) and white mold (caused by *Sclerotinia sclerotiorum* (Lib.) de Bary). High levels of resistance to these and other desirable traits exist in the relatives and gene pools of *P. vulgaris*. Early maturity, adaptation to higher altitude, upright plant type, high pod quality and seed yield, and resistance to bean common mosaic virus (a potyvirus) and/or rust (*Uromyces appendiculatus*) have been bred into cultivars. However, most of the genetic variability available in the common bean races, gene pools, and wild relatives remains to be utilized. To maximize and sustain bean production, high-yielding and high-quality cultivars having less dependence on various inputs should be developed. Singh (2001) suggested a three-tiered breeding approach for the broadening of the genetic base of the common bean involving: (1) gene introgression from alien germplasm, (2) pyramiding favorable alleles from different sources, and (3) simultaneous improvement of multiple traits for common bean cultivars.

Examples of trait transfer in interracial crosses include (1) upright plant type introduced into race Durango materials from race Mesoamerica (Singh 1999; Kelly 2004); (2) transfer of *Apion godmani* resistance from race Jalisco to race Mesoamerica; and (3) introduction of the recessive *bgm-1* gene for resistance to the *bean golden mosaic virus* (BGYMV) from race Durango to race Mesoamerica (Singh 1999). *I* gene for resistance to BCMV and *Co-2* gene for resistance to anthracnose are of Mesoamerican origin (Johnson et al. 1997), yet these have been introduced into numerous cultivars of Andean gene pool. Gonzalez-Alonso et al. (1995) showed that photosynthetic rates of Mesoamerican and Andean beans are quite distinct. Mesoamerican accession have higher specific leaf mass, high chlorophyll content, high CO_2 exchange rate, higher stomatal conductance, higher stomatal density, and lower N content than Andean accessions. Andean large-seeded bean varieties achieve higher yield by maturing earlier, and Mesoamerican beans in contrast achieve higher yield by maturing later but producing a large number of pods and a higher biomass. Beaver and Kelly (1994) used recurrent selection with early-generation (F_2) and late-generation (F_5) testing to combine the medium–large seed type characteristic of Andean genotypes and indeterminate growth habit of Mesoamerican genotypes to increase the yield potential as high as 30 % to that of the determinate parent. Wild beans are a potential source of genetic diversity for arcelin seed protein which provides resistance to bean weevils (*Zabrotes subfasciatus*). Several expressed allelic variants for the arcelin locus have been identified in wild common bean populations. No expressed allelic variant was found in cultivated common bean (Osborn et al. 1988). The *Arl* gene has been successfully transferred into cultivated background (Kornegay et al. 1993).

The transfer of quantitative traits between Mesoamerican and Andean gene pools appears to be problematic. Attempts by breeders to combine traits from Andean and Mesoamerican gene pools such as larger seed size from the Andean gene pool and high yield potential from the Mesoamerican gene pool generally failed (Singh 1989) with a few exceptions (Beaver and Kelly 1994). The average performance of the progeny derived from intergene pool crosses is below that of the parents (Welsh et al. 1995). Divergence between the two gene pools may have led to the independent suites of genes controlling the overall performance in Mesoamerican and Andean gene pools. Hybridization between gene pools breaks up these suites which are difficult to reconstitute in the progeny because of recombination and lack of adaptation. Many of the concerns raised about the introgression between Mesoamerican and Andean cultivated gene pools are also valid for the introgression between wild and cultivated gene pools. There are additional concerns that wild beans lack adaptation because of their photoperiod x temperature requirements and to the field cultivation, due to lack of domestication syndrome traits (Koinange et al. 1996). The selection of determinacy and insensitivity to photoperiod, either directly through phenotypic means or indirectly by molecular markers, could accelerate the introgression from wild beans to cultivated ones. These two traits are of prime importance in any germplasm conversion program aimed to broaden the genetic basis of the cultivated common beans.

2.7 Interspecific Hybridization in Crop Species

Interspecific hybridization among *Phaseolus* species: The genus *Phaseolus* is Neotropical in origin. Although we know clearly what a bean is, it is less certain how many *Phaseolus* species exist. A reasonable estimate would be 50–60 species, pending additional germplasm explorations in Central America (Debouck 2000). Understanding the relationships between the species is a question of practical importance to explore the variability aspect. Recent phylogenetic studies that included both wild and domesticated species of *Phaseolus* using morphological, biochemical, and molecular data (seed proteins, isozymes and nuclear, chloroplastic and mitochondrial DNA, etc.) have confirmed that the genus is monophyletic (Debouck 1999). Two to nine subclades may exist at the subgeneric level (Baudoin et al. 1998; Delgado-Salinas et al. 1999). One lineage includes the common bean, while another encompasses *P. lunatus* (Fofana et al. 1999, 2001; Maquet et al. 1999). Three species, *P. coccineus*, *P. polyanthus*, and *P. vulgaris*, belong to the same evolutionary branch (Schmit et al. 1993, 1995). However, differences emerge between the number, kind of taxa, and type of DNA examined.

Of the 50–60 wild *Phaseolus* species of American origin, only five, namely, common (*P. vulgaris*), yearlong (*P. polyanthus*), scarlet runner (*P. coccineus*), tepary (*P. acutifolius*), and lima bean (*P. lunatus*), have been domesticated. Each domesticated species constitutes a primary gene pool with its wild ancestral form. Secondary and tertiary gene pools may exist for all the domesticated species, depending on the phylogenetic events that lead to the formation of the biological species (Debouck 1999). Recently, a novel wild species, *P. costaricensis*, was shown to belong to the secondary gene pool of *P. vulgaris*. *P. costaricensis* is only known from Costa Rica and Panama. There are over 29,000 domesticated and more than 1,300 wild accessions of *P. vulgaris* housed in the germplasm bank at CIAT, Cali, Colombia, and elsewhere. At CIAT, the numbers of accessions belonging to the secondary and tertiary gene pools, respectively, are 1,049 and 335. In spite of this diversity, the genetic base of commercial cultivars of specific market classes is narrow. In fact, only a small portion (<5 % of the available genetic diversity) has been used globally despite nearly a century of organized bean improvement.

Systematic evaluation of wild common bean as well as wild and domesticated germplasm of alien species for resistance to pests, diseases, and other useful traits has been limited. Nevertheless, alien germplasm seems to be a promising source of common bean improvement as resistance to bruchids was found in wild *P. vulgaris*. *P. polyanthus* is well known for its resistance to ascochyta blight as well as to BGYMV. *P. coccineus* is a source of resistance to anthracnose as well as root rots, white mold, BYMV (*bean yellow mosaic virus*), and BGYMV. Tolerance to leafhoppers exists in *P. acutifolius*, and high levels of resistance to CBB (Common Bacterial Blight) and bruchids are found in some accessions of tepary bean (Baudoin 2001; Debouck 1999; Schmit and Baudoin 1992; Singh 1999).

Thus, major production constraints include the lack of resistance to diseases/pests; the slow progress in identifying useful genes in related species has led to the adoption of intraspecific hybridizations among *Phaseolus* species. However, *P. vulgaris* and *P. coccineus* were frequently intercrossed between 1940 and 1985. It was observed that in reciprocal crosses using *P. coccineus* as the female parent, segregants naturally reverted to the cytoplasm donor parent after a few generations (Baudoin et al. 1995). Major genes have established a barrier between these two species, and chromosome pairing is not perfect. Since reproductive isolation may be due to domestication, attempts were made to cross *P. vulgaris* with wild variants of *P. coccineus*. Nevertheless, few commercial cultivars have been created this way. *P. polyanthus* crosses more easily with *P. coccineus* and related forms than with *P. vulgaris*, particularly if the latter is the pollen donor (Baudoin et al. 2001). *P. polyanthus* belongs to the *P. vulgaris* clade, but its nuclear genome has been introgressed with *P. coccineus* genes and this limits its use in interspecific hybridizations. Especially when *P. vulgaris* is used as the female parent, crosses between *P. vulgaris* and *P. costaricensis* are simple to perform without embryo rescue, but it is not clear whether *P. coccineus* genes have contaminated the nuclear genome of *P. costaricensis* (Debouck 1999). *P. coccineus* and its allies may thus be the reservoir of diversity with the greatest potential once the primary gene pool and the *P. vulgaris* phylum have been fully exploited. An unrealized dream of combining the potential of lima bean (part of the quaternary gene pool) that is well adapted to tropical conditions with the genome of the common bean has failed to produce fertile hybrids.

2.8 Barriers to Interspecific Hybridization

The major reproductive barrier in interspecific hybridization among the genus *Phaseolus* occurs mainly during early embryo development after fertilization (Baudoin et al. 1995). When maintained in vivo, embryos resulting from *P. polyanthus* (female) × *P. vulgaris* crosses develop poorly despite the close phylogenetic relationship among species. Infertility in *P. polyanthus* × *P. vulgaris* crosses results from early nutritional barriers that are related to a deficient endosperm tissue development, while in reciprocal crosses, endothelium proliferation and, to some extent, hypertrophy of the vascular elements are causes of early embryo abortion (Geerts 2001; Geerts et al. 1999, 2002; Lecomte et al. 1998). To a large extent, the importance of these abnormalities depends on the compatibility between the parent genotypes. Although several hybrids between *P. vulgaris* and species belonging to its tertiary gene pool can only be obtained by embryo rescue, most infertility results from male sterility caused by incomplete chromosomal pairing in Metaphase I (Baudoin et al. 1995). Where sterility of hybrids precludes any form of introgression, traditional chromosome doubling yields weak semi-fertile amphidiploids.

Both inter- and intraspecies barriers to hybridization and genetic recombination are of great consequence for genetic and breeding studies of *P. vulgaris*. Marechal (1971) and Pratt et al. (1985) reported that genic factors, in addition to varying degrees of chromosomal differences among

Phaseolus species, constitute an important barrier to interspecific hybridization and exploitation of genetic variability. Hucl and Scoles (1985) have reviewed interspecific hybridization among different species of the genus *Phaseolus*. Smartt (1970) attempted interspecific hybridization between the species *Phaseolus vulgaris*, *P. coccineus*, *P. acutifolius*, and *P. lunatus* and found a close genetic relationship between *P. vulgaris* and *P. coccineus*. Viable hybrids of varying fertility can be made reciprocally between these species. Affinity is indicated with *P. acutifolius* by the formation of viable but sterile hybrids. Less affinity is apparent between these three species and *P. lunatus* with which they do not produce viable interspecific hybrids. The effects of cytoplasmic constitution on both fertility and viability were shown in the F_2 and backcross progenies of *P. vulgaris* × *P. coccineus* hybrids and on viability in the *P. vulgaris* × *P. acutifolius* and reciprocal hybrids. In slightly fertile *P. vulgaris* × *P. coccineus* hybrids, the distortion in F_2 segregations was due to differential loss of *P. coccineus* genes in sporogenesis and embryogenesis occurring in a *P. vulgaris* plasmon. Bemis and Kedar (1961) proposed a three-allele model to account for the occurrence of differential crossability between particular *P. vulgaris* and *P. coccineus* genotypes. Dimorphic segregation for normal versus abnormal plants showing dwarfing was also observed in the cross of *P. vulgaris* × *P. coccineus* by Zagorcheva and Poriazov (1983). Various degrees of incompatibility ranging in the expression from pale green foliage to virescent, variegated, crippled, or stunted growth, semilethal chlorosis, and death of stunted plants have been reported in intervarietal crosses within cultivated common bean. The occurrence of abnormal F_1 weakness or dwarfism within *P. vulgaris* crosses have been reviewed (Gepts and Bliss 1985; Singh and Gutierrez 1984). Shii et al. (1980) and van Reheenen et al. (1979) reported that F_1 abnormality was controlled by two complementary dominant genes which suppress the growth and root development. The allelic dosage of *Dl1* in the root and *Dl2* in the shoot rather than the genotype of the whole plant determines the severity of expression. Gepts and Bliss (1985) found that a small-seeded parent carrying *Dl1* gene possessed "S" phaseolin type, which is predominant in Middle American forms. Medium- and large-seeded parents carrying *Dl2* gene had either "T" or "C" phaseolin pattern, a characteristic of beans originated in Andean South America. Noncarriers of *Dl1* and *DL2* produced a higher frequency of compatible hybrids between *P. vulgaris* and *P. acutifolius* crosses (Parker and Michaels 1986). Thus, perhaps the same loci are responsible for at least some of the intra- and interspecific genetic barriers to hybridization and gene exchange.

2.9 Conventional and Contemporary Approaches of Interspecific Gene Transfer

Plant breeding programs have used different crossing methods to integrate economically important traits from wild and related species into cultivated crops. Some approaches which have been used by various workers to transfer genes across the *Phaseolus* spp. are described below.

2.9.1 Backcrossing

In distant crosses, often useful recombinants are not obtained in early generations due to delayed segregation, and populations are required to be advanced up to F_9–F_{10} generations to get desirable recombinants. This problem can be overcome by backcrossing F_1 hybrids with cultivated species in early generations. With recurrent backcrossing, the direction of the initial cross uses the cultivated species as the maternal parent; the F_1 primary hybrid and every subsequent generation are backcrossed to the cultivated parent species. The goal of such a mating scheme is to incorporate the economically important trait from the wild species into the cultigen without any of the traits from the wild species that would be detrimental or unsuitable in cultivation. When *Phaseolus vulgaris* is crossed with *P. acutifolius* and *P. parvifolius*, one or more backcrosses to the recurrent parent are required to restore the fertility of the hybrids. Interspecific crosses to introgress high iron from related species appears to hold promise, especially for the Mesoamerican beans where it has been difficult to reach levels of iron of 90^+ ppm.

CIAT has made interspecific crosses with high iron accessions of *P. dumosus* (*P. polyanthus*) and *P. coccineus* which have expressed as high as 127 ppm iron in grain harvested under greenhouse conditions (although field-harvested grain is often lower in iron). One backcross of the interspecific F_1 has been performed, and so far, a small number of F3 families average nearly 20 ppm more than high-iron checks, while the range among families has been as high as 98 ppm. Although interspecific crosses with an Andean bean (CAL 96) were also created, these did not show evidence of significant introgression of the high-iron trait (Blair et al. 2009b).

2.9.2 Congruity Backcrossing

A slight modification to backcrossing or recurrent backcrossing is congruity backcrossing. Recurrent backcrossing has only infrequently been capable of such a task; frequent problems are the loss of the important trait(s) from the nonrecurrent parent, difficulties in transferring traits that are quantitatively inherited, chromosome or genome elimination, and large linkage groups that are difficult to break. Haghighi and Ascher (1988) were the first to report the use of congruity backcrossing as a useful means to create an interspecific hybrid between *P. vulgaris* and *P. acutifolius*. In this crossing scheme, the hybrids are backcrossed with each of the parent species in alternate generations. This eliminates the problems noted above with the many forms of recurrent backcrossing as both genomes are constantly incorporated and less, if any, traits are lost. In interspecific crosses between the common bean (*P. vulgaris*) and tepary bean (*P. acutifolius*), congruity backcrossing is a better approach as compared to recurrent backcrossing (Mejia-Jimenez et al. 1994).

2.9.3 Advanced Backcross-QTL Strategy

Since the mid-1990s, convincing evidence at morphological and molecular levels has accumulated for the utility of wild progenitors and related species as donors of productivity alleles. Nobel breeding strategies such as AB-QTL have been deployed to exploit the worth of progenitor and related species as this helps to minimize the linkage drag associated with gene introgression (Tanksley and Nelson 1996). QTL analysis of yield traits in an advanced backcross population derived from a cultivated Andean and wild common bean has been done by Blair et al. (2006b).

2.9.4 Reciprocal Crossing

In wide crosses, disharmony between the nucleus of one species and cytoplasm of other species is often a reason for postfertilization barrier, and under such conditions, reciprocal crosses can be exploited in developing hybrids. Reciprocal differences are very common in wide crosses. Mejia- Jimenez et al. (1994) reported that when *P. acutifolius* is used as female parent in a cross with *P. vulgaris* and or first backcrossing of *P. vulgaris* × *P. acutifolius* hybrid on to *P. acutifolius,* is often more difficult than *P. vulgaris* used as female parent.

2.9.5 Bridge Species

In the common bean, genes for various biotic and abiotic stresses are available in secondary and tertiary gene pools which are difficult to transfer into cultivated species. So when direct hybridization between cultivated and wild species does not result in fertile hybrids, the involvement of a third species as a bridge species has often been used for the introgression of genes across the species. Parker and Michaels (1986) noted that a bridge crossing scheme would facilitate the introgression of *P. acutifolius* germplasm into incompatible *P. vulgaris* lines.

2.9.6 Gamete Selection and Recurrent Selection

Gamete selection using multiple-parent crosses (Singh et al. 1998; Asensio et al. 2005, 2006;

Terán and Singh 2009) and recurrent selection (Singh et al. 1999; Terán and Singh 2010a, b), respectively, may be more effective. But the use of these methods is not common in the common bean because of the large number of pollinations required and other demands on resources.

2.9.7 Growth Hormones

In distant crosses, if hybrid seeds die when their embryos are too small to be cultured, post-pollination application of growth regulators such as gibberellic acid, naphthalene acetic acid, kinetin, or 2, 4-D dimethylamine, singly or in combination, may be helpful in maintaining the developing seeds by facilitating the division of the hybrid zygote and endosperm. Some interspecific crosses have been successful in *Phaseolus* (Stalker 1980) by the application of growth regulators after pollination. This suggests that further breakthroughs in wide crossing may be possible through the exploitation of growth regulators followed by embryo rescue.

2.9.8 Embryo Culture

A range of in vitro methods have been developed to overcome postfertilization barriers in a number of plant species. When the mismatch between embryo and endosperm development starts very early and ovary culture and/or ovary-slice culture fails, ovules can be cultured in vitro. If young fruits can stay for a longer time on the mother plant, embryo rescue can be applied by different methods, i.e., *in ovulo* rescue and embryo rescue. Embryo culture can be applied successfully in crosses in which pollinated flowers can stay on the plant for a notable time, before natural abscission occurs and the technique has been applied in a large number of crops. Ovary culture and ovary-slice culture has been used in *Phaseolus* by Sabja et al. (1990) during interspecific hybridization of *Phaseolus vulgaris* × *P. acutifolius* by using pods 10 days after pollination (DAP) in a modified liquid Murashige and Skoog (MS) medium (Murashige and Skoog 1962). Ninety percent of the ovary culture generated embryos though survived, but viable hybrid plants were produced only from the ovaries collected 14 DAP and onward (Sabja et al. 1990). In most cases, ovary culture is applied during the growth of a small undifferentiated embryo inside the ovule, and the differentiated embryo is used for culture. *P. parvifolius* crosses easily with *P. acutifolius* and *P. vulgaris* using the embryo rescue technique. Other species such as *P. filiformis* and *P. angustissimus* have also been crossed with the common bean, but embryo-rescued hybrid plants were completely sterile (Baudoin 2001). Although several hybrids between *P. vulgaris* and species belonging to its tertiary gene pool can only be obtained by embryo rescue, most infertility results from male sterility which is caused by incomplete chromosomal pairing in metaphase I (Baudoin et al. 1995). Where sterility of hybrids precludes any form of introgression, traditional chromosome doubling yields weak semi-fertile amphidiploids.

2.9.9 Androgenesis and Chromosome Doubling

Peters et al. (1977) reported an equal amount of haploid and diploid callus cells with less than 3 % cells showing polyploidy. Tai and Cheng (1990) cultured the common bean anthers on B5 medium with 2 mg/l 2, 4-D and 1 mg/l kinetin. Experiments on bean anther culture with modification in MS base medium have also been conducted by other workers (Munoz and Baudoin 2002; Munoz et al. 1993; Munoz-Florez and Baudoin 1994). Callus cells during early growth stages were predominantly haploid, but with age, ploidy levels increased, indicating spontaneous chromosome doubling. Interspecific F_1 hybrids may display sterility owing to lack of chromosome pairing during meiosis and hampers further breeding. However, the somatic (mitotic) chromosome doubling may induce homologous pairing of chromosomes and restores fertility (Hermsen 1984). Colchicine was used successfully to produce fertile allotetraploids in *Phaseolus* (Weilenmann et al. 1986). Allopolyploids may function as fertile bridges for gene introgression into the cultivar assortment (Hermsen 1984; Nandakumar and Shivanna 1993). Chromosome doubling has been attempted to overcome incompatibility barriers but may not be very useful, given the difficulty of exploiting amphidiploids (Broughton et al. 2003).

2.9.10 Transformation

Conventional breeding has contributed significantly to the trait improvement of *P. vulgaris*. However, breeding cannot add certain genes that do not exist naturally in the *P. vulgaris* gene pool. Due to this limitation of plant breeding, new trait improvement approaches such as interspecific horizontal gene transfer via genetic engineering need to be utilized in order to complement the limitations encountered by conventional breeding of this crop (Aragão et al. 2002). The common bean has been transformed using two main approaches: particle bombardment or biolistics and *Agrobacterium* transformation. The applicability of the particle bombardment to introduce and transiently express genes into dry bean was demonstrated in the beginning of the 1990s (Genga et al. 1992; Aragão et al. 1992, 1993). The bombardment of meristematic cells of embryonic axes revealed that foreign genes could efficiently reach the superficial cell layers, demonstrating the feasibility of producing transgenic bean using this method (Aragão et al. 1993). Russell et al. (1993) reported the production of transgenic navy bean plants using an electrical particle acceleration device. However, the protocol was time-consuming and the frequency of transformation was low (0.03 %) and variety-limited. Later, Kim and Minamikawa (1996) reported the recovery of transgenic bean plants (cv. Goldstar) by bombarding embryonic axes. A transformation system was developed for the regeneration of transgenic bean plants based on the bombardment of apical meristematic regions associated with an efficient tissue culture protocol for multiple shoot induction, elongation, and rooting (Aragão et al. 1996). The frequency of transformation ranged from 0.2 to 0.9 % using linear or circular plasmid vectors, respectively (Aragão et al. 1996; Vianna et al. 2004). One important constraint in the transformation system based on the bombardment of meristematic tissue of embryonic axes is the difficulty of having efficient selection of transformed cells. Using the selective agent imazapyr, an herbicidal molecule of the imidazolinone class capable of systemically translocating and concentrating in the apical meristematic region of the plant, we have been able to increase the recovery of fertile soybean (Aragão et al. 2000) and, more recently, dry bean plants (Bonfim et al. 2007). Since the first report on *P. vulgaris* transformation in 1993, a few groups have transformed the bean to introduce agronomic traits. Russell et al. (1993) introduced into the bean the *bar* gene, which encodes the phosphinothricin acetyltransferase (PAT) and confers resistance to both phosphinothricin and the herbicide glufosinate ammonium, and the coat protein gene from the bean golden mosaic geminivirus (BGMV) in an attempt to produce virus-resistant plants. The introduced *bar* gene was shown to confer resistance to the herbicide under glasshouse conditions. However, transgenic bean plants expressing the BGMV coat protein gene did not exhibit virus resistance (D. Maxwell, unpublished results). Faria et al. (2006) reported highly BGMV-resistant plants by the expression of a mutated *AC1* viral gene and using interfering RNA (RNAi). Transgenic bean lines containing the *bar* gene and resistant to the herbicide glufosinate ammonium were tested in field (Aragão et al. 2002). This was the first field release of a transgenic line of *P. vulgaris*.

Agrobacterium tumefaciens-assisted transformation is a standard protocol successfully applied to generate transgenic legume plants of some *Phaseolus* species (Zambre et al. 2005), although the frequency of transformation for *Phaseolus vulgaris* (common bean) has been very low (Zhang et al. 1997). The only report where *A. tumefaciens* was used to achieve expression of a *lea* gene that conferred abiotic tolerance in *P. vulgaris* is a very recent protocol based on sonication and vacuum assistance (Liu et al. 2005). Although *A. rhizogenes*-mediated root transformation has been described for a number of legumes, the induction of composite plants has not been previously reported for the common bean (Díaz et al. 2000; Boisson-Dernier et al. 2001; and Van de Velde et al. 2003). Transgenic roots in the genus *Phaseolus* offers a new strategy for overexpressing or suppressing endogenous genes. This method can be readily scaled up to perform functional genomics focused on root biology and root–microbe interactions.

It is thus not surprising that attempts to transfer polygenetic traits from related species to *P. vulgaris* met with limited success. Yet less than 5 % of the *Phaseolus* germplasm has been used in hybridization programs. New ways of enhancing introgression (congruity backcrossing, single seed descent, and recurrent selection) promise to restore fertility, as well as to augment the frequency of desirable genes in the breeding population. The Phaseomics program of CGIAR has envisaged that the availability of molecular markers, developing more detailed linkage maps, coupled with marker-assisted selection will facilitate the retention of desirable genes and the elimination of harmful ones and remove restrictions on interspecific hybridizations.

2.10 Molecular Markers, Genome Mapping and Genomics as an Adjunct to Breeding

In the last 25 years, an impressive array of DNA markers has been/are being developed that display an ever-increasing ease of application. In the common bean, markers have been used not only to understand the organization of genetic diversity but also increasingly to map and tag genes of agronomic interest, particularly genes for disease resistance (Gepts et al. 1999; Freyre et al. 1998; Faleiro et al. 2004). Mapping and tagging of genes for resistance has facilitated our understanding of the inheritance of several diseases, previously considered to be very difficult to deal with including white mold (Miklas et al. 2001), bean gold mosaic (Urrea et al. 1996), and common bacterial blight (Kelly 2004). Markers have also been very useful to circumvent epistatic interactions occurring in genotypes with multiple resistance genes as is the case with the landrace G2333 (Young et al. 1998; Alzate-Marin et al. 2001; Pathania et al. 2006).

The most favored markers currently are microsatellite markers (SSRs) and single-nucleotide polymorphisms (SNPs) because they require little DNA and can be automated for high-throughput analysis. A number of SSR markers have been developed (Yu et al. 2000; Gaitán-Solís et al. 2002; Blair et al. 2003) and mapped on common bean chromosomes (Miklas et al. 2006b). Further efforts are now necessary to increase the number of microsatellite markers located on the common bean linkage map. Recently, a large set of expressed sequence tags (ESTs) has been published (Ramírez et al. 2005). Because the bean genotypes involved originate in the Mesoamerican and Andean gene pools, respectively, a sequence comparison will reveal SNPs.

Most genes to date have been tagged with RAPD markers which have been converted to SCAR markers (Miklas 2005) with few exceptions, for example, the *Ur-13* gene tagged with an AFLP marker (Mienie et al. 2005) and a QTL conferring resistance to white mold linked with the phaseolin seed protein locus (*Phs*) (Miklas et al. 2001). A few codominant RAPD markers have been generated as for *bgm-1* (Urrea et al. 1996), *bc-3* (Johnson et al. 1997), and $Co\text{-}4^2$ (Awale and Kelly 2001), but most markers identified have been dominant and in coupling phase linkage (*cis*) with the R genes and QTL. The predominance of dominant markers reduces the efficiency of marker-assisted selection. The use of dominant markers in repulsion phase (*trans*) linkage alone or in combination with markers in coupling can be done to improve selection efficiency (Kelly and Miklas 1998). Codominant markers like SSRs accelerate the identification of homozygous individuals without the necessity for progeny tests, but still too few exist to be effective for gene tagging studies. Ideally, the selection of markers flanking the resistance gene would be used to improve marker-assisted selection (MAS) efficiency, but such flanking markers are generally not available for the genes and QTL that have been tagged thus far in the common bean. For marker-assisted backcrossing, the lack of flanking markers can be partially overcome by also selecting for the recurrent parent genomic background outside the target locus using random markers (Hagiwara et al. 2001). Other gene sequence-based marker systems like RGA (López et al. 2003), resistance gene analog polymorphism (Mutlu et al. 2006), and TRAP (Miklas et al. 2006a) show promise for tagging targeted traits but have not yet led to MAS. As EST databases expand for the common bean, concurrently STS and SNP markers will be generated, leading to additional and more direct marker systems for tagging genes. The utility of tightly linked markers for MAS has been evaluated by surveying an array of accessions with and without the gene for the presence and absence of the marker. These surveys have revealed the utility for many markers to be limited to specific gene pools (Miklas et al. 1993). Once limitations of a marker are understood, MAS can be implemented more effectively as a breeding tool. The utility of QTL for MAS is determined by the expression of the QTL in multiple environments and in different populations and by the effectiveness of MAS itself.

Multiple disease-resistant Mesoamerican bean lines possessing *Co-4*, *Co-6*, *Co-10*, *Ur-OuroNegro*, and *Phg-1*R genes with "carioca-type" grains have been developed through MAS (Ragagnin et al. 2005; Alzate-Marin et al. 2005). Similarly, MAS was used to develop black and red bean cultivars with multiple disease resistance (Costa et al. 2006). Ultimately, sequencing the bean genome and anchoring of the sequence contigs on genetic and physical maps will be the major source of markers for genetic analysis and marker-assisted selection in the common bean. The first sequence contig from a bacterial artificial chromosome (BAC) clone in the common bean was obtained by Melotto et al. (2004). Finally, the use of bioinformatic tools applied to genomic sequence information from other species, especially legumes such *Medicago truncatula*, *Lotus japonicus*, and *Glycine max* but also nonlegumes such as *Arabidopsis thaliana*, will no doubt also provide additional sequence information and, hence, PCR-based molecular markers.

2.10.1 Genetic Mapping

Linkage mapping in common beans as in other crops has benefited from a range of molecular technologies that have greatly supplemented the number of genetic markers used in genetic maps for the species. As a result, a large number of markers are in use today for common bean mapping compared to early linkage maps that were based almost entirely on a limited number of morphological markers such as those for flower or seed color or certain pod traits and growth habit characteristics (Bassett 1991). Isozymes and seed proteins, both analyzed based on biochemical assays, were among the first molecular markers to be used in genetic maps but suffered from the limitation of requiring multiple protocols and different source tissues (Gepts 1988; Arndt and Gepts 1989; Koenig and Gepts 1989a; Vallejos and Chase 1991). With the advent of DNA technologies, RFLP markers emerged as the first DNA-based markers in common beans to be used on a large scale (Vallejos et al. 1992; Nodari et al. 1993a; Adam-Blondon et al. 1994). New mapping populations were created at UC Davis and the University of Florida (UF) based on the crosses BAT93 × Jalo EEP558 (F_2) and XR-235-1-1 × Calima (BC_1) to determine the linkage relationships of large numbers of RFLP markers with totals of 224Bng and 108D series clones mapped in these studies. As single-copy markers, RFLPs were especially useful for comparative mapping and map integration and led to the creation of the core linkage map of Freyre et al. (1998). In that study, RFLP probes from the initial UC Davis and UF maps plus those of Adam-Blondon et al. (1994) were analyzed to create a core map for the BAT93 × Jalo EEP558 (RIL) population that combined RFLPs with additional marker types including RAPDs, allozymes, and known genes. RFLPs were also useful for comparative mapping across legume species as they hybridized well to genomic DNA of soybean and mung bean (Boutin et al. 1995). The sequencing of the Bng probes used to establish the UF map showed them to be rich in gene sequences (Murray et al. 2002).

Polymerase chain reaction (PCR)-based markers added greatly to the efficiency of genetic mapping in common beans by increasing marker throughput, initially with multi-locus marker systems such as those of the RAPD and then AFLP techniques and more recently with single-copy marker systems such as sequence-tagged sites/sequence-characterized amplified regions (STSs/SCARs) and SSR primer pairs. RAPD and AFLP markers were especially useful for saturating RFLP-anchored genetic maps and for creating genetic maps for additional populations (Freyre et al. 1998; Johnson and Gepts 2002; Tar'an et al. 2002). As genetic maps became more saturated and based on a wide range of molecular marker types, the number of linkage groups coalesced to equal the haploid complement of chromosomes for beans ($n = 11$). The cumulative genetic distance represented by linkage maps also increased as a greater number of markers was used. Comparative mapping based on RAPD bands was frequently useful for linking genetic maps by correlating the presence of fragments with the evolutionary origin of parental genotypes in either the Mesoamerican or Andean gene pools. The advent of single-copy PCR-based markers such as SSR and STS/SCAR markers has further aided in map comparison and integrating genetic maps. Over 30 STS or SCAR markers developed for many different disease or insect resistance genes and seed color traits have been placed on the core genetic map (McClean et al. 2002; Blair et al. 2006c; Miklas et al. 2006b) (Fig. 2.2).

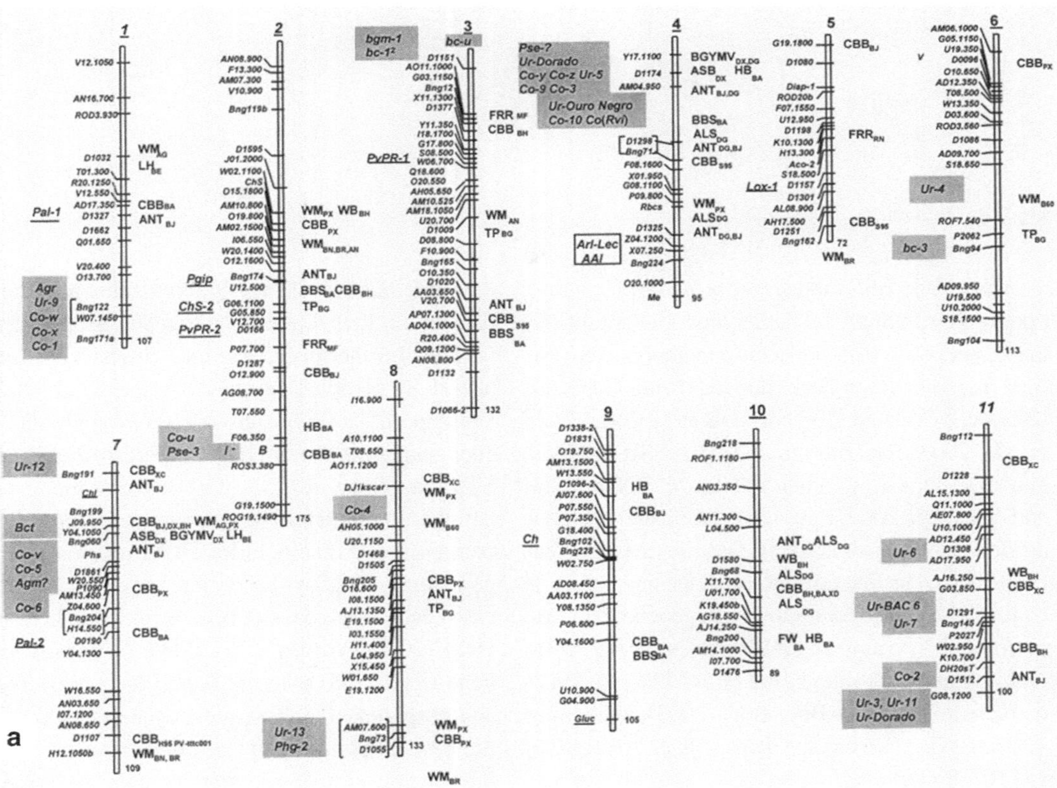

Fig. 2.2 Comprehensive genomic map of the genes of known function for external phenotypic traits, and QTL in common bean, including genes with a biochemical function (mostly disease response genes) (**a**), the domestication genes and color genes (**b**) in common bean [Abstracted from Kelly et al. 2003]. Each linkage group is represented by its core map version (Freyre et al. 1998). To the left of each linkage group, are the framework molecular markers (smaller font) and the biochemical genes (larger font) and major phenotypic trait genes (*shaded boxes*) (Source: Miklas et al. 2006a, b) (**a & b**: Two maps to fit all the genes and QTL on the map properly)

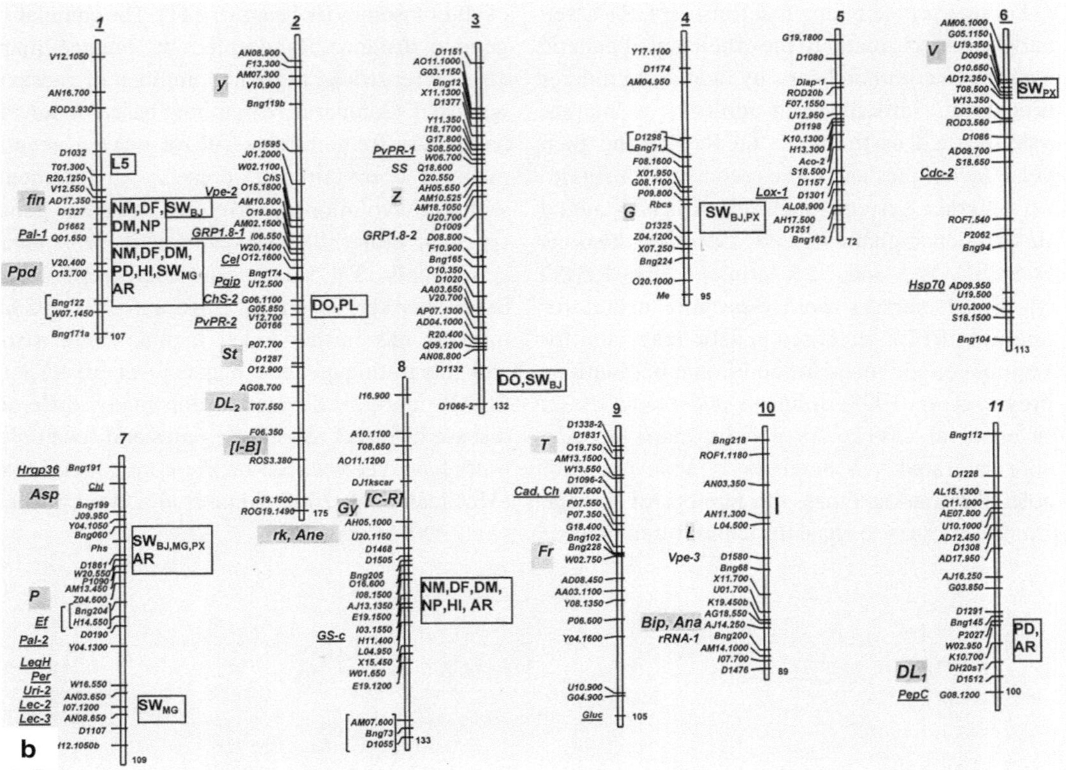

Fig. 2.2 (continued)

Meanwhile, SSR markers are proving informative due to their multiallelic nature, codominant inheritance, and wide distribution in the genome. So far, SSR markers have been derived from GenBank sequences (Yu et al. 2000; Métais et al. 2002; Blair et al. 2003; Guerra-Sanz 2004), SSR-enriched genomic libraries (Gaitán-Solís et al. 2002; Yaish and de la Vega 2003; Buso et al. 2006), and bacterial artificial chromosome (BAC) sequences (Caixeta et al. 2005). The first effort towards the integration of SSR markers into a common bean genetic map was performed by Yu et al. (2000), mapping 15 SSRs onto a framework map based on RAPD and RFLP markers, followed by Blair et al. (2003), integrating 100 SSRs into two linkage maps with AFLP, RAPD, and RFLP markers.

Additional integration of new SSR loci is ongoing at Empresa Brasileira de Pesquisas Agropecuarias Arroz e Feijão (Santo Antônio de Goiás, GO, Brazil) (for BAT93 × Jalo EEP558) and CIAT (for DOR364 × G19833) and will allow the identification of more polymorphic and transferable markers that can be mapped across populations. Furthermore, existing SSRs have been useful in mapping studies (Ochoa et al. 2006; Blair et al. 2006c) showing their potential for anchoring new genetic maps. Apart from SSR and STS markers, a new set of gene-based markers is being implemented for genetic mapping in the common bean that includes TRAP and resistance-gene analog (RGA)-based markers targeting disease resistance genes (López et al. 2003; Mutlu et al. 2006; Miklas et al. 2006a). Meanwhile, gene-based SNP markers being mapped at CIAT and mapping of EST sequences at North Dakota State University and at the University of Saskatchewan hold promise for the development of a transcriptional map for beans that could help in the establishment of correlations between candidate genes and specific QTLs. For the future, the use of reference populations, anchor markers, and comparative mapping between populations will continue to assist in the placement of new markers onto genetic maps and the comparison of genetic distances between markers. Comparative mapping has become important for correlating locations of QTL for biotic or abiotic stress resistance, nitrogen fixation, seed characteristics, and root traits (Nodari et al. 1993b; Tar'an et al. 2002; Beebe et al. 2006; Miklas et al. 2006b; Ochoa et al. 2006). In addition, the use of anchor markers, especially those that are highly polymorphic such as SSRs, will allow the genetic mapping of more narrow crosses that are often of interest to plant breeders.

2.10.2 Expressed Sequence Tags

Partial sequencing of cDNA inserts or expressed ESTs, obtained from many tissues and organs, has been used as an effective method for gene discovery, molecular marker generation, and transcription pattern characterization. It is an efficient approach for identifying a large number of plant genes expressed during the different developmental stages and in response to a variety of environmental conditions. Contributions to the development of ESTs come from collaborative projects performed within the frame of Phaseomics among scientists from academic institutions in Brazil, Colombia, Mexico, and the USA.

The complete EST data set, generated by Ramírez et al. (2005) and Melotto et al. (2005), has been assembled into contigs to provide a single *P. vulgaris* gene index containing 20,578 ESTs (Graham et al. 2006). The *P. vulgaris* EST sequence resource developed has provided tools for projects oriented to characterize the transcript profile of different bean organs and/or growth conditions. For instance, macroarray technology was used for transcriptome analysis of bean mature nodules elicited by the nitrogen-fixing bacteria: *Rhizobium tropici* (Ramírez et al. 2005). This study led to the identification of genes overexpressed in nodules as compared to other plant organs such as roots, leaves, stems, and pods. Nodule transcript profile showed that genes related to nitrogen and carbon metabolism are integrated for ureide production (Ramírez et al. 2005).

Because genetic and genomic resources of the common bean are limited, a comparative genomic approach using bean ESTs and those of other plant species is important to determine the usefulness of sequence information from other plants to study the common bean. Bean unigenes share more orthologous sequences with soybean and *Arabidopsisthaliana* unigenes than with *Orzya sativa* and *Zea mays* unigenes (Melotto et al. 2005). However, the degree of conservation among genes of bean, soybean, and *A. thaliana* varies between functional categories (Melotto et al. 2005). Single-linkage clustering analysis of homologous sequences among five different plant species (*P. vulgaris*, *Lupinus albus*, soybean, *M. truncatula*, and *A. thaliana*) has been used to identify candidate genes that are relevant for adaptation to phosphorus deficiency in the bean and are shared with other species (Graham et al. 2006).

2.10.3 TILLING

Targeted induced local lesions in genomes (TILLING) is a powerful reverse genetic approach that uses gene-specific primers for the identification of mutants of a gene of interest from a large mutagenesis population (McCallum et al. 2000). Theoretically, given a sufficient population size, genome saturation can be achieved and mutants for any gene can be identified. Tagged mutagenesis, such as T-DNA or transposon-based systems, generally requires an effective transformation system. TILLING, however, does not rely on transformation and thus allows for reverse genetic approaches in transformation recalcitrant species, such as the common bean. Significant advances have been made in the development of a TILLING platform in the common bean. The common bean genotype BAT93, from the Mesoamerican gene pool, was selected for TILLING by the common bean research community because it is one of the parents of the core genetic map (Freyre et al. 1998), it has been used in the generation of a large BAC library (Kami et al. 2006), has broad adaptation, and has desirable characteristics, such as disease resistance (Broughton et al. 2003). Based on genome saturation in other species, such as *Lotus japonicus* (Perry et al. 2003), it was estimated that about 5,000 mutagenesis lines would be required for genome saturation in BAT93.

The TILLING consortium has currently produced about 1,500 M2 families and has advanced about 900 families to the M3 generation. DNA has been extracted from all M2 families, and these DNAs will be used to confirm appropriate mutagen concentration and for TILLING analysis. Once the sequences of *P. vulgaris* loci of interest are available, gene-specific primers will be designed using codons optimized to detect deleterious lesions (CODDLE), developed for the discovery of deleterious lesions within coding sequences. The consortium expects to identify allelic series, which will allow for the study of gene function, as has been the case in previous TILLING efforts (Henikoff et al. 2004; McCallum et al. 2000). These initial studies have shown that the common bean EMS mutagenesis protocol is effective at generating mutants and that mutants can be identified using appropriate screening protocols. TILLING in the common bean will have wide applications for both basic and applied research, as requests for mutants have already been received.

The genome size of *P. vulgaris* (580 Mbp/haploid genome) is comparable to that of rice (490 Mbp/haploid genome; Bennett and Leitch 2005). As the genome of the common bean has been sequenced, the sequence information will be powerful tools to improve agronomic and nutritional traits and facilitate marker-assisted selection, an important genetic alternative to improve the economic competitive of the crop, and comparative gene discovery in legumes and to study the function of genes within an economically important group of legumes, namely, soybean, cowpea, and pigeon pea.

2.11 Conclusions

Common bean, a versatile food legume originated in the New World, adapted to many niches, agronomically as well as consumer preferences point of view. Primary centers of origin as well as domestication are located in Mesoamerica and the Andes based on which this crop has been separated into two major gene pools, namely, Mesoamerican and Andean based on morpho-molecular markers. Both the gene pools are known to harbor different races. Out of 50–60 spp. in the genus *Phaseolus*, only five (*Phaseolus vulgaris* L., *Phaseolus lunatus*, *Phaseolus acutifolius* A. Gray, *Phaseolus coccineus* L., and *Phaseolus dumosus* (=*polyanthus*) L.) are cultivated; however, *Phaseolus vulgaris* L. is the most widely grown and exploited. This crop suffers from various biotic and abiotic stresses like diseases, insect pests, nematodes, drought, and nutritional deficiencies. A number of resistance sources have been identified against few diseases; however, this aspect needs exploration for other stresses. Bean improvement programs across the world are mainly based on intraspecific hybridization; hence, efforts are needed to address the problems associated with interspecific gene transfer. Attempts have been made to explore approaches like congruity backcrossing, inbred backcross breeding, and recurrent and gamete selection to facilitate interspecific gene transfer which needs to be prioritized for the effective and rapid introgression of useful genes from alien species. The identification of different molecular markers, particularly microsatellites in the common bean, and unveiling of whole genome sequence will facilitate the marker-assisted selection-based bean improvement. Molecular markers are increasingly being used for indirect selection for resistance to bacterial, fungal, and viral diseases in the common bean (Singh and Schwartz 2010). Currently, the limiting factor for MAS is the number of markers of known map location, primarily because of the lack of DNA sequence information available. The transformation of common bean, although very difficult to popularize at the moment, is no doubt possible (Aragão et al. 1998, 2002; Faria et al. 2006). Its use to introgress pest and pesticide resistance alleles from distantly related *Phaseolus* (e.g., quarternary gene pool and beyond) and non-Phaseolus species may still be an impractical goal due to barriers to acceptance and costs involved. For sustained development of improved bean cultivars, researchers need to continue to gain knowledge about the biotic and abiotic stresses of economic importance in production regions; identify, share, and preserve sources of resistance to the important stresses; develop faster and more reliable screening procedures for both direct selection (phenotypic) and MAS of resistance traits; gain a better understanding of the inheritance and mechanisms of resistance especially for complex stresses; conduct molecular genetic and genomic studies relevant to gaining a better understanding of the genetics and physiology of resistance; translate molecular and genomic information obtained into tools useful for marker-aided breeding; and integrate marker-aided breeding to compliment classical breeding.

References

Abawi GS, Widmer TL (2000) Impact of soil health management practices on soil borne pathogens, nematodes and root diseases of vegetable crops. Appl Soil Ecol 15:37–47

Acosta-Gallegos JA, Adams MW (1991) Plant traits and yield stability of dry bean (*Phaseolus vulgaris*) cultivars under drought stress. J Agric Sci (Cambridge) 117:213–219

Acosta-Gallegos JA, Ocoa-Marquez R, Arrieta-Montiel MP, Ibarra-Perez F, Pajarito-Ravelero A, Sanchez-Valdez I (1995) Registration of Pinto Villa common bean. Crop Sci 35:1211

Acosta-Gallegos JA, Acosta E, Padilla S, Goytia MA, Rosales R, López E (1999) Mejoramiento de la resistencia a la sequía del frijol común en México. Agron Mesoam 10:83–90

Acosta-Gallejos JA, Ibarra-Perez FJ, Cazares-Enriqez B, Castillo-Rosales A, Rosales-Serna R, Kelly JD, Singh SP (2004) Registration of 'Negro Vizcaya' shiny black bean. Crop Sci 44:1866–1867

Adam-Blondon A, Sévignac M, Bannerot H, Dron M (1994) SCAR, RAPD and RFLP markers tightly linked to a dominant gene (*Are*) conferring resistance to anthracnose in common bean. Theor Appl Genet 88:865–870

Aggarwal VD, Pastor-Corrales MA, Chirwa RM, Buruchara RA (2004) Andean beans (*Phaseolus vulgaris* L.) with resistance to the angular leaf spot pathogen *Phaeoisariopsis griseola*) in southern and eastern Africa. Euphytica 136:201–210

Allen DJ, Ampofo JKO, Wortmann CS (1998) Pests diseases and nutritional disorders of the common bean in Africa. CIAT, Cali, p 131

Alzate-Marin AL, Menatim H, Baia GS, Paula TJ, de Souza KA, da Costa MR, de Barros E, Moreira MA (2001) Inheritance of anthracnose resistance in the common bean differential cultivar G 2333 and identification of a new molecular marker linked to the *Co-4(2)* gene. J Phytopathol- Phytopathologische Zeitschrift 149:259–264

Alzate-Marin AL, Cervigni GDL, Moreira MA, Barros EG (2005) Seleção assistida por marcadores moleculares visando ao desenvolvimento de plantas resistentes a doenças, com ênfase em feijoeiro e soja. Fitopatol Bras 30:333–342

Aragão FJL, Desa MFG, Almeida ER, Gander ES, Rech EL (1992) Particle bombardment mediated transient expression of a Brazil nut methionine-rich albumin in bean (*Phaseolus vulgaris* L). Plant Mol Biol 20:357–359

Aragão FJL, Desa MFG, Davey MR, Brasileiro ACM, Faria JC, Rech EL (1993) Factors influencing transient gene-expression in bean (*Phaseolus vulgaris* L) using an electrical particle acceleration device. Plant Cell Rep 12:483–490

Aragão F, Barros L, Brasileiro A, Ribeiro S, Smith F, Sanford J, Faria J, Rech E (1996) Inheritance of foreign genes in transgenic bean (*Phaseolus vulgaris* L) co-transformed via particle bombardment. Theor Appl Genet 93:142–150

Aragão FJL, Ribeiro SG, Barros LMG (1998) Transgenic beans (*Phaseolus vulgaris* L.) engineered to express viral antisense RNAs show delayed and attenuated symptoms to bean golden mosaic geminivirus. Mol Breed 4(6):491–499

Aragão FJL, Sarokin L, Vianna GR, Rech EL (2000) Selection of transgenic meristematic cells utilizing a herbicidal molecule results in the recovery of fertile transgenic soybean [*Glycine max* (L.) Merril] plants at a high frequency. Theor Appl Genet 101:1–6

Aragão FJL, Vianna GRM, Albino MC, Rech EL (2002) Transgenic dry bean tolerant to the herbicide glufosinate ammonium. Crop Sci 42(4):1298–1302

Arndt GC, Gepts P (1989) Segregation and linkage for morphological and biochemical markers in a wide cross in common bean (*Phaseolus vulgaris*). Ann Rep Bean Improv Coop 32:68–69

Asensio S, Manzanera MC, Asensio C, Singh SP (2005) Introgressing resistance to bacterial and viral diseases from the Middle American to Andean common bean. Euphytica 143:223–228

Asensio S, Manzanera MC, Asensio C, Singh SP (2006) Gamete selection for resistance to common and halo bacterial blights in dry bean intergene pool populations. Crop Sci 46:131–135

Awale HE, Kelly JD (2001) Development of SCAR markers linked to $Co\text{-}4^2$ gene in common bean. Annu Rep Bean Improv Coop 44:119–120

Baggett JR (1956) The inheritance of resistance to strains of bean yellow mosaic virus in the interspecific cross *Phaseolus vulgaris* × *P. coccineus*. Plant Dis Rep 40:702–707

Balardin RS, Kelly JD (1998) Interaction among races of *Colletotrichum lindemuthianum* and diversity in *Phaseolus vulgaris*. J Am Soc Hortic Sci 123:1038–1047

Barron JE, Pasini RS, Davis DW, Stuthman DD, Graham PH (1999) Response to selection for seed yield and nitrogen (N2) fixation in common bean (*Phaseolus vulgaris* L.). Field Crop Res 62:119–128

Bassett MJ (1991) A revised linkage map of common bean. Hortic Sci 26:834–836

Baudoin JP (2001) Contribution des ressources phytogénétiques à la selection variétale de légumineuses alimentaires tropicales. Biotechnol Agron Soc Environ 5(4):221–230

Baudoin JP, Camarena F, Lobo M (1995) Amélioration de quatre espèces de légumineuses alimentaires tropicales *Phaseolus vulgaris, P. coccineus, P. polyanthus* et *P. lunatus*. Sélection intra- et interspécifique. In: Quel Avenir Pour l'amélioration des Plantes? Quatrièmes Journées Scientifiques du Réseau Biotechnologie Végétale de l'UREF, Namur, 18–21 Octobre 1993. Ed. par J Dubois, Y Demarly, AUPELF-UREF. John Libbey Eurotext, Paris, pp 31–49

Baudoin JP, Fofana B, du Jardin P, Vekemans X (1998) Diversity and genetic organization of the genus *Phaseolus*. Analysis of the global and chloroplastic genome. In: International symposium on breeding of protein and oil crops. EUCARPIA, Fundación Pedro Barrié De La Maza, pp 75–76

Baudoin JP, Camarena F, Lobo M, Mergeai G (2001) Breeding *Phaseolus* for intercrop combinations in Andean highlands. In: Cooper HD, Spillane C, Hodgkin T (eds) Broadening the genetic basis of the crop. CAB International, Wallingford, pp 373–384

Beaver JS, Kelly JD (1994) Comparison of selection methods for dry bean populations derived from crosses between gene pools. Crop Sci 34:34–37

Beaver JS, Rosas JC (1998) Heritability of the length of reproductive period and rate of seed mass accumulation in common bean. J Am Soc Hortic Sci 123:407–411

Beaver JS, Osorno JM, Ferwerda FH, Muñoz Perea CG (2005) Registration of bean golden yellow mosaic virus resistant germplasms PR9771-3-1, PR0247-49 and PR0157-4-1. Crop Sci 45:2126

Becerra-Velásquez VL, Gepts P (1994) RFLP diversity in common bean (*Phaseolus vulgaris* L.). Genome 37:256–263

Beebe SE, Bliss FA (1981) Root rot resistance in common bean germplasm of Latin American origin. Plant Dis 65:485–489

Beebe SE, Pastor-Corrales M (1991) Breeding for disease resistance. In: van Schoonhoven A, Voysest O (eds) Common beans: research for crop improvement. CAB International/CIAT, Wallingford/Cali, pp 561–617

Beebe S, Lynch J, Galwey N, Tohme J, Ochoa I (1997) A geographical approach to identify phosphorus-efficient genotypes among landraces and wild ancestors of common bean. Euphytica 95:325–336

Beebe S, Skroch P, Tohme J, Duque MC, Pedraza F, Nienhhuis J (2000) Structure of genetic diversity among common bean landraces of middle American origin based on correspondence analysis of RAPD. Crop Sci 40:264–273

Beebe S, Rengifo J, Gaita´n-Solý´s E, Duque MC, Tohme J (2001) Diversity and origin of Andean landraces of common bean. Crop Sci 41:854–862

Beebe S, Rao I, Teran H, Cajiao C (2004) Breeding concepts and approaches in food legumes: the example of the common bean. Abstract of paper presented at the "Second national workshop on food and forage legumes" Addis Ababa, Ethiopia, 22–26 Sept 2003

Beebe SE, Rojas-Pierce M, Yan X, Blair MW, Pedraza F et al (2006) Quantitative trait loci for root architecture traits correlated with phosphorus acquisition in common bean. Crop Sci 46:413–423

Bemis WP, Kedar N (1961) Inheritance of morphological abnormalities in seedlings of two species of *Phaseolus*. J Hered 52:171–178

Bennett M, Leitch I (2005) Angiosperm DNA C-Values database. Release 4.0

Berglund-Brucher O, Brucher H (1976) The South American wild bean (*Phaseolus aborigineus* Burk.) an ancestor of common bean. Econ Bot 30:257–272

Bianchini A, Moda-Cirino V, Fonseca NS, Toledo FF (1994) Genética de resistência ao v´ýrus do mosaico dourado do feijoeiro. (Abstr) Fitopatol Bras 19:329

Bitocchi E, Nanni L, Bellucci E, Rossi M, Giardini A et al (2012) Mesoamerican origin of the common bean (*Phaseolus vulgaris* L.) is revealed by sequence data. Proc Natl Acad Sci 109:788–796

Blair MW, Muñoz MC, Beebe SE (2002) QTL analysis of drought and abiotic stress tolerance in common bean RIL populations. In: Annual report, Biotechnology Research Project. CIAT, Cali, Colombia, pp 68–72

Blair MW, Pedraza F, Buendia HF, Gaita n-Solý´s E, Beebe SE, Gepts P, Tohme J (2003) Development of a genome-wide anchored microsatellite map for common bean (*Phaseolus vulgaris* L.). Theor Appl Genet 107:1362–1374

Blair MW, Giraldo MC, Buendía HF, Tovar E, Duque MC, Beebe SE (2006a) Microsatellite marker diversity in common bean (*Phaseolus vulgaris* L.). Theor Appl Genet 113:100–109

Blair MW, Iriarte G, Beebe S (2006b) QTL analysis of yield traits in an advanced backcross population derived from a cultivated Andean and wild common bean (*Phaseolus vulgaris* L.) cross. Theor Appl Genet 112:1149–1163

Blair MW, Muñoz C, Garza R, Cardona C (2006c) Molecular mapping of genes for resistance to the bean pod weevil (*Apion godmani* Wagner) in common bean. Theor Appl Genet 112:913–923

Blair MW, Dìaz JM, Hidalgo R, Dìaz LM, Duque MC (2007) Microsatellite characterization of Andean races of common bean (*Phaseolus vulgaris* L.). Theor Appl Genet 116:29–43

Blair M, Diaz LM, Buendia HF, Duque MC (2009a) Genetic diversity, seed size associations and population structure of a core collection of common beans (*Phaseolus vulgaris* L.). Theor Appl Genet 119:955–972

Blair MW, Astudillo C, Grusak M, Graham R, Beebe S (2009b) Inheritance of seed iron and zinc content in common bean (*Phaseolus vulgaris* L.). Mol Breed 23:197–207

Blair MW, González LF, Kimani PM, Butare L (2010) Genetic diversity, inter-gene pool introgression and nutritional quality of common beans (*Phaseolus vulgaris* L.) from Central Africa. Theor Appl Genet 121:237–248

Blair MW, Hurtado N, Sharma PN (2012a) New gene-derived simple sequence repeat markers for common bean (*Phaseolus vulgaris* L.). Mol Ecol Res 12:661–668

Blair MW, Soler A, Cortes AJ (2012b) Diversification and population structure in common beans (*Phaseolus vulgaris* L.). PLoS One 7(11):e49488

Bliss FA (1993) Breeding common bean for improved nitrogen fixation. Plant Soil 152:71–79

Boisson-Dernier A, Chabaud M, Garcia F, Becard G, Rosenberg C et al (2001) Agrobacterium rhizogenes-transformed roots of *Medicago truncatula* for the study of nitrogen-fixing and endomycorrhizal symbiotic associations. Mol Plant-Microbe Inter 14:695–700

Bonfim K, Faria JC, Nogueira EOPL, Mendes EA, Aragão FJL (2007) RNAi-mediated resistance to Bean golden mosaic virus in genetically engineered common bean (*Phaseolus vulgaris*). Mol Plant-Microbe Inter 20:717–726

Boutin S, Young N, Olson T, Yu Z, Shoemaker R et al (1995) Genome conservation among three legume genera detected with DNA markers. Genetics 38:928–937

Broughton WJ, Hernandez G, Blair MW, Beebe SE, Gepts P, Vanderleyden J (2003) Beans (*Phaseolus* spp.): model food legumes. Plant Soil 252:55–128

Brucher H (1988) The wild ancestor of *Phaseolus vulgaris* in South America. In: Gepts P (ed) Genetic resources of Phaseolus beans: their maintenance, domestication, evolution and utilization. Kluwer, Dordrecht, pp 185–214

Burkart A (1941) Sobre la existencia de razas silvestres de *Phaseolus vulgaris*. Resol. Y Trab. I Ren. Argent. Agron., Buenos Aires, p 52

Burkart A, Brucher H (1953) Phaseolus aborigineus Burkart, die mutmassliche andine Stammform der Kulturbohne. Zuchter 23:65–72

Buso GSC, Amaral ZPS, Brondani RPV, Ferreira ME (2006) Microsatellite markers for the common bean – *Phaseolus vulgaris*. Mol Ecol Notes 6:252–254

Cabral PDS, Soares TCB, Lima ABP, de Miranda FD, Souza FB, Gonçalves LSA (2011) Genetic diversity in local and commercial dry bean (*Phaseolus vulgaris*) accessions based on microsatellite markers. Genet Mol Res 10(1):140–149

Caixeta ET, Borém A, Kelly JD (2005) Development of microsatellite markers based on BAC common bean clones. Crop Breed Appl Biotechnol 5:125–133

Camarena F, Baudoin JP (1987) Obtention des premiers hybrids interspecifiques entre *Phaseolus vulgaris* et *Phaseolus polyanthus* avec le cytoplasme de cette derniere forme. Bull Rech Agron Gembloux 22:43–55

Campa A, Giraldez R, Ferreira JJ (2009) Genetic dissection of the resistance to nine anthracnose races in the common bean differential cultivars MDRK and TU. Theor Appl Genet 119:1–11

Cardoso JMK (2009) Estimativa da Diversidade Genética entre Acessos do Tipo Carioca de Feijão Comum com Base em Marcadores Moleculares. Dissertação de Mestrado, Instituto Agronômico de Campinas, Campinas

Castellanos JZ, Peña-Cabrales JJ, Acosta-Gallegos JA (1996) 15 N-determined dinitrogen fixation capacity of common bean (*Phaseolus vulgaris*) cultivars underwater stress. J Agric Sci 126:327–333

Chacon MI, Gonzalez AV, Gutierrez JP, Beebe S, Debouck DG (1996) Increased evidence for common bean (*Phaseolus vulgaris*) domestication in Colombia. Ann Rep Bean Improv Coop 39:201–202

Chacón MI, Pickersgill S, Debouck DG (2005) Domestication patterns in common bean (Phaseolus vulgaris L.) and the origin of Mesoamerican and Andean cultivated races. Theor Appl Genet 110:432–444

Chacon MI, Pickersgill B, Debouck DG, Arias JS (2007) Phylogeographic analysis of the chloroplast DNA variation in wild common bean (*Phaseolus vulgaris* L.) in the Americas. Plant Syst Evol 266:175–195

Chowdhury MA, Yu K, Park SJ (2002) Molecular mapping of root rot resistance in common bean. Ann Rep Bean Improv Coop 45:96–97

CIAT (1975) Annual report. http://www.ciat.cgiar.org/beans/drought_tolerance05.htm

CIAT (2002) Annual report. http://www.ciat.cgiar.org/beans/drought_tolerance05.htm

Costa MR, Tanure JPM, Arruda KMA, Carneiro JES, Moreira MA et al (2006) Pyramiding of anthracnose, angular leaf spot and rust resistance genes in black and red bean cultivars. Ann Rep Bean Improv Coop 49:187–188

Debouck DG (1991) Systematics and morphology. In: van Schoonhoven A, Voyasset O (eds) Common beans: research for crop improvement. CAB International/CIAT, Wallingford/Cali, pp 55–117

Debouck DG (1999) Diversity in *Phaseolus* species in relation to the common bean. In: Singh SR (ed) Common bean improvement in the 21st century. Kluwer Academic Publisher, Dordrecht, pp 25–52

Debouck DG (2000) Biodiversity, ecology and genetic resources of *Phaseolus* beans – seven answered and unanswered questions. In: The seventh MAFF international workshop on genetic resources. Wild legumes. National Institute of Agrobiological Resources, Tsukuba, Ibaraki, Japan 1:95–123

Debouck DG, Smartt J (1995) Beans. In: Smartt J, Simmonds NW (eds) Evolution of crop plants, 2nd edn. Longman, Harlow, pp 287–294

Debouck DG, Toro O, Paredes OM, Johnson WC, Gepts P (1993) Genetic diversity and ecological distribution of *Phaseolus vulgaris* (*Fabaceae*) in northwestern South America. Econ Bot 47:408–423

Delgado-Salinas A (1985) Systematics of the genus *Phaseolus* (Leguminosae) in North and Central America. Ph.D. dissertation. University of Texas, Austin, p 363

Delgado-Salinas A, Bruneau A, Doyle JJ (1993) Chloroplast DNA phylogenetic studies in New World *Phaseolinae* (*Leguminosae*; *Papilionoideae*; *Phaseoleae*). Syst Bot 18:6–17

Delgado-Salinas A, Turley T, Richman A, Lavin M (1999) Phylogenetic analysis of the cultivated and wild species of *Phaseolus* (Fabaceae). Syst Bot 24:438–460

Díaz LM, Blair MW (2006) Race structure within the Mesoamerican gene pool of common bean (Phaseolus vulgaris L.) as determined by microsatellite markers. Theor Appl Genet 114:143–154

Díaz CL, Spaink HP, Kijne JW (2000) Heterologous rhizobial lipochitin oligosaccharides and chitin oligomers induce cortical cell divisions in red clover roots, transformed with the pea lectin gene. Mol Plant Microbe Int 13:268–276

Dickson MH, Boettger MA (1976) Selection for seed quality in white seeded snap bean. Ann Rep Bean Improv Coop 19:24–25

Dickson MA, Boettger MA (1977) Applied selection for mechanical damage resistance in snap bean using the mechanical damage simulator. Ann Rep Bean Improv Coop 20:38–39

Ellis MA, Galvez GE, Sinclair JB (1976) Hongos internalmente portados por la semilla y calidad de la semilla de frijol (*Phaseolus vulgaris* L.) cosechado en fincas de pequenos agricultores en cuatro departamentos de Colombia. Not Fiopatol 5:79–82

Ellstrand N, Prentice H, Hancock J (1999) Gene flow and introgression from domesticated plants into their wild relatives. Ann Rev Ecol Syst 30:539–563

Escalante AM, Coello G, Eguiarte LE, Piñero D (1994) Genetic structure and mating systems in wild *Phaseolus coccineus* and *P. vulgaris* (Fabaceae). Am J Bot 81:1096–1103

Evan AM (1976) Beans. In: Simmonds NW (ed) Evolution of crop plants. Longman, London, pp 168–172

Fairbairn JN (1993) Evaluation of soil, climate and land use information at three scales: the case study of low income bean farming in Latin America. University of Reading, UK

Faleiro FG, Ragagnin VA, Moreira MA, Barros EG (2004) Use of molecular markers to accelerate the breeding of common bean lines resistant to rust and anthracnose. Euphytica 183:213–218

FAO (2012) FAOSTAT. Available from http://faostat.fao.org/faostat/collections?subset=agriculture

FAO (2013) FAOSTAT. Available from http://faostat.fao.org/faostat/collections?subset=agriculture

Faria JC, Albino MMC, Dias BBA, Cancado LJ, da Cunha NB, Silva LD, Vianna GR, Aragão FJL (2006) Partial resistance to bean golden mosaic virus in a transgenic common bean (*Phaseolus vulgaris* L.) line expressing a mutated *rep* gene. Plant Sci 171:565–571

Federici CT, Waines JG (1988) Interspecific hybrid compatibility of selected *Phaseolus vulgaris* L. lines with *Phaseolus acutifolius* A. Gray, *Phaseolus lunatus* L., *Phaseolus filiformis* Bentham. Ann Rep Bean Improv Coop 31:201–202

Ferreira JJ, Alveraze E, Fueyo MA, Roca A, Giraldez R (2000) Determination of outcrossing rate in *Phaseolus vulgaris* L. using seed protein markers. Euphytica 113:259–263

Ferreira JL, de Souza C, Eustáquio J, Teixeira Alexsandro L, de Lanes FF, Roberto CP, Aluízio B (2007) Gene flow in common bean (*Phaseolus vulgaris* L.). Euphytica 153:165–170

Fofana B, Baudoin JP, Vekemans X, Debouck DG, du Jardin P (1999) Molecular evidence for an Andean origin and a secondary gene pool for the Lima bean (*Phaseolus lunatus*) using chloroplast DNA. Theor Appl Genet 98:202–212

Fofana B, du Jardin P, Baudoin JP (2001) Genetic diversity in the Lima bean (*Phaseolus lunatus* L.) as revealed by chloroplast DNA (cpDNA) variations. Genet Resour Crop Evol 48:437–445

Foster EF, Pajarito A, Acosta- Gallegos J (1995) Moisture stress impact on N partitioning, N-remobilization and N-use efficiency in beans (*Phaseolus vulgaris* L). J Agric Sci 124:27–37

Frahm MA, Rosas JC, Mayek-Pérez N, López-Salinas E, Acosta-Gallegos JA, Kelly JD (2004) Breeding beans for resistance to terminal drought in the Lowland tropics. Euphytica 136:223–232

Freyre R, Ríos R, Guzmán L, Debouck D, Gepts P (1996) Ecogeographic distribution of *Phaseolus spp.* (*Fabaceae*) in Bolivia. Econ Bot 50:195–215

Freyre R, Skroch P, Geffroy V, Adam-Blondon AF, Shirmohamadali A et al (1998) Towards an integrated linkage map of common bean. 4. Development of a core map and alignment of RFLP maps. Theor Appl Genet 97:847–856

Freytag GF, Debouck DG (2002) Taxonomy, distribution and ecology of genus *Phaseolus* (*Leguminosae-Papilionoideae*) in North America, Mexico and Central America. Sida, Botanical Miscellany 23. Botanical Research Institute of Texas [BRIT], Forth Worth, p 298

Gaitán-Solís E, Duque MC, Edwards KJ, Tohme J (2002) Microsatellite repeats in common bean (*Phaseolus vulgaris*): isolation, characterization, and cross-species amplification in *Phaseolus* ssp. Crop Sci 42:2128–2136

Galindo L, Blair MW, Beebe S, Ishitani M, Tohme J (2003) Development of a molecular marker: DREB for drought tolerance in common bean. In: Annual report, Biotechnology Research Project, CIAT, Cali, Colombia, pp 177–179

Geerts P (2001) Study of embryo development in *Phaseolus* in order to obtain interspecific hybrids. Ph.D. thesis. Gembloux Agricultural University, Belgium, p 183

Geerts P, Mergai G, Baudoin JP (1999) Rescue of early heart-shaped embryos and plant regeneration of *Phaseolus polyanthus* Greenm. and *P. vulgaris* L. Biotechnol Agron Soc Environ 3(3):141–148

Geerts P, Toussaint A, Mergeai G, Baudoin JP (2002) Study of the early abortion in reciprocal crosses between *Phaseolus vulgaris* and *Phaseolus polyanthus* Greenn. Biotechnol Agron Soc Environ 6:109–119

Geffroy V, Sicard D, Oliveira JCF, Sevignac M, Cohen S, Gepts P, Neema C, Langin T, Dron M (1999) Identification of an ancestral resistance gene cluster involved in the coevolution process between *Phaseolus vulgaris* and its fungal pathogen *Colletotrichum lindemuthianum*. Mol Plant-Microbe Inter 12:774–784

Genga AM, Allavena A, Ceriotti A, Bollini R (1992) Genetic transformation in *Phaseolus* by high-velocity particle microprojectiles. Acta Hortic 300:309–313

Gentry HS (1969) Origin of common bean *Phaseolus vulgaris*. Econ Bot 23:55–69

Gepts P (1988) A middle American and an Andean common bean gene pool. In: Gepts P (ed) Genetic resources of Phaseolus beans. Kluwer Academic Publishers, Dordrecht, pp 375–390

Gepts P (1998) Origin and evolution of common bean: past events and recent trends. Hortic Sci 33(7):1124–1130

Gepts P (2004) Domestication as a long-term selection experiment. Plant Breed Rev 24:1–44

Gepts P, Bliss FA (1985) F1 hybrid weakness in the common bean: differential geographic origin suggests two gene pools in cultivated bean germplasm. J Hered 76:447–450

Gepts P, Bliss FA (1988) Dissemination pathways of common bean (*Phaseolus vulgaris* L.) deduced from phaseolin electrophoretic variability, II: Europe and Africa. Econ Bot 42(1):86–104

Gepts P, Debouck D (1991) Origin, domestication and evolution of the common bean (*Phaseolus vulgaris L.*). In: Van Shoonhoven A, Voysest O (eds) Common beans: research for crop improvement. Commonwealth Agricultural Bureau, Wallingford, pp 7–53

Gepts P, Osborn TC, Rashka K, Bliss FA (1986) Electrophoretic analysis of phaseolin protein variability in wild form and landraces of common bean, *Phaseolus vulgaris*: evidence for multiple centers of domestication. Econ Bot 40:451–468

Gepts P, Papa R, González A, Acosta J, Delgado-Salinas A (1999) Human effects on *Phaseolus vulgaris* adaptation during and after domestication. In: Van Raamsdonk LWD, den Nijs JCM (eds) Plant evolution in man-made habitats. Hugo de Vries Laboratory, Amsterdam, pp 161–181

Gepts P, Aragão-Francisco JL, de Barros E, Matthew WB et al (2008) Genomics of *Phaseolus* beans, a major source of dietary protein and micronutrients in the Tropics. In: Moore PH, Ming R (eds) Genomics of tropical crop plants. Springer, Berlin, pp 113–143

González A, Wong A, Delgado-Salinas A, Papa R, Gepts P (2005) Assessment of inter simple sequence repeat markers to differentiate sympatric wild and domesticated populations of common bean (*Phaseolus vulgaris* L.). Crop Sci 45:606–615

Gonzalez-Alonso J, Mora-Rodrúguez R, Below PR, Coyle EF (1995) Dehydration reduces cardiac output and increases systemic and cutaneous vascular resistance during exercise. J Appl Physiol 79:1487–1496

Gonzalez-Torres RI (2004) Estimacion de flujo de genes en Phaseolus vulgaris L. mediante marcadores moleculares; Microsatellites y polymorphismo de AND de chloroplasto. Master's thesis. Universidad Nacional de Colombia

Graham PH, Ranalli P (1997) Common bean (*Phaseolus vulgaris* L.). Field Crop Res 53:131–146

Graham A, Ramírez M, Valdés-López O, Lara M, Tesfaye M et al (2006) Identification of phosphorus stress induced genes in *Phaseolus vulgaris* L. through clustering analysis across several species. Funct Plant Biol 33:789–797

Guerra-Sanz JM (2004) New SSR markers of *Phaseolus vulgaris* from sequence databases. Plant Breed 123:87–89

Guzmán P, Gilbertson RA, Nodari R, Johnson WC, Temple SR, Mandala D, Mkandawire ABC, Gepts P (1995) Characterization of variability in the fungus *Phaeoisariopsis griseola* suggests coevolution with the common bean (*Phaseolus vulgaris*). Phytopathology 85:600–607

Haghighi KR, Ascher PD (1988) Fertile intermediate hybrid between *Phaseolus vulgaris* L. and *P. acutifolius* from congruity backcrossing. Sex Plant Reprod 1:51–58

Hagiwara WE, dos Santos JB, do Carmo LM (2001) Use of RAPD to aid selection in common bean backcross breeding programs. Crop Breed Appl Biotechnol 1:355–362

Hamann A, Zink D, Nagl W (1995) Microsatellite fingerprinting in genus *Phaseolus*. Genome 38:507–515

Hammer K (1984) Das domestication syndrome. Kulturpflanze 32:11–34

Hardy O, Dubois S, Zoro Bi I, Baudoin JP (1997) Gene dispersal and its consequences on genetic structure of wild populations of Lima bean (*Phaseolus lunatus*) in Costa Rica. Plant Genet Resour Newsl 109:1–6

Harlan JR (1971) Agricultural origins. Centers and noncenters. Science 174:468–474

Hedrick UP (1931) The vegetable of New York: the beans of New York. Agric Exp Stn Rep 1:1–110, New York (Geneva)

Henikoff S, Till BJ, Comai L (2004) Tilling. Traditional mutagenesis meets functional genomics. Plant Physiol 135:630–636

Hermsen JGT (1984) Nature, evolution and breeding of polyploids. Iowa State J Res 58:411–412

Honma S (1956) A bean interspecific hybrid. J Hered 47:217–220

Hoover EE, Brenner ML, Ascher PD (1985) Comparison of development of two bean crosses. Hortic Sci 20:84–86

Hornakova O, Závodná M, Žáková M, Kraic J, Debre F (2003) Diversity of common bean landraces collected in the Western and Eastern Carpatien. Czech J Genet Plant Breed 39(3):73–83. http://www.rbgkew.org.uk/cval/database1.html

Hucl P, Scoles GJ (1985) Interspecific hybridization in common bean. A review. Hortic Sci 20:352–357

Ibarra-Pérez F, Ehdaie B, Waines G (1997) Estimation of out-crossing rate in common bean. Crop Sci 37:60–65

Ibrahim AM, Coyne DP (1975) Genetics of stigma shape, cotyledon position and flower color in reciprocal crosses between *Phaseolus vulgaris* L. × *Phaseolus coccineus* (Lam.) and implications in breeding. J Am Soc Hortic Sci 100:622–626

Ijani ASM, Mabagala RB, Nchimba-Msolla S (2000) Root-knot nematode species associated with beans and weeds in the Morogoro region, Tanzania. Afr Plant Prot 6:37–41

Ishitani M, Rao I, Wenzl P, Beebe S, Tohme J (2004) Integration of genomics approach with traditional breeding towards improving abiotic stress adaptation: drought and aluminum toxicity as case studies. Field Crop Res 90:35–45

Islam FMA, Basford KE, Redden RJ, Jara C, Beebe S (2002) Patterns of resistance to angular leaf spot, anthracnose and common bacterial blight in common bean germplasm. Aust J Exp Agric 42:481–490

Islam FM, Beebe S, Muñoz M, Tohme J, Redden RJ, Basford KE (2004) Using molecular markers to assess the effect of introgression on quantitative attributes of common bean in the Andean gene pool. Theor Appl Genet 108:243–252

Jaaska V (1996) Isozyme diversity and phylogenetic affinities among the *Phaseolus* beans (*Fabaceae*). Plant Syst Evol 200:233–252

Johnson WC, Gepts P (2002) The role of epistasis in controlling seed yield and other agronomic traits in an Andean × Mesoamerican cross of common bean (*Phaseolus vulgaris* L.). Euphytica 125:69–79

Johnson WC, Guzmán P, Mandala D, Mkandawire A, Temple S et al (1997) Molecular tagging of the *bc-3* gene for introgression into Andean common bean. Crop Sci 37:248–254

Kairaillah MM, Adams MW, Sears BB (1990) Mitochondrial polymorphisms of Malawian bean lines: further evidence for two major gene pools. Theor Appl Genet 80:753–761

Kami J, Becerra V, Debouck DG, Gepts P (1995) Identification of presumed ancestral DNA sequences of phaseolin in *Phaseolus vulgaris*. Proc Natl Acad Sci 92:1101–1104

Kami J, Poncet V, Geffroy V, Gepts P (2006) Development of four phylogenetically-arrayed BAC libraries and sequence of the APA locus in *Phaseolus vulgaris*. Theor Appl Genet 112:987–998

Kelly JD (2004) Advances in common bean improvement: some case histories with broader applications. Acta Hortic (ISHS) 637:99–122

Kelly JD, Miklas PN (1998) The role of RAPD markers in breeding for disease resistance in common bean. Mol Breed 4:1–11

Kelly JD, Vallejo VA (2004) A comprehensive review of the major genes conditioning resistance to anthracnose in common bean. Hortic Sci 39:1196–1207

Kelly JD, Hosfield GL, Varner GV, Uebersax MA, Taylor J (1999) Registration of 'Matterhorn' great northern bean. Crop Sci 39:589–599

Kelly JD, Gepts P, Miklas PN, Coyne DP (2003) Tagging and mapping of genes and QTL and molecular-marker assisted selection for traits of economic importance in bean and cowpea. Field Crop Res 82:135–154

Khairallah M, Sears B, Adams M (1992) Mitochondrial restriction fragment length polymorphisms in wild *Phaseolus vulgaris* L.: insights on the domestication of the common bean. Theor Appl Genet 84:915–922

Kim JW, Minamikawa T (1996) Transformation and regeneration of French bean plants by the particle bombardment process. Plant Sci 117:131–138

Kipe-Nolt JA, Vargas H, Giller KE (1993) Nitrogen fixation in breeding lines of *Phaseolus vulgaris* L. Plant Soil 152:103–106

Koenig R, Gepts P (1989a) Allozyme diversity in wild *Phaseolus vulgaris*: further evidence for two major centers of genetic diversity. Theor Appl Genet 78:809–817

Koenig R, Gepts P (1989b) Segregation and linkage of genes for seed proteins, isozymes, and morphological traits in common bean (*Phaseolus vulgaris*). J Hered 80:455–459

Koenig R, Singh SP, Gepts P (1990) Novel phaseolin types in wild and cultivated common bean (*Phaseolus vulgaris*, Fabaceae). Econ Bot 44:50–60

Koinange EMK, Gepts P (1992) Hybrid weakness in wild *Phaseolus vulgaris* L. J Hered 83:135–139

Koinange EMK, Singh SP, Gepts P (1996) Genetic control of the domestication syndrome in common bean. Crop Sci 36:1037–1045

Kornegay J, White JW, Ortiz de la Cruz O (1992) Growth habit and gene pool effects on inheritance of yield in common bean. Euphtyica 62:171–180

Kornegay J, Cardona C, Posso CE (1993) Inheritance of resistance to Mexican bean weevil in common bean, determined by bioassay and biochemical tests. Crop Sci 33:589–594

Kwak M, Gepts P (2009) Structure of genetic diversity in the two major gene pools of common bean (*Phaseolus vulgaris* L., Fabaceae). Theor Appl Genet 118:979–992

Kwak M, Kami JA, Gepts P (2009) The putative Mesoamerican domestication center of *Phaseolus vulgaris* is located in the Lerma-Santiago basin of Mexico. Crop Sci 49:554–563

Kwapata K, Thang N, Mariam S (2012) Genetic transformation of common bean (*Phaseolus vulgaris* L.) with the gus color marker, the bar herbicide resistance, and the barley (*Hordeum vulgare*) HVA1 drought tolerance genes. Int J Agron. doi:10.1155/2012/198960

Lackey JA (1983) A review of genetic concepts in American *Phaseolinae* (*Fabaceae, Faboideae*). Iselya 2:21–64

Leakey CLA (1988) Genotypic and phenotypic markers in common beans. In: Gepts P (ed) Genetic resources of *Phaseolus* beans. Kluwer, Dordrecht

Lecomte B, Longly B, Crabbe J, Baudoin JP (1998) Etude comparative du développement de l'ovule chez deux espèces de *Phaseolus*: *P. polyanthus* et *P. vulgaris*. Dévelopement de l'ovule dans le genre *Phaseolus*. Biotechnol Agron Soc Environ 2(1):77–84

Lentini Z, Debouck DG, Espinoza AM, Arya R (2006) Gene flow analysis into wild/weedy relatives from crops with center of origin/diversity in tropical America. In: Proceedings of the 9th International Symposium on

Biosafety of Genetically Modified Organisms (ISBGMO), Jeju Island, South Korea, 24–29 Sept 2006 (International Society for Biosafety Research [ISBR] n.p.), pp 152–156. www.isbr.info/isbgmo/docs/9th_isbgmo_program.pdf

Liaca V, Delgado SA, Gepts P (1994) Chloroplast DNA as evolutionary marker in the *Phaseolus vulgaris* complex. Theor Appl Genet 88:642–652

Liao H, Yan X, Rubio G, Beebe SE, Blair MW, Lynch JP (2004) Basal root gravitropism and phosphorus acquisition efficiency in common bean. Funct Plant Biol 31:1–12

Liebman M, Roew RJ, Corson S, Marra MC, Honeycutt CW, Murphy BA (1993) Agronomic and economic performance of conventional vs reduced input bean cropping systems. J Prod Agric 6:369–378

Liebman M, Corson S, Rowe RJ, Halteman WA (1995) Dry bean response to nitrogen fertilizer in two tillage and residue management systems. Agron J 87:538–546

Lioi L (1989) Geographical variation of phaseolin patterns in an old world collection of Phaseolus vulgaris. Seed Sci Technol 17:317–324

Liu ZC, Park BJ, Kanno A, Kameya T (2005) The novel use of a combination of sonication and vacuum infiltration in Agrobacterium-mediated transformation of kidney bean (*Phaseolus vulgaris* L) with lea gene. Mol Breed 16:189–197

López CE, Acosta IF, Jara C, Pedraza F, Gaitán-Solís E et al (2003) Identifying resistance gene analogs associated with resistances to different pathogens in common bean. Phytopathology 93:88–95

MacBryde FW (1947) Cultural and historical geography of southwest Guatemala, Publication no. 4. Institute of Social Anthropology, Smithsonian Institution, Washington, DC, p 25

Mahuku GS, Riascos JJ (2004) Virulence and molecular diversity within *Colletotrichum lindemuthianum* isolates from Andean and Mesoamerican bean varieties and regions. Eur J Plant Pathol 110:253–263

Manshardt RM, Bassett MJ (1984) Inheritance of stigma position in *Phaseolus vulgaris* × *Phaseolus coccineus* hybrid populations. J Hered 75:45–50

Maquet A, Vekemans X, Baudoin JP (1999) Phylogenetic study on wild allies of Lima bean and implication on its origin. Plant Syst Evol 218:43–54

Maquet A, Masumboko B, Ouedraogo M, Zoro Bi I, Badoin JP (2001) Estimation of gene flow among wild populations of *Phaseolus lunatus* L. using isozyme markers. Ann Rep Bean Improv Coop 44:27–28

Marechal R (1971) Observations sur quelques hybrids dans le genre Phaseolus, II : les phenomenes meiotiques. Bull Rech Agron Gembloux 6:461–489

Marechal R, Mascherpa J, Stanier F (1978) Etude taxonomique d'un groupe complexe d'especes des genres *Phaseolus* et *Vigna* (*Papilionaceae*) sur la base de donnees morphologiques et polliniques, traitees par l/analyse informatique. Memories des Conservatoire et Jardin Botaniques de la Ville de Geneve. Boissiera, Geneva 28:1–273

Martinez-Castillo J, Zuzumbo- Villarreal D, Gepts P, Colunga Garcia-Marin P (2007) Gene flow and genetic structure in the wild-weedy –domesticated complexes of *Phaseolus lunatus* in its Mesoamerican centre of domestication and diversity. Crop Sci 47:58–66

McCallum CM, Comai L, Greene EA, Henikoff S (2000) Targeting induced local lesions in genomes (TILLING) for plant functional genomics. Plant Physiol 123:439–442

McClean PE, Lee RK, Otto C, Gepts P, Bassett MJ (2002) Molecular and phenotypic mapping of genes controlling seed coat pattern and color in common bean (*Phaseolus vulgaris* L.). J Hered 93:148–152

McClean PE, Terpstra J, McConnell M, White C, Lee R, Mamidi S (2012) Population structure and genetic differentiation among the USDA common bean (*Phaseolus vulgaris* L.) core collection. Genet Resour Crop Evol 59:499–515

Melotto M, Coelho MF, Pedrosa-Harand A, Kelly JD, Camargo LEA (2004) The anthracnose resistance locus *Co-4* of common bean is located on chromosome 3 and contains putative disease resistance-related genes. Theor Appl Genet 109:690–699

Melotto M, Monteiro-Vitorello CB, Bruschi AG, Camargo LEA (2005) Comparative bioinformatics analysis of genes expressed in common bean (*Phaseolus vulgaris* L.) seedlings. Genome 48:562–570

Menjia-Jimenez A, Munoz C, Jacobsen HJ, Roca WM, Singh SP (1994) Interspecific hybridization between common and tepary beans: increased hybrid embryo growth, fertility and efficiency of hybridization through recurrent and congruity back crossing. Theor Appl Genet 88:324–331

Métais I, Hamon B, Jalouzot R, Peltier D (2002) Structure and level of genetic diversity in various bean types evidenced with microsatellite markers isolated from a genomic enriched library. Theor Appl Genet 104:1346–1352

Mienie C, Liebenberg M, Pretorius Z, Miklas P (2005) SCAR markers linked to the common bean rust resistance gene *Ur-13*. Theor Appl Genet 111:972–979

Miklas P (2000) Use of Phaseolus germplasm in breeding pinto great northern and red bean for pacific, north west and inter mountain regions. In: Bean research products and utilization. Proceedings of Idaho bean workshop, Idaho University, Russia

Miklas PN (2005) List of SCAR marker for disease resistance: Aug 2005 update http://www.ars.usda.gov/SP2UserFiles/Place/53540000/miklas/SCARtable.pdf

Miklas PN, Singh SP (2007) Common bean. In: Kole C (ed) Genome mapping and molecular breeding in plants. Berlin: Pulses, sugar and tuber crops. Springer 1:1–31

Miklas PN, Stavely JR, Kelly JD (1993) Identification and potential use of a molecular marker for rust resistance in common bean. Theor Appl Genet 85:745–749

Miklas PN, Johnson WC, Delorme R, Gepts P (2001) QTL conditioning physiological resistance and avoidance to white mold in dry bean. Crop Sci 41:309–315

Miklas PN, Hu J, Grünwald NJ, Larsen KM (2006a) Potential application of TRAP (Targeted Region Amplified Polymorphism) markers for mapping and tagging disease resistance traits in common bean. Crop Sci 46:910–916

Miklas PN, Kelly JD, Beebe SE, Blair MW (2006b) Common bean breeding for resistance against biotic and abiotic stresses: from classical to MAS breeding. Euphytica 147:105–131

Mkandawire ABC, Mabagala RB, Guzman P, Gepts P, Gilbertson RL (2004) Genetic diversity and pathogenic variation of common blight bacteria (*Xanthomonas campestris* pv. *Phaseoli* and *X. campestris* pv. *phaseoli* var. *fuscans*) suggests coevolution with the common bean. Phytopathology 94:593–603

Mmbaga M, Steadman JR, Stavely JR (1996) The use of host resistance in disease management of rust in common bean. Integr Pest Manag Rev 1:191–200

Mumba LE, Galwey NW (1998) Compatibility of crosses between gene pools and evolutionary classes in common bean (*Phaseolus vulgaris* L.). Genet Resour Crop Evol 45:69–80

Mumba LE, Galwey NW (1999) Compatibility between wild and cultivated common bean (*Phaseolus vulgaris* L.) genotypes of the Mesoamerican and Andean gene pools: evidence from the inheritance of quantitative characters. Euphytica 108:105–119

Munoz LC, Baudoin JP (1994) Influence of cold pretreatment and carbon source on callus induction from anthers in *Phaseolus*. Ann Rep Bean Improv Coop 37:129–130

Munoz LC, Baudoin JP (2002) Improvement of *in vitro* induction of androgenesis in *Phaseolus* beans (*P. vulgaris* L. and *P. coccineus* L.). Acta Agron 51:81–87

Munoz LC, Baudoin JP, Bradfer C (1993) In-vitro induction of androgenesis in *Phaseolus*. Ann Rep Bean Improv Coop 36:18–19

Munoz-Florez LC, Baudoin JP (1994) Anther culture in some *Phaseolus* species. In: Roca WM, Mayer JE, Pastor CMA, Tohme MJ (eds) Proceedings of the international scientific meeting of the Phaseolus Bean Advanced Biotechnology Research Network, Cali, Colombia, February 1993. Centro Internacional de Agricultura Tropical (CIAT), Colombia, pp 205–212

Murashige T, Skoog F (1962) A revised medium for rapid growth and bioassays with tobacco tissue cultures. Physiol Plant 15:473–497

Murray J, Larsen J, Michaels TE, Schaafsma A, Vallejos CE et al (2002) Identification of putative genes in bean (*Phaseolus vulgaris*) genomic (Bng) RFLP clones and their conversion to STSs. Genome 45:1013–1024

Mutlu N, Miklas PN, Coyne DP (2006) Resistance gene analog polymorphism (RGAP) markers co-localize with disease resistance genes and QTL in common bean. Mol Breed 17:127–135

Nandakumar PBA, Shivanna KR (1993) Intergeneric hybridization between *Diplotaxis siettiana* and crop brassicas for production of alloplasmic lines. Theor Appl Genet 85:770–776

Navarro F, Sass M, Nienhuis J (2003) Identification and mapping bean root rot resistance in a population of Mesoamerican × Andean origin. Ann Rep Bean Improv Coop 46:213–214

Negri V, Maxted N, Vetelainen M (2009) European landrace conservation: an introduction. In: Vetelainen M, Negri V, Maxted N (eds) European landraces: on-farm conservation, management and use. Bioversity International, Rome, pp 1–22

Nodari RO, Tsai SM, Gilbertson RL, Gepts P (1993a) Towards an integrated linkage map of common bean. II. Development of an RFLP-based linkage map. Theor Appl Genet 85:513–520

Nodari RO, Tsai SM, Guzmán P, Gilbertson RL, Gepts P (1993b) Towards an integrated linkage map of common bean. 3. Mapping genetic factors controlling host-bacteria interactions. Genetics 134:341–350

Ocampo CH, Martin JP, Sanchez- Yelamo MD, Ortiz JM, Toro O (2005) Tracing the origin of spanish common bean cultivars sing biochemical and molecular markers. Genet Resour Crop Evol 52:33–40

Ochoa IE, Blair MW, Lynch JP (2006) QTL analysis of adventitious root formation in common bean under contrasting phosphorus availability. Crop Sci 46:1609–1621. doi:10.2135/cropsci2005.12-0446

Ockendon DJ, Currah L, Taylor JD (1982) Transfer of resistance to halo blight (*Pseudomonas phaseolicola*) from *Phaseolus vulgaris* to *Phaseolus coccineus*. Ann Rep Bean Improv Coop 25:84–85

Ortwin-Sauer C (1966) The early Spanish man. University of California Press, Berkeley

Osborn TC, Alexander DC, Sun SSM, Cardona C, Bliss FA (1988) Insecticidal activity and lectin homology of arcelin seed protein. Science (Wash) 240:207–210

Ouedraogo M, Baudoin JP (2002) Comparative analysis of genetic structure and diversity of wild lima bean populations from Central Valley of Costa Rica, using microsatellite and isozyme markers. Ann Rep Bean Improv Coop 45:240–241

Papa R, Gepts P (2004) Asymmetric gene flow and introgression between wild and domesticated populations. In: dens Nijz HCM, Bartsch D, Sweet J (eds) Introgression from genetically modified plants to wild relatives. CAB International, Wallingford, pp 125–138

Papa R, Gepts P (2003) Asymmetry of gene flow and differential geographical structure of molecular diversity in wild and domesticated common bean (*Phaseolus vulgaris* L.) from Mesoamerica. Theor Appl Genet 106:239–250

Papa R, Acosta J, Delgado-Salinas A, Gepts P (2005) A genome-wide analysis of differentiation between wild and domesticated Phaseolus vulgaris from Mesoamerica. Theor Appl Genet 111:1147–1158

Parker JP, Michaels TE (1986) Simple genetic control of hybrid plant development in interspecific crosses between *Phaseolus vulgaris* and *Phaseolus acutifolius* A. Gray. Plant Breed 97:315–323

Pastor-Corrales MA, Jara C (1995) La evolución de *Phaeoisariopsis griseola* con el frijol común en América Latina. Fitopatol Colomb 19:15–24

Pathania A, Sharma PN, Sharma OP, Chahota RK, Ahmad B, Sharma P (2006) Evaluation of resistance sources and genetics of resistance in kidney bean to Indian virulences of *Colletotrichum lindemuthianum*. Euphytica 149:97–103

Payro de la Cruz E, Gepts P, Colunga Garcia-Marin P, Zizumbo Villareal D (2005) Spatial distribution of genetic diversity in wild populations of *Phaseolus vulgaris* L. from Guanajuato and Michoacán, Mexico. Genet Res Crop Evol 52:589–599

Perry JA, Wang TL, Welham TJ, Gardner S, Pike JM (2003) A TILLING reverse genetics tool and a web-accessible collection of mutants of the legume *Lotus japonicus*. Plant Physiol 131:866–871

Peters JE, Crocomo OJ, Sharp WR, Paddock EF, Tegenkamp I, Tegenkamp T (1977) Haploid callus cells from anthers of *Phaseolus vulgaris*. Phytomorphology 27:79–85

Polegri L, Negri V (2010) Molecular markers for promoting agrobiodiversity conservation: a case study from Italy. How cowpea landraces were saved from extinction. Genet Resour Crop Evol 57:867–880

Pratt RC, Bresssan RA, Hasegawa PM (1985) Genotypic diversity enhance recovery of hybrids and fertile backcrosses of *Phaseolus vulgaris* L. × *P. acutifolius* A. Gray. Euphytica 34:329–344

Prendota K, Baudoin JP, Marechal R (1982) Fertile allopolyploids from cross the *Phaseolus acutifolius* × *Phaseolus vulgaris*. Bull Rech Agron Gembloux 17:177–189

Pueyo JJ, Delgado-Salinas A (1997) Presence of alpha-amylase inhibitor in some members of the subtribe *Phaseolinae* (*Phaseoleae* = *Fabaceae*). Am J Bot 84:79–84

Purseglove JW (1976) The origin and migrations of crops in tropical Africa. In: Harlan JR, de Wet JMJ, Stemler ABL (eds) Origin of African plant domestication. Mouton, The Hague, pp 293–309

Ragagnin VA, Alzate-Marin AL, Souza TLPO, Sanglard DA, Moreira MA et al (2005) Use of molecular markers to pyramiding multiple genes for resistance to rust, anthracnose and angular leaf spot in the common bean. Ann Rep Bean Improv Coop 48:94–95

Raggi L, Barbara T, Negri V (2013) Italian common bean landraces: diversity and population structure. Genet Resour Crop Evol 60:1515–1530

Ramírez M, Graham MA, Blanco-López L, Silvente S, Medrano-Soto A et al (2005) Sequencing and analysis of common bean ESTs. Building a foundation for functional genomics. Plant Physiol 137:1211–1227

Ramirez-Vallejo P, Kelly JD (1998) Traits related to drought resistance in common bean. Euphytica 99:127–136

Rao IM (2001) Role of physiology in improving crop adaptation to abiotic stresses in the tropics: the case of common bean and tropical forages. In: Pessarakli M (ed) Handbook of plant and crop physiology. Marcel Dekker, New York, pp 583–613

Rodriguez-Guerra R, Ramirez-Rueda MT, Martinez de laVega O, Simpson J (2003) Variation in genotype, pathotype and anastomosis groups of *Colletotrichum lindemuthianum* isolates from Mexico. Plant Pathol 52:228–235

Rosales-Serna R, Kohashi-Shibata J, Acosta-Gallegos JA, Trejo-Lopez C, Ortiz- Cereceres J, Kelly JD (2004) Biomass distribution, maturity acceleration and yield in drought stressed common bean cultivars. Field Crop Res 85:203–211

Rossi M, Bitocchi E, Bellucci E, Nanni L, Rau D et al (2009) Linkage disequilibrium and population structure in wild and domesticated populations of *Phaseolus vulgaris* L. Evol Appl 2:504–522

Russell DR, Wallace KM, Bathe JH, Martinell BJ (1993) Stable transformation of *Phaseolus vulgaris* via electric discharge mediated particle acceleration. Plant Cell Rep 12:165–169

Sabja AM, Mok DWS, Mok MC (1990) Seed and embryo growth in pod cultures of *Phaseolus vulgaris* and *P. vulgaris* × *P. acutifolius*. Hortic Sci 25(10):1288–1291

Sandlin CM, Steadman JR, Araya CM, Coyne DP (1999) Isolates of *Uromyces appendiculatus* with specific virulence to landraces of *Phaseolus vulgaris* of Andean origin. Plant Dis 83:108–113

Santalla M, Rodino AP, de Ron AM (2002) Allozyme evidence supporting Southwestern Europe as a secondary center of genetic diversity for the common bean. Theor Appl Genet 104:34–944

Schmit V, Baudoin JP (1992) Screening for resistance to *Ascochyta* blight in populations of *P. coccineus* L. and *P. polyanthus* Greenman. Field Crop Res 30:155–165

Schmit V, du Jardin P, Baudoin JP, Debouck DG (1993) Use of chloroplast DNA polymorphism for the phylogenetic study of seven *Phaseolus* taxa including *P. vulgaris* and *P. coccineus*. Theor Appl Genet 87:506–516

Schmit V, Muñoz JE, du Jardin P, Baudoin JP, Debouck DG (1995) Phylogenetic studies of some *Phaseolus* taxa on the basis of chloroplast DNA polymorphisms. In: Roca WM, Mayer JE, Pastor-Corrales MA, Tohme J (eds) *Phaseolus* beans advanced biotechnology research network Proceedings of the second international scientific meeting, CIAT, Cali, Colombia, 7–10 Sept 1993, pp 69–75

Schneider KA, Rosales-Serna R, Ibarra-Pérez FB, Cazares-Enriquez JA, Gallegos A, Ramirez-Vallejo P, Wassimi N, Kelly JD (1997) Improving common bean performance under drought stress. Crop Sci 37:43–50

Schneider KA, Grafton KF, Kelly JD (2001) QTL analysis of resistance to *Fusarium* root rot in bean. Crop Sci 41:535–542

Schwartz HF, Pastor-Corraies MA (1989) Bean production problems in the tropics, 2nd edn. CIAT, Cali

Schwartz HF, Peairs FB, Hall R (1999) Integrated pest management. Common bean improvement in the twenty-first century. Kluwer Academic Press, Dordrecht, pp 371–388

Schwartz HF, Steadman JR, Hall R, Forster RL (2005) Compendium of bean diseases, 2nd edn. APS Press, St. Paul, p 109

Sharma PN, Kumar A, Sharma OP, Sud D, Tyagi PD (1999) Pathogenic variability in *Colletotrichum lindemuthianum* and evaluation of resistance in *Phaseolus vulgaris* in the north-western Himalayan region of India. J Phytopathol 147:41–45

Sharma TR, Rana JC, Sharma R, Rathour R, Sharma PN (2006) Genetic diversity analysis of exotic and Indian accessions of common bean (*Phaseolus vulgaris L.*) using RAPD markers. Indian J Genet 66(4): 275–278

Sharma PN, Bilal A, Sharma OP, Pathania A, Sharma P (2007) Pathological and molecular diversity in *Colletotrichum lindemuthianum* (bean anthracnose) across Himachal Pradesh- a north-western Himalayan state of India. Australas Plant Pathol 36:191–197

Sharma PN, Diaz LM, Blair MW (2013) Genetic diversity of two Indian common bean germplasm collections based on morphological and microsatellite markers. Plant Genet Resour 11:121–130

Shii CT, Mok MC, Temple SR, Mok DWS (1980) Expression of developmental abnormalities in hybrid of *Phaseolus vulgaris* L. interaction between temperature and allelic dosage. J Hered 71(4):218–222

Singh SP (1982) A key for identification of different growth habits of frijol *Phaseolus vulgaris* L. Ann Rep Bean Improv Coop 25:92–95

Singh SP (1989) Pattern of variation in cultivated common beans (*Phaseolus vulgaris* Fabaceae). Econ Bot 43:39–57

Singh SP (1991) Bean genetics. In: van Schoonhoven A, Voysest O (eds) Common beans: research for crop improvement. CAB International, Wallingford, pp 199–286

Singh SP (1999) Integrated genetic improvement. In: Singh SP (ed) Common bean improvement in the 21st century. Kluwer Academic Publishers, Dordrecht, pp 133–165

Singh SP (2001) Broadening the genetic base of common bean cultivars. Crop Sci 41:1659–1675

Singh SP, Gutierrez JA (1984) Geographical distribution of DL1 and DL2 genes causing hybrid dwarfism in *Phaseolus vulgaris* L. their association with seed size and their significance in breeding. Euphytica 33(2):337–345

Singh SP, Molina A (1996) Inheritance of crippled trifoliate leaves occurring in interracial crosses of common bean and its relationship with hybrid dwarfism. J Hered 87(6):464–469

Singh SP, Schwartz HF (2010) Breeding common bean for resistance to diseases: a review. Crop Sci 50:2199–2223

Singh SP, Schwartz HF (2011) Review: breeding common bean for resistance to insect pests and nematodes. Can J Plant Sci 91:239–250

Singh SP, Gepts P, Debouck DG (1991a) Races of common bean (*Phaseolus vulgaris* L., Fabaceae). Econ Bot 45:379–396

Singh SP, Gutierrez A, Molina A, Urrea C, Gepts P (1991b) Genetic diversity in cultivated common bean: II Marker- based analysis of morphological and agronomic traits. Crop Sci 31:23–29

Singh SP, Nodari R, Gepts P (1991c) Genetic diversity in cultivated common beans. I. Allozymes. Crop Sci 31:19–23

Singh SP, Debouck DG, Roca WM (1997) Successful interspecific hybridization between *Phaseolus vulgaris* L. and *Phaseolus costaricensis* Freytag & Debouck. Ann Rep Bean Improv Coop 40:40–41

Singh SP, Cardona C, Morales FJ, Pastor-Corrales MA, Voysest O (1998) Gamete selection for upright carioca bean with resistance to five diseases and a leafhopper. Crop Sci 38:666–672

Singh SP, Terán H, Muñoz CG, Takegami JC (1999) Two cycles of recurrent selection for seed yield in common bean. Crop Sci 39:391–397

Singh SP, Muñoz CG, Terán H (2001a) Registration of common bacterial blight resistant dry bean germplasm VAX 1, VAX 3, and VAX 4. Crop Sci 41:275–276

Singh SP, Terán H, Gutiérrez JA (2001b) Registration of SEA 5 and SEA 13 drought tolerant dry bean germplasm. Crop Sci 41:276–277

Smartt J (1970) Interspecific hybridization between cultivated American species of the genus Phaseolus. Euphytica 19:480–488

Smartt J (1979) Interspecific hybridization in grain legumes: a review. Econ Bot 33:329–337

Smartt J (1990) Grain legumes. Evolution and genetic resources. Cambridge University Press, Cambridge, p 379

Smartt J, Haq N (1972) Fertility and segregation of amphidiploid *Phaseolus vulgaris* L. × *Phaseolus coccineus* L. and its behaviour in backcrosses. Euphytica 21:496–591

Snoeck C, Vanderleyden J, Beebe S (2003) Strategies for genetic improvement of common bean and rhizobia towards efficient interactions. Plant Breed Rev 23:21–72

Sonnante G, Stockton T, Nodari RO, Becerra Velásquez VL, Gepts P (1994) Evolution of genetic diversity during the domestication of common-bean (*Phaseolus vulgaris* L.). Theor Appl Genet 89:629–635

Sponchiado BN, White JW, Castillo JA, Jones PG (1989) Root growth of four common bean cultivars in relation to drought tolerance in environments with contrasting soil types. Exp Agric 25:249–257

Stalker HT (1980) Utilization of wild species for crop improvement. Adv Agron 33:111–147

Sullivan JG, Freytag G (1986) Predicting interspecific compatibilities in beans (*Phaseolus*) by seed protein electrophoresis. Euphytica 35:201–209

Tai KS, Cheng SH (1990) Tissue culture of cultivated *Leguminosae*. J Agric Assoc China 149:49–52

Tanksley SD, Nelson JC (1996) Advanced back cross QTL analysis, a method for the simultaneously discovery and transfer of valuable QTLs from unadapted germplasm into elite breeding lines. Theor Appl Genet 92:191–203

Tapia H, Comacho A (1988) Manejo Integrado de la Produccion de Frijol Basado en Labranza Zero. GTZ, Eschborn

Tar'an B, Michaels TE, Pauls KP (2002) Genetic mapping of agronomic traits in common bean. Crop Sci 42:544–556

Terán H, Singh SP (2009) Gamete selection for improving physiological resistance to white mold in common bean. Euphytica 167:271–280

Terán H, Singh SP (2010a) Recurrent selection for physiological resistance to white mold in dry bean. Plant Breed 129:327–333

Terán H, Singh SP (2010b) Gamete and recurrent selection for improving physiological resistance to white mold in common bean. Can J Plant Sci 90:153–162

Thomas CV, Waines JG (1984) Fertile backcross and allotetraploid plants from crosses between tepary beans and common beans (*Phaseolus acutifolius*, *Phaseolus vulgaris*). J Hered 75:93–98

Thung M (1990) Phosphorus: a limiting nutrient in bean (Phaseolus vulgaris L.) production in Latin America and field screening for efficiency and response. In: El Bassam N, Dambrooth M, Loughman BG (eds) Genetic aspects of plant mineral nutrition. Kluwer, Dordrecht, pp 501–521

Thung MDT, Rao I (1999) Integrated management of abiotic stresses. In: Singh SP (ed) Common bean improvement in twenty first century. Kluwer, Dordrecht, pp 331–370

Thurston HD (1992) Sustainable practices for plant disease management in traditional farming systems. Westview Press, Boulder

Tohme J, González O, Beebe S, Duque MC (1996) AFLP analysis of gene pools of a wild bean core collection. Crop Sci 36:1375–1384

Toro O, Tohme J, Debouck DG (1990) Wild bean (Phaseolus vulgaris L): description and distribution. Centro Internacional de Agricultura Tropical, Cali

Tsai SM, Nodari RO, Moon DH, Camargo LEA, Vencovsky R, Gepts P (1998) QTL mapping for nodule number and common bacterial blight in *Phaseolus vulgaris* L. Plant Soil 204:135–145

Tseng TC, Tu JC, Sot LC (1995a) Comparison of the profiles of seed-borne fungi and the occurrence of aflatoxins in mold-damaged beans and soybeans. Microbios 84:105–116

Tseng TC, Tu JC, Tsean SS (1995b) Mycoflora and mycotoxins in dry bean (*Phaseolus vulgaris*) produced in Taiwan and in Ontario, Canada. Bot Bull Acad Sin 36:229–234

Urrea C, Miklas P, Beaver J, Riley R (1996) A codominant randomly amplified polymorphic DNA (RAPD) marker useful for indirect selection of bean golden mosaic virus resistance in common bean. J Am Soc Hortic Sci 121:1035–1039

Vadez V, Lasso JH, Beck DP, Drevon JJ (1999) Variability of N2-fixation in common bean (*Phaseolus vulgaris* L.) under P deficiency is related to P use efficiency: N2-fixation tolerant to P deficiency. Euphytica 106:231–242

Vallejos CE, Chase CD (1991) Linkage between isozyme markers and a locus affecting seed size in *Phaseolus vulgaris* L. Theor Appl Genet 81:413–419

Vallejos CE, Sakiyama NS, Chase CD (1992) A molecular marker-based linkage map of *Phaseolus vulgaris* L. Genetics 131:733–740

Van de Velde W, Mergeay J, Holsters M, Goormachtig S (2003) *Agrobacterium rhizogenes*-mediated transformation of *Sesbania rostrata*. Plant Sci 165:1281–1288

van Reheenen HA, Muigai SGS, Kitivo DK (1979) Male sterility in beans (*Phaseolus vulgaris* L.). Euphytica 28:761–763

van Schoonhoven A, Pastor-Corrales MA (1987) Standard system for the evaluation of bean germplasm. CIAT, Cali, p 53

Vekemans X, Hardy O, Berken B, Fofana B, Baudoin JP (1998) Use of PCR- RFLP on chloroplast DNA to investigate the phylogenetic relationships in the genus *Phaseolus*. Biotechnol Agro Soc Environ 2:128–134

Verdecourt B (1970) Studies in the *Leguminosae-Papilionoideae* for the flora of tropical East Africa. IV. Kew Bull 24:507–569

Vianna GR, Albino MMC, Dias BBA, Silva LDM, Rech EL, Aragão FJL (2004) Fragment DNA as vector for genetic transformation of bean (*Phaseolus vulgaris* L.). Sci Hortic 99:371–378, Amsterdam

Vieira AL, Patto Ramalho MA, dos Santos JB (1989) Crossing incompatibility in some bean cultivars utilized in Brazil. Rev Brasil Genet 12:169–171

Wallace DH, Wilkinson RE (1975) Breeding for resistance in dicotyledonous plants to root rot fungi. In: Bruehll GW (ed) Biology and control of soil-borne plant pathogens. American Phytopathological Society, St. Paul

Weilenmann E, Baudoin JP, Maréchal R (1986) Obtention d'allopolyploides fertiles chez le croisement entre *Phaseolus vulgaris* et *Phaseolus filiformis*. Bull Rech Agron Gembloux 21(1):35–46

Welsh W, Bushuk W, Roca W, Singh SP (1995) Characterization of agronomic traits and markers of recombinant inbred lines from intraracial and interracial populations of *Phaseolus vulgaris*. Theor Appl Genet 91:169–177

White JW (1987) Preliminary results of the Bean International Drought Yield Trial (BIDYT). In: Proceedings of the international bean drought workshop, Cali, Colombia, 19–21 Oct 1987, pp 126–145

White JW, Singh SP (1991) Breeding for adaptation to drought. In: van Schoonhoven A, Voysest O (eds) Common beans: research for crop improvement. CAB International/CIAT, Wallingford/Cali, pp 501–551

White JW, Ochoa R, Ibarra F, Singh SP (1994) Inheritance of seed yield, maturity and seed weight of common bean (*Phaseolus vulgaris* L.) under semi-arid rainfed conditions. J Agric Sci 122:265–273

Wortmann CS, Kirkby RA, Eledu CA, Allen DJ (1998) Atlas of common bean (*Phaseolus vulgaris* L.) production in Africa. CIAT Pan-African Bean Research Alliance, Cali

Yaish MWF, de la Vega MP (2003) Isolation of (GA)(n) microsatellite sequences and description of a predicted

MADS-box sequence isolated from common bean (*Phaseolus vulgaris* L.). Genet Mol Biol 26:337–342

Yan X, Beebe SE, Lynch JP (1995a) Genetic variation for phosphorus efficiency of common bean in contrasting soil types. II. Yield response. Crop Sci 35:1094–1099

Yan X, Lynch JP, Beebe SE (1995b) Genetic variation for phosphorus efficiency of common bean in contrasting soil types. I. Vegetative response. Crop Sci 35:1086–1093

Yan X, Lynch JP, Beebe SE (1996) Utilization of phosphorus substrates contrasting common bean genotypes. Crop Sci 36:936–941

Yan X, Liao H, Trull MC, Beebe SE, Lynch JP (2001) Induction of a major leaf acid phosphatase does not confer adaptation to low phosphorus availability in common bean. Plant Phys 125:1–11

Young R, Melotto M, Nodari R, Kelly JD (1998) Marker-assisted dissection of the oligogenic anthracnose resistance in the common bean cultivars, 'G 2333'. Theor Appl Genet 96:87–94

Yu K, Park S, Poysa V, Gepts P (2000) Integration of simple sequence repeat (SSR) markers into a molecular linkage map of common bean (*Phaseolus vulgaris* L.). J Hered 91:429–434

Zagorcheva L, Poriazov I (1983) Hybrid development and fertility in *Phaseolus vulgaris* × *Phaseolus coccineus*. Ann Rep Bean Improv Coop 26:92–93

Zambre M, Goossens A, Cardona C, Van Montagu M, Terryn N et al (2005) A reproducible genetic transformation system for cultivated *Phaseolus acutifolius* (tepary bean) and its use to assess the role of arcelins in resistance to the Mexican bean weevil. Theor Appl Genet 110:914–924

Zhang ZY, Coyne DP, Mitra A (1997) Factors affecting *Agrobacterium*-mediated transformation of common bean. J Am Soc Hortic Sci 122:300–305

Zhang X, Blair MW, Wang S (2008) Genetic diversity of Chinese common bean (*Phaseolus vulgaris* L.) landraces assessed with simple sequence repeat markers. Theor Appl Genet 117:629–640

Zizumbo-Villarreal D, Colunga-Garcý́a Marý́n P, Payró de la Cruz E, Delgado-Valerio P, Gepts P (2005) Population structure and evolutionary dynamics of wild–weedy–domesticated complexes of common bean in a Mesoamerican region. Crop Sci 45:1073–1083

Chickpea

3

Sarvjeet Singh, Inderjit Singh, Karan Kapoor, P.M. Gaur, S.K. Chaturvedi, N.P. Singh, and J.S. Sandhu

Abstract

The narrow genetic base of cultivated chickpea warrants systematic collection, documentation and evaluation of chickpea germplasm and particularly wild *Cicer* species for effective and efficient use in chickpea breeding programmes. Limiting factors to crop production, possible solutions and ways to overcome them, importance of wild relatives and barriers to alien gene introgression and strategies to overcome them and traits for base broadening have been discussed. It has been clearly demonstrated that resistance to major biotic and abiotic stresses can be successfully introgressed from the primary gene pool comprising progenitor species. However, many desirable traits including high degree of resistance to multiple stresses that are present in the species belonging to secondary and tertiary gene pools can also be introgressed by using special techniques to overcome pre- and post-fertilization barriers. Besides resistance to various biotic and abiotic stresses, the yield QTLs have also been introgressed from wild *Cicer* species to cultivated varieties. Status and importance of molecular markers, genome mapping and genomic tools for chickpea improvement are elaborated. Because of major genes for various biotic and abiotic stresses, the transfer of agronomically important traits into elite cultivars has been made easy and practical through marker-assisted selection and marker-assisted backcross. The usefulness of molecular markers such as SSR and SNP for the construction of high-density genetic maps of chickpea and for the identification of genes/QTLs for stress resistance, quality and yield contributing traits has also been discussed.

S. Singh (✉) • I. Singh • K. Kapoor
Department of Plant Breeding and Genetics, Punjab Agricultural University, Ludhiana 141004, Punjab, India
e-mail: sarvjeetm@rediffmail.com; inderjitpau@rediffmail.com; karanpb09@ymail.com

P.M. Gaur
International Crops Research Institute for the Semi-Arid Tropics, Patancheru, 502324 Hyderabad, Andhra Pradesh, India
e-mail: p.gaur@cgiar.org

S.K. Chaturvedi • N.P. Singh
Indian Institute of Pulses Research, Kanpur 208024, Uttar Pradesh, India
e-mail: sushilk.chaturvedi@gmail.com; narendrasingh@scientist.com

J.S. Sandhu
Department of Agriculture and Cooperation, Ministry of Agriculture, Krishi Bhawan, New Delhi 110001, India
e-mail: js_sandhu@hotmail.com

M. Singh et al. (eds.), *Broadening the Genetic Base of Grain Legumes*,
DOI 10.1007/978-81-322-2023-7_3,

3.1 Introduction

Chickpea (*Cicer arietinum* L.) is cultivated in almost all parts of the world covering Asia, Africa, Europe, Australia, North America and South America continents. It is known by various common or local names in different countries like *Hamas*, *Hommos*, *Humz*, *Nakhi* and *Melanch* in Arabian countries; *Keker* in the Netherlands; *Kichererbse* in Germany and Belgium; *Ceseror* and *Cicerolle* in France, *Ceci* in Vatican City and Switzerland, *Simbra* in Ethiopia; *Lablabi* in Turkey; *Garbanzo* or *Garbanzobean* in Spain; *Gravanço* in Portugal; and *Ovetichie* in Russia. Similarly, in India, chickpea is known by various names like *Chana* or *Gram* or *Bengal gram* or *Chani* in Haryana, Rajasthan, Uttarakhand, Uttar Pradesh, Madhya Pradesh, Chhattisgarh, Bihar, Jharkhand, etc.; *Chhole* in Punjab, Jammu and Kashmir and Delhi; *Chola* in West Bengal; *Harbara* in Maharashtra; *Boot* in Orissa; *Sanagulu* in Andhra Pradesh; *Kadale* in Karnataka; *Kadalai* in Tamil Nadu; and *Kadala* in Kerala, indicating its widespread cultivation and knowledge of utilization.

Chickpea, a member of Fabaceae, is a self-pollinated true diploid ($2n = 2x = 16$) with genome size of 738 Mbp (Varshney et al. 2013a). It is an ancient cool season food legume crop cultivated by man and has been found in Middle Eastern archaeological sites dated 7500–6800 BC (Zohary and Hopf 2000). Its cultivation is mainly concentrated in semiarid environments (Saxena 1990). Chickpea is the second most important food legume crop after common bean (FAOSTAT 2011). It is grown in more than 50 countries on an area of 13.2 m ha, producing approximately 11.62 m tonnes annually. India ranks first in the world's production and area by contributing around 70.7 % to the world's total production (FAOSTAT 2011). It is one of the most important food legume plants in sustainable agriculture system because of its low production cost, wider adaptation, ability to fix atmospheric nitrogen and fit in various crop rotations (Singh 1997) and presence of prolific tap root system. Chickpea can fix atmospheric nitrogen up to 140 kg/ha through its symbiotic association with *Rhizobium* and meets its 80 % requirement (Saraf et al. 1998). It also helps in enhancing the soil quality for subsequent cereal crop cultivation by adding organic matter for the maintenance of soil health and ecosystem. Deep and tap root system of chickpea is known to help in opening up of the soil to the deeper strata, ensuring better texture and aeration of the soil for next crop.

It is a rich source of quality protein (20–22 %) to the predominantly vegetarian population in Indian subcontinent, other South Asian countries and the Middle East. It has the highest nutritional compositions and free from anti-nutritive components compared to any other dry edible grain legumes, and thus, it is considered a functional food or nutraceutical. Besides proteins, it is rich in fibre and minerals (phosphorus, calcium, magnesium, iron and zinc), and its lipid fraction is high in unsaturated fatty acids (Williams and Singh 1987). It has no anti-nutritional factors (Mallikarjuna et al. 2007) and contains higher amounts of carotenoids like β-carotene than genetically engineered 'golden rice' (Abbo et al. 2005). This plant holds a good repute in 'Ayurvedic' and 'Unani' systems of medicine. In India, acid exudates from the leaves were used medicinally for aphrodisiac, bronchitis, cholera, constipation, diarrhoea, dysentery, snakebite, sunstroke and warts. It also has the property to act as hypo-cholesteremic agent; germinating chickpea is believed to reduce the blood cholesterol level. Sprouted seeds are eaten as a vegetable or salad. Young leaves and stems and green pods are eaten like vegetables. Leaves yield an indigo-like dye. The dried seeds may be used in soups or after grinding as flour. Grain husks, stems and leaves may be used in livestock feed. In the USA and Europe, chickpeas are marketed dried, canned or in various vegetable mixtures. Mashed chickpea mixed with oils and spices (hummus) is a popular hors d'oeuvre in the Mediterranean Middle East. Vavilov (1926) supported the idea of Southwest Asia and the Mediterranean region being the primary centres of origin, with Ethiopia as the secondary centre. van der Maesen (1987) suggested that Anatolia in Turkey was the area where chickpea was believed to have originated. Two types of chickpea cultivars are recognized globally – *kabuli* and *desi*. The *kabuli* types are generally grown in the Mediterranean region including Southern Europe, Western Asia and Northern Africa, and the *desi* types are grown mainly in Ethiopia and Indian subcontinent. *Desi* chickpeas are characterized by flowers of varying colours, angular to round seeds with dark seed coat, anthocyanin pigmentation on stem or other plant parts, rough seed surface and with erect, semierect or semi-spreading growth habit, whereas *kabuli* types generally have owl- or ram-shaped beige-coloured seeds, white flowers, smooth seed surface, lack of anthocyanin pigmentation and semi-spreading to erect growth habit (Pundir et al. 1985). Of the total production, the *desi* and *kabuli* chickpeas contribute around 80 % and 20 %, respectively. *Kabuli* type is mainly grown in temperate regions, while the *desi* type chickpea is grown mostly in the semiarid tropics (Malhotra et al. 1987; Muehlbauer and Singh 1987).

3.2 Systematics, Genetic Relationships and Crop Gene Pool

3.2.1 Systematics

The *Cicer* genus belongs to family Leguminosae, subfamily Papilionaceae and tribe Cicereae Alef. The *Cicer* genus currently comprises 43 species, out of which 9 are annual and 34 are perennial species (Muehlbauer et al. 1994). Most of these species are found in West Asia and North Africa, covering Turkey in the north to Ethiopia in the south and Pakistan in the east to Morocco in the west. Of the 9 annual *Cicer* species, *C. arietinum* is the only cultivated species. The eight other annual *Cicer* species are *C. reticulatum*, *C. echinospermum*, *C. pinnatifidum*, *C. judaicum*, *C. bijugum*, *C. cuneatum*, *C. chorassanicum* and *C. yamashitae*. The wild annual progenitor of chickpea has been identified as *C. reticulatum* L. (Ladizinsky and Adler 1976), and the perennial progenitor is proposed as *C. anatolicum* (Tayyar and Waines 1996). The *Cicer* species, including cultivated and wild, have been classified into four sections based on their geographical distribution, life cycle and morphological characteristics (van der Maesen 1987). The 8 annual species, namely, *C. arietinum*, *C. reticulatum*, *C. echinospermum*, *C. pinnatifidum*, *C. bijugum*, *C. judaicum*, *C. yamashitae* and *C. cuneatum*, were grouped in *Monocicer* section, *C. chorassanicum* and *C. incisum* (perennial species) in *Chamaecicer* section, 23 perennial species in *Polycicer* section and the remaining 7 woody perennial species in *Acanthocicer* section. The distribution of different *Cicer* species is given in Table 3.1.

3.2.2 Genetic Relationships

The knowledge of genetic relationships between the cultivated and its wild relatives is a prerequisite to exploit related species for the introgression of useful traits from wild to cultivated background, to track the evolution of cultivated species and also to establish the relatedness among the species within the genus. Before making use of wild forms in a better way, there is need to understand crossability relationships, chemotaxonomic relationships and cytogenetical affinities among the wild species and cultigens (Hawkes 1977). Interspecific hybridization, seed storage protein profiles, isozymes, karyotype and molecular markers have been used as criteria to study species relationships in the genus *Cicer* (Kaur et al. 2010a).

3.2.3 Crop Gene Pool

Harlan and de Wet (1971) proposed the concept of gene pools and their uses in crop improvement. The genus *Cicer* is classified into three gene pools (primary, secondary and tertiary) based on crossability with cultigens. The experimental evidences permitted to define the gene pool of chickpea (Ladizinsky and Adler 1976; Ahmad et al. 1987). The accumulated evidence of experimental hybridization to the gene pool approach would save the misdirected efforts in attempting wide crosses. Various researchers proposed different pools for different species. van der Maesen et al. (2007) proposed recent classification in which primary gene pool consists of cultivated species and landraces. The secondary gene pool consists of the progenitor species, *C. reticulatum* and *C. echinospermum*, the species that are crossable with *C. arietinum* but with reduced fertility of the resulting hybrids and progenies; nevertheless, both are cross-compatible with the cultigen and do not need in vitro interventions to produce hybrids. The tertiary gene pool

Table 3.1 Distribution of different annual and perennial *Cicer* species

Species	Distribution
Annual species	
C. arietinum	Mediterranean region to Burma, Ethiopia, Mexico, Chile
C. chorassanicum	N and C Afghanistan, N and NE Iran
C. bijugum	SE Turkey, N Syria, N Iraq
C. cuneatum	Ethiopia, SE Egypt, NE Sudan, Saudi Arabia
C. echinospermum	Turkey, E Anatolia, N Iraq
C. judaicum	Palestine, Lebanon
C. pinnatifidum	Cyprus, N Iraq, Syria, Turkey, USSR
C. reticulatum	E Turkey
C. yamashitae	Afghanistan
Perennial species	
C. acanthophyllum	Afghanistan, Pakistan, USSR
C. anatolicum	Turkey, Iran, Iraq, Armenia
C. atlanticum	Morocco
C. balcaricum	Caucasus (USSR)
C. baldshuanicum, C. flexuosum, C. grande, C. incanum, C. korshinskyi, C. laetum, C. mogoltavicum, C. paucijugum, C. rassuloviae, C. songaricum	USSR
C. canariense	Canary islands, Tenerife, La Palma
C. fedtschenkoi	USSR, N and NE Afghanistan
C. floribundum, C. heterophyllum, C. isauricum	Turkey
C. graecum	Greece
C. incisum	Greece, Turkey, Iran, Lebanon, USSR
C. kermanense	SE Iran
C. macrocanthum	Afghanistan, India, Pakistan, USSR
C. microphyllum	E Afghanistan, Tibet, India, Pakistan, USSR
C. montbretti	Albania, Bulgaria, Turkey
C. multijugum, C. rechingeri	Afghanistan
C. nuristanicum	Afghanistan, India, Pakistan
C. oxyodon	Iran, Afghanistan, N Iraq
C. pungens	Afghanistan, USSR
C. spyroceras, C. stapfianum, C. subaphyllum	Iran
C. tragacanthoides	Iran, USSR

consists of all the annual and perennial *Cicer* species that are not crossable with cultivated species.

The *Cicer* species possess wealth of useful genes for biotic and abiotic stresses and hold promise for enhancing seed yield through introgression of wild genes into cultivated species (Singh et al. 1982a, b, 2005, 2014; Singh and Ocampo 1993). Large inter-accessions as well as intra-accessions variation has been observed in different *Cicer* species for resistance to *Botrytis* grey mould (Kaur et al. 2007). Verma et al. (1990) studied the crossability of cultivated chickpea, used as female, with *C. echinospermum, C. judaicum* and *C. bijugum,* used as males. All the F_1s were fertile and were successfully advanced to later generations. All the crosses were developed under field conditions using mixture of plant growth hormones at the time of emasculation and pollinations. The success was attributed to the use of growth hormones and large number of pollinations and, thus, indicated that there is a need to reclassify *Cicer* species. The recent studies (Sandhu et al. 2005; Kaur et al. 2013) indicated that *C. pinnatifidum*, a valuable source for several biotic and abiotic stresses, can be crossed successfully with cultivated chickpea for the transfer of high level of resistance to *Botrytis* grey mould and *Ascochyta* blight (Kaur et al. 2013). These studies further confirm findings of Verma et al. (1990).

3.3 Assessment of Gene Flow for Crop Improvement

Due to long evolutionary process, high-yielding genes might have been eroded from populations or become silent resulting in average grain yield of most of the present-day cultivars. It is also true that in most of the cases, high-yielding varieties developed through hybridization in the last 40–50 years have utilized limited variability as the derivatives of already utilized plant genetic resources were considered in making crosses for the improvement of targeted traits. Therefore, the assessment of gene flow and constraints in flow has been discussed below.

Due to domestication bottlenecks, the genetic base of various legume crops including chickpea became narrow (Spillane and Gepts 2001). The bottleneck is perpetuated further by various reproductive isolation factors preventing gene flow. The domesticated plants may be carried by the cultivators to sites far removed from its original habitat. During transfer between latitudes, there may be further narrowing of the genetic base, because the population would not be well adapted to the new day-length conditions, and so only a small number of the genotypes would survive. There may be other chance occurrences that narrow the genetic base such as disease epidemics, which may decimate populations. Post-domestication, the crops evolved under human selection but continued to possess a breadth of genetic variation in order to overcome challenges from changes in biotic and abiotic milieu. The advent of modern plant breeding resulted in the creation of plant varieties that optimized adaptation at the cost of adaptability. The narrowed genetic base of germplasm is often evident from plateau in yield gains as have been observed in several other crop plants. Historically, instances of more disastrous consequences have also been observed. Often quoted instances include the *Ascochyta* blight epidemics caused by *Ascochyta rabiei* in North India in 1980 and 1982 (Singh et al. 1982b, 1984b). During the process of evolution, chickpea like other crops are subjected to genetic bottlenecks and subsequent founder effect that resulted in narrow genetic base. The progenitor species *C. reticulatum* (Ladizinsky and Adler 1976) is narrowly distributed in Southeastern Turkey and harbours limited adapted variation compared to wheat and barley (Berger et al. 2003). During domestication, it is likely that only a small proportion of diversity of wild population is sampled. Subsequent gene flow between new cultivars and its wild progenitors might be restricted by breeding barriers and nature of domestication event (Cooper et al. 2001). Shifting from autumn to spring season probably to avoid *Ascochyta* blight has reduced the genetic diversity and selection to suit post-rainy season cropping has further narrowed the genetic base. The replacement of landraces by elite cultivars developed through hybridization using closely related parents caused another bottleneck. With changing climatic conditions and evolution of new pathogens, selection will be rigorous for chickpea germplasm to withhold biotic and abiotic stresses, which will further narrow down the genetic base. The pedigree analysis of 86 chickpea varieties released in India through hybridization and selection revealed that top ten ancestors contributed more than 35 % to their genetic base and about 41 % varieties have 'Pb7' as one of the ancestors in their pedigree (Kumar et al. 2004, 2008, 2009). Because of the limited genetic variability within the primary gene pool, the genetic improvement of chickpea by classical breeding involving intervarietal crosses has met with limited success (Singh and Ocampo 1993). Thus, the situation warrants an urgent need to broaden the genetic base of cultivated varieties through genetic enhancement by involving unadapted germplasm, exotic germplasm and landraces in hybridization (Duvick 1995). Since the genetic variability within the primary gene pool is limited, there is a need for introgression of alien genes through pre-breeding efforts for widening the base of cultivated gene pool (Verma et al. 1990; van Rheenen et al. 1993; Nadarajan and Chaturvedi 2010).

3.4 Level of Diversity in Crop Germplasm

The sum total of hereditary materials present in a crop species and its wild relatives is referred to as germplasm. This is also known as genetic resources or gene pool. In other words, germplasm is a collection of genetic resources for an organism which include inbred lines, landraces, open pollinated varieties, exotic accessions, wild species, cultivars and breeding stocks. These types of germplasm can carry unidentified variation that may be a valuable resource for breeders and other researchers. Germplasm can be collected from centres of diversity, gene banks, gene sanctuaries, farmer's field, markets and seed companies. Genetic pool represents the entire genetic variability or diversity within a crop species. Agricultural practices have gradually dispersed the local traditional varieties and crop wild relatives (CWRs) leading to a loss of indigenous diversity. However, CWR and landraces (LR) are the two major components of agro-biodiversity that offer the widest range of diversity for breeders in crop improvement programmes. The CWRs and locally adapted traditional crop varieties contain vital sources of useful genes. These invaluable resources are threatened by the climate change as well as by a range of other human-induced pressures and socio-economic changes, while the value of CWR and LR for food security is widely recognized. A systematic strategy for the conservation of the highest priority CWR and LR resources is required at global level.

3.4.1 Crop Wild Relatives

The crop wild relatives (CWRs) provide the broadest range of genetic diversity in grain legumes, including chickpea, and have the ability to provide the wide range of germplasm resources for the incorporation of various traits of agro-economic importance (Singh et al. 2013). Many species are being extinct because natural habitats are being lost due to increased human pressure and ecological threats. There is an urgent need for systematic exploration and sample of the genetic diversity in wild relatives that was partially captured during the domestication. The *ex situ* conservation was pioneered by Vavilov (1926), and subsequent explorations resulted in large collections in gene banks. However, these gene banks are dominated by cultivated forms of crops. Crop wild relatives are also important from phylogenetic perspective, applied in interspecific crosses to increase the diversity, as the majority of allelic variation is predicted to occur outside of the crop itself. In spite of the existence of vast collections of CWR, their use for crop improvement has been limited as assessing the genetic diversity was a challenge until now in most of the legumes including chickpea. Selections for higher yield and quality characteristics during domestication have resulted in narrowing of the genetic variation in cultivated chickpea. The wild *Cicer* species do not consist of useful variation for morphological characteristics and protein content, but they are rich sources of resistance to various biotic and abiotic stresses (Singh et al. 1998; Croser et al. 2003; Sandhu et al. 2006; Mallikarjuna et al. 2007; Kaur et al. 2013), yield QTLs (Singh and Ocampo 1997; Singh et al. 2005) and biochemical traits (Kaur et al. 2010b).

3.4.2 Landraces

A variable population, which is identifiable and usually has a local name, is designated as landrace. It lacks 'formal' crop improvement and characterized by a specific adaptation to the environmental conditions of the area of cultivation and is associated with the traditional uses, knowledge, habits, dialects and celebrations of the people who developed and continue to grow it (Lorenzetti and Negri 2009). The LR belongs to the people who developed it and feel to be its owner (in a specific human context). They are maintained because of their better quality than commercial varieties and better yield performance/persistence under difficult pedo-climatic conditions. It is estimated that less than one third of them is already marketed as niche, typical product (Negri 2003). However, most of them are highly threatened because they are cultivated primarily by ageing farmers (Negri 2003; Galluzzi et al. 2010). The lack of traditional information severely hampers the possibility of conserving and using these LR effectively. There is an urgent need to make an inventory and focus should be given to the priority species. Landraces as an agro-biodiversity resource is not only critical for future food security but is a vital component of our biodiversity and cultural heritage. The extent of loss of crop genetic diversity is associated with the loss of landraces which is very difficult to quantify accurately. This erosion of our agro-biodiversity resources is likely to be critical for future food security. It has been recognized in a number of international legal platforms, including the Convention on Biological Diversity and the International Treaty on Plant Genetic Resources for Food and Agriculture. In general, it has been noticed that the average landrace maintainer was found to be 65 years old and there was little evidence of the next generation being willing or able to continue the family role as maintainers. Hence the need for research in this area and the production of a corresponding inventory is necessary. Any new variation will not replace the old varieties that have been lost. Rich diversity in the form of landraces of chickpea has been reported in Bundelkhand and part of Chhattisgarh in India.

3.4.3 Commercial Cultivars

In the 1970s, most of the commercial varieties were developed through selection from landraces. The major emphasis was given on increasing yield potential. During the 1980s, the focus was laid on breeding for disease resistance. Consequently, several varieties (Avarodhi, JG 315, Pusa 209, GNG 16, Pusa 212, Pusa 240, Pusa 244, Pusa 256, Pusa 413, ICCV 10, ICCV 37, Phule G 5, Phule G 12, etc.) resistant/tolerant to *Fusarium* wilt were developed and released for their cultivation in different regions of the country (Chaturvedi et al. 2003). Similarly, genotypes tolerant to *Ascochyta* blight, namely, C235, Gaurav, H 75-35, BG 261, PBG 1, GNG 146 and PBG 5, were developed for North West Plain Zone (Delhi, Punjab, Haryana, North Western Rajasthan and Western UP) and G 543 and PBG 5 for the state of Punjab (Sandhu et al. 2004). However, during the 1990s, the major thrust was given on breeding for multiple disease resistance and high-input responsive varieties. Sources for drought tolerance (RSG 44, RSG 963, RSG 888, ICC 4958, ICCV 10, Vijay, GL 769, GPF 2, PDG 3, PDG 4, Phule G 5), cold tolerance (ICCV 88506, ICCV 88503, Phule G 96006, ICC 8923, PDG 84-10, GL 28008, GL 28028) and salt tolerance (CSG 88101, CSG8962) were identified for their use in breeding programme. As a result, multiple disease-resistant varieties, namely, Bharati, Pusa 372, Pusa 362, BG 391, KWR 108 and GNG 1581 against wilt and root rot and GNG 469 against *Ascochyta* blight and root rot, and high-input responsive variety like DCP 92-3 were released for cultivation. Rice fallows (about 11.0 m ha) in Eastern India (eastern UP, Bihar, West Bengal, Orissa, Jharkhand and Assam) and Central India (eastern MP and Chhattisgarh) provide opportunities for horizontal expansion of area under chickpea. This requires the development of varieties amenable for late planting to popularize rice-chickpea sequential cropping system. As a result of concerted breeding efforts, varieties like KPG 59, Pant G 186, BGM 547, RSG 963, Rajas and Pusa 372 were developed for late sown condition in Eastern India. Similarly, recently developed varieties like JG 14, Vaibhav, JSC 55 and JSC 56 have great potential for adaptation in late sown condition and rice fallow of Central India due to short duration, drought and heat tolerance traits.

The development of short-duration varieties like ICCV 2, JG 74, Vijay, JG 11, JG 16, JAKI 9218 and KAK 2 was the major catalyst for the expansion of chickpea area in Southern and Central India. In spite of reduction in duration, the yield potential of these early maturing varieties remains almost unaffected, thus improving per day the productivity of the crop. Presently emphasis has been laid on the development of extra-large seeded *kabuli* chickpea varieties (>50 g/100 seed weight). Some of the promising varieties, viz. Phule G 0517, IPCK 02, MNK 1 and PKV Kabuli 4-1, have been released for Maharashtra, Madhya Pradesh, Andhra Pradesh and Karnataka, and a medium bold seeded variety L 552 has been released for Punjab state. A major breakthrough has been witnessed in developing large seeded *kabuli* varieties with high-yield potential such as KAK 2, BG 1003, BG 1053, JGK 1, Phule G 95311, IPCK 2002-29, IPCK 2004-29, L 555 and HK 05-169 (Chaturvedi et al. 2010). Similarly, prominent large seeded *desi* varieties, viz. BG 256, Phule G 5, BGM 391, K 850, Radhey, Gujarat Gram 2 and L 556, were also developed.

3.5 Production-Related Problems

3.5.1 *Fusarium* Wilt

Chickpea wilt occurs in 32 countries across 6 continents in the world (Nene et al. 1991; Singh and Sharma 2002). It is caused by *Fusarium oxysporum* f. sp. *ciceris*. The yield losses caused by it vary from 10 to 90 % (Jimenez-Diaz et al. 1989; Singh and Reddy 1991). At present there are eight distinct physiological races of *Fusarium oxysporum*, viz. 0, 1A, 1B/C, 2, 3, 4, 5 and 6, out of which four races that have been identified as 1, 2, 3 and 4 are prevalent in India (Haware and Nene 1982) and races 0, 5 and 6 are reported from Spain (Jimenez-Diaz et al. 1989). Breeding for *Fusarium* wilt is of prime importance because of the nature of the pathogen – it can persist in soil year after year even in the absence of the host (Haware et al. 1996). Due to the presence of such a number of races, it is very difficult to develop a cultivar which shows stability to the disease across different regions. Most of the resistance against *Fusarium* wilt is of vertical type (Sharma et al. 2005) as is mostly governed by a single major gene. Chickpea genotypes differ in the development of initial symptoms of wilt, indicating different degrees of resistance which is controlled by the segregation of a single gene (Upadhyaya et al. 1983). Such individual genes which are part of oligogenic resistance mechanism delay the onset of disease symptoms, and such a phenomenon is called as late wilting. The yield losses are significantly less in the case of genotypes which show the phenomenon of late wilting due to the combination of two recessive genes for resistance. Resistance to race 0 is due to two independent genes (Rubio et al. 2003), while resistance is digenic or trigenic for race 1A, 2 and 4. Resistance to race 3 and 5 is controlled by a single gene (Sharma et al. 2005). Singh et al. (2012a) have also reported resistance against *Fusarium* wilt in the indigenous chickpea germplasm.

3.5.2 *Ascochyta* Blight

Ascochyta blight is the most important foliar disease of chickpea in many parts of the world including India. It is caused by *Ascochyta rabiei* and has a devastating effect on the chickpea production by causing yield losses ranging from 10 to 100 % (Nene and Reddy 1987; Singh 1990). It is reported in 37 countries all over the world. Among the most affected regions are the Indian subcontinent and Mediterranean region as their prevailing climatic conditions are conducive for the pathogens. *Ascochyta* disease epidemics are common and had occurred in India, Pakistan, the USA, Northwest Pacific, Australia and Syria in the past (Malhotra et al. 2003). *Ascochyta rabiei* isolates have been classified into either a two- or three-pathotype system (I, II and III) according to their levels of virulence (Udupa et al. 1998; Chen et al. 2004; Jayakumar et al. 2005). The main reason for the epidemics is the nature of pathogen which is unstable and continuously evolves new races which break down the host resistance and, thus, reduces the life of cultivars under production systems. The development of more virulent pathotypes has been reported in Syria (Reddy and Kabbabeh 1985), Italy (Stamigna et al. 2000) and Pakistan (Jamil et al. 2000). The genetics of resistance to *Ascochyta* blight is mainly digenic in nature, and both recessive and dominant genes are involved in its control (Bhardwaj et al. 2010). This has forced the breeders to devise a different approach for breeding cultivars with durable resistance. Under new breeding strategy, plant breeders have shifted to gene pyramiding in elite lines instead of incorporating vertical resistance. An alternative strategy of deploying different lines possessing resistance against different races of the pathogen prevalent in different regions can also be effective for minimizing yield losses.

3.5.3 *Botrytis* Grey Mould

Botrytis grey mould (BGM) is a second major foliar disease of chickpea and is prevalent in 15 countries including India, Bangladesh, Nepal, Pakistan, Australia, Argentina, Myanmar, Canada, Columbia, Hungary, Mexico, Spain, Turkey, the USA and Vietnam. Earlier there was no reliable source of resistance to BGM in India (Singh and Reddy 1991), but in recent years, derivative lines from the interspecific crosses of *C. arietinum* and *C. pinnatifidum* had shown high level of genetic resistance to the BGM (Kaur et al. 2013). This resistance can be incorporated into elite lines to develop high-yielding chickpea cultivars with durable resistance. According to Mallikarjuna (unpublished results), the resistance introgressed from wild *C. echinospermum* to cultivated chickpea was found to be monogenic in nature, indicating that resistance to BGM can be easily incorporated to elite lines from the interspecific derivatives.

3.5.4 Pod Borer

Among the various insects that attack chickpea crop, pod borer (*Helicoverpa armigera)* is the most damaging insect. It is quite prevalent in Asia, Africa, Australia and some other chickpea-growing regions. As pod borer is a polyphagous insect that attacks on more than 182 plant species, the development of cultivars resistant or tolerant to *H. armigera* could be integrated in the pest management strategy particularly in the developing countries (Fitt 1989; Sharma and Ortiz 2002). More than 14,000 chickpea germplasm accessions were screened under field conditions at ICRISAT for resistance towards *H. armigera* (Lateef and Sachan 1990). This resulted in the identification and release of moderately resistant/tolerant chickpea cultivars (Gowda et al. 1983; Lateef 1985; Lateef and Pimbert 1990). Still a complete resistance against pod borer is far from reach as different chickpea cultivars show differential inhibition activity of gut proteinases of *H. armigera*, which indicate that *H. armigera* is adapted to a wide range of host protein inhibitors (Singh et al. 2008).

3.5.5 Bruchids

Apart from field, chickpea is also damaged in the storage and one such insect is bruchids (*Callosobruchus chinensis*). It causes loss of grains in the Mediterranean region and in India, where infestation levels approach 13 % (Mookherjee et al. 1970; Dias and Yadav 1988) to total loss (Weigand and Tahhan 1990). Till now there is no report of resistance in the cultivated chickpea, though wild chickpea accessions have shown some resistance to bruchids (Singh et al. 1994, 1998). Owing to crossing barrier, it has not been possible to transfer this trait to the cultivated background. Thus, chemical methods are advised for the control of bruchids (Duke 1981).

3.5.6 Cold Tolerance

If temperature ranges between 0 and 12 °C, then this stress can be defined as chilling without snow cover (Wery et al. 1993). If below 10 °C, the susceptibility of reproductive phase of chickpea to chilling temperatures increases (Sandhu et al. 2005; Srinivasan et al. 1999; Nayyar et al. 2005b). When crop is sown in autumn or in early spring, it is exposed to freezing stress during vegetative growth in WANA region, Europe and Central Asia (Singh et al. 1994). Usually chickpea grown in winter season is more productive than the traditionally grown spring season in the Mediterranean region (Singh and Hawtin 1979). This is due to long growing season and better moisture availability. Problem encountered by winter season crop is flower drop and pod abortion, which leads to major yield loss, when mean temperature of the day falls below 15 °C (Savithri et al. 1980; Srinivasan et al. 1999; Clarke and Siddique 2004; Nayyar et al. 2005a). Chaturvedi et al. (2009) reviewed the work on cold tolerance in chickpea extensively and highlighted the importance of cold tolerance at reproductive stage in chickpea. The deployment of genes for cold tolerance will protect crop from getting damaged and reduce yield losses. Cold tolerance was found to be controlled by at least five genes with both additive and nonadditive gene effects and was dominant over susceptibility (Malhotra and Singh 1990, 1991). Selfing generations would result in reduced dominance and epistatic effects, which would be better for the selection of cold tolerance in chickpea. However, Bhardwaj and Sandhu (2009) reported that cold tolerance is under the control of a single recessive gene. Screening of wild *Cicer* species showed promising traits for cold tolerance (Berger et al. 2012), but till now there are no reports of introgression of cold tolerance from wild to the cultivated chickpea.

3.5.7 Drought Tolerance

Drought is the second most important abiotic stress that contributes immensely to the losses in chickpea production globally. Most of the time, it is terminal drought that has an adverse effect on the crop productivity (Khanna-Chopra and Sinha 1987). In order to counter this drought stress, development of early maturing cultivars will make judicious use of the available soil moisture efficiently and produce relatively higher yields. Critical review made by Upadhyaya et al. (2012) illustrates efficient methods for phenotyping with respect to drought in chickpea and pigeon pea. Drought tolerance may be under polygenic or oligogenic control with additive and nonadditive gene effects. Root traits have been given most importance in recent years because lines with longer root systems have shown better drought tolerance. As chickpea is grown under receding moisture conditions, the long root trait plays an important role in countering drought. Apart from this, early flowering is considered to be very important for drought escape. There a single recessive locus involved in controlling early flowering (Kumar and van Rheenen 2000). Genotypes with recessive allele in homozygous condition escape drought and give a good yield, and such alleles could be easily transferred to the drought-susceptible lines using suitable breeding method for developing drought-tolerant cultivars. Apart from this, wild *Cicer* species have been screened, and few accessions of *C. pinnatifidum* and *C. reticulatum* were found to be resistant against drought (Toker et al. 2007). In the case of cultivated chickpea, the line ICC 4958 is currently considered as potential donor for drought tolerance. Some chickpea cultivars with improved drought tolerance have been released using ICC 4958 as one of the donors in South India and Kenya (Gaur et al. 2012a).

3.5.8 Heat Tolerance

Most of the time, heat stress occurs in combination with/overlapping with drought stress (Toker and Canci 2009). It is very difficult to make distinction between whether the crop is under heat or drought stress; thus, less progress has been made on heat tolerance (Malhotra and Saxena 1993). Chickpea is grown in post-rainy period in South Asia, which results in the exposure of crop to drought along with high temperature. It is predicted that climate change would result in temperature rise by 3–4 °C over current levels by 2050 (Basu et al. 2009). Chickpea is usually grown in winter season in Northern India. But it experiences a high temperature (>35 °C) during the reproductive phase. Most sensitive organs of plant to heat are flowers (Wery et al. 1993; Toker and Canci 2006). During the flowering or reproductive period, if temperature rises above the threshold level, it would adversely affect the pod formation and seed set and, thus, results in reduced grain yield (Summerfield et al. 1984; Wang et al. 2006; Basu et al. 2009; Kumar et al. 2013). Adverse effects of high temperature occur in seed germination, respiration, membrane stability, photosynthesis, hormone level, nutrient absorption, protoplasmic movement, quality of seeds, fruit maturation, fertilization, materials transport, withering, burning of lower leaves, desiccation of poorly developed plants, stunting flower and pod abortion, reduced root nodulation, nitrogen fixation and seed yield (Saxena et al. 1988; Kurdali 1996; Chen et al. 1982; Wahid and Close 2007). In comparison to other cool season legume crops, chickpea is more tolerant to heat stress (Summerfield et al. 1984; Erskine et al. 1994; McDonald and Paulsen 1997; Patrick and Stoddard 2010). However, acute heat stress in chickpea could lead to high-yield losses and crop failure (Devasirvatham et al. 2012). Large genetic variations have been observed for heat tolerance in chickpea when reference set was screened against heat stress at several locations in India (Krishnamurthy et al. 2010). A field screening technique for heat tolerance has been developed and several sources of heat tolerance were identified (Gaur et al. 2014a).

3.6 Traits of Importance for Base Broadening

All studies on assessing genetic diversity in chickpea using molecular markers indicate that chickpea has a narrow genetic base. A large variability is seen in chickpea germplasm for morphological traits, but it could be a reflection in expression of a limited number of mutant genes, as a single mutant gene may cause marked changes in the appearance of the plant (Gaur and Gour 2003). The narrow genetic base of chickpea is a major concern for plant breeding programmes as the genetic variability is a key factor that contributes to genetic gain from selections. Thus, broadening the genetic base of chickpea is very much needed for improving effectiveness of breeding efforts. The genetic variability in the cultivated chickpea can be enhanced by gene introgression from the wild species and by inducing genetic changes through induced mutagenesis. The wild species of chickpea constitutes a valuable genetic resource, particularly for resistance to biotic and abiotic stresses and the nutritional quality traits. Sources of resistance are available in the cultivated species for several biotic (*Fusarium* wilt, *Ascochyta* blight) and abiotic (drought, heat) stresses (reviewed by Gaur et al. 2010). There is a need to diversify sources of resistance being used in the breeding programmes by bringing new sources of resistance from the wild species. There are several biotic stresses, such as dry root rot, *Botrytis* grey mould, phytophthora blight and pod borer, for which high levels of resistance are not available in the germplasm of cultivated species and can be introgressed from the wild species or induced through mutagenesis. Over a dozen mutants have been directly released as varieties (Gaur et al. 2007), while many others have been used as parents in crossing programmes. Mutants with novel traits, such as cymose inflorescence with more than three flowers per node (Gaur and Gour 2002), brachytic growth habit (Gaur et al. 2008) and determinate growth habit (Hegde 2011), have been identified in chickpea and have potential for developing new plant types in chickpea. Singh et al. (2014) have also identified several useful agro-morphological traits and important biotic parameters in various wild annual *Cicer* species and suggested their introgressions for widening the genetic base of cultivated gene pool.

In present day, when farm labourers are becoming expansive day by day, the farmers are demanding chickpea varieties with traits such as suitability to mechanical harvesting and tolerance to herbicides to enhance mechanization of chickpea cultivation (Sandhu et al. 2010; Gaur et al. 2012a). Large genetic variations have been observed for postemergence herbicide tolerance in chickpea (Gaur et al. 2013a) which can be utilized for the development of herbicide-tolerant cultivars. Several diseases, such as dry root rot, collar rot, wet root rot and stem rot, were minor hitherto and are becoming potential threat to chickpea cultivation in many parts of the world including India. There is a need to strengthen research efforts on identifying useful sources of resistance and breeding for enhancing resistance to these diseases. The other traits which have not received much attention in the past, but are important under current scenario of growing environments and consumer requirements, include nutrient use efficiency, especially phosphorus, and nutritional quality traits (protein, iron, zinc, β-carotene, oligosaccharides, etc.).

3.7 Interspecific Hybridization in Crop Species

The crop wild relatives are species closely related to crops, including crop progenitors, identified as critical resources that are vital for wealth creation, food security and environmental stability (Meilleur and Hodgkin 2004; Stolton et al. 2006; Maxted et al. 2008). Historically, the commercial use of wild relatives started in the late nineteenth century (Prescott-Allen and Prescott-Allen 1988), and in the middle of twentieth century, the value of CWR was widely recognized and breeding efforts to explore the potential of wild relatives were initiated in many crop species. The use of wild relatives increased in the 1970s and 1980s (Hodgkin and Hajjar 2008) and in the mid-1980s. Prescott-Allen and Prescott-Allen (1988) asserted that the achievements were substantial enough to recognize the potential of wild relatives.

The proportion of wild or weedy relatives in gene bank holdings has significantly increased in a span of 21 years starting from 1983 (Plucknett et al. 1987) to 2004 (http://singer.grinfo.net/). Summarizing the use of wild relatives for the improvement of major crop species in the last 20 years, Hajjar and Hodgkin (2007) have listed the number of traits. They showed that the extent of utilization varies from crop to crop. Tomato takes lead with 55 traits followed by rice and potato with 12 traits each. Using CWR, wheat was improved for nine traits and sunflower for seven traits. Millet was featured on the list with three traits and maize and chickpea with two traits each.

Wild species have contributed substantially to crop improvement for many characters including seed yield in different crop plants (Stalker 1980). Similarly in chickpea, the productivity genes/alleles have been introgressed from *C. echinospermum*, *C. reticulatum* (Singh and Ocampo 1997; Singh et al. 2005) and *C. pinnatifidum* (Sandhu et al. 2006; Singh et al. 2012a, b, c). However, the transfer of specific genes is frequently associated with the transfer of large alien chromosome segments having undesirable traits (Tanksley and Nelson 1996). Owing to linkage drag, the genes for primitive or wild traits are often introduced along with desirable traits. Breaking linkages with unwanted type and restoring the genotype associated with accepted agronomic background may take a long time. The first report on interspecific crosses involving *C. arietinum* on one hand and *C. reticulatum* and *C. cuneatum* on the other hand was published by Ladizinsky and Adler (1976). The cross between *C. arietinum* and *C. reticulatum* was achieved successfully. Subsequently, wide hybridization was attempted between *C. arietinum* and *C. echinospermum* by various workers (Pundir and Mengesha 1995; Singh and Ocampo 1993). van Dorrestein et al. (1998) attempted crosses involving *C. arietinum, C. bijugum* and *C. judaicum*. Due to the use of in vitro technique, success has been made in achieving hybrids between *C. arietinum* and *C. bijugum* and *C. arietinum* and *C. judaicum*. Badami et al. (1997) also reported successful hybridization between *C. arietinum* and *C. pinnatifidum* using embryo rescue technique. No success has been reported in hybridization between *C. arietinum* and *C. microphyllum*. The crossability and further generation advance studies showed that the species like *C. pinnatifidum and C. judaicum* can be crossed with cultivated chickpea (Verma et al. 1990, 1995; Sandhu et al. 2006, 2007). The derived populations showed wide variation for different agro-morphological traits. Further, F_5 population of an interspecific cross with *C. pinnatifidum* showed that some of the derived lines were superior in yield performance than the recommended check variety and some of the lines also showed resistance against *Botrytis* grey mould.

3.8 Barriers to Interspecific Hybridization

Cross-incompatibility, inviability of F1 hybrid and its progenies are the most common barriers to wide hybridization. Cross-incompatibility between parent species arises when pollen grain does not germinate or pollen tube does not reach to the ovary or male gamete does not fuse with female gamete (Chowdhury and Chowdhury 1978). Post-fertilization barriers in legumes such as production of shrivelled hybrid seed with reduced germination (hybrid inviability), production of dwarf and weak F_1 plants (hybrid weakness) and death of F_1 plants at critical stage of development (hybrid lethality) have been observed by various workers in *Cicer* (Ahmad et al. 1988). This may be attributed to disharmonies between genomes of parental species between genome(s) of one species and cytoplasm of other, or between genotype of F_1 zygote and the genotype of maternal tissues. These barriers were found in varying degrees in most of the interspecific crosses (Al-Yasiri and Coyne 1966; Biswas and Dana 1976; Chowdhury and Chowdhury 1977; Machado et al. 1982; Chen et al. 1983; Gopinathan et al. 1986). Reciprocal crosses should be attempted, if there is disharmony between genome of one species and cytoplasm of the other. The application of flower and fruit setting hormones to pollinated buds (GA_3, indole-3-acetic acid, 6-benzyl amino purine, etc.) and pre-pollination foliar spray of x-aminocaproic acid (immunosuppressant) were also found useful in overcoming inviability of wide crosses. With the advent of in vitro techniques, viz. embryo rescue and ovule culture, ambit of crossable species is greatly enhanced (van Rheenen 1992). In many instances, the hybrid zygote from wide cross died after few days. Growth regulators in such cases increase the embryo survival. These embryos further rescued by culturing them on artificial medium. Murashige (1977) and Raghvan (1980) have presented a good discussion on embryo culture for crop breeding. The procedure of embryo culture has been detailed by Hadley and Opeshaw (1980). This technique has facilitated the production of many viable hybrids in different legume crops (Ahn and Hartmann 1978; Cohen et al. 1984; Chen et al. 1990; Sharma and Satija 1996; Gomathinayagam et al. 1998).

Hybrid sterility generally arises when chromosomes do not pair in F_1, and therefore, gametes receive different number of chromosomes leading to sterility. Studies have shown that in some of the interspecific hybrids, sterility was of segregational type (as per Stebbins 1966 classification) and was mainly due to interchange, inversion and possible duplication and deficiency type of structural heterozygosities in the F_1 individuals (Biswas and Dana 1976; Karmarkar and Dana 1987). Chromosome doubling of the parental species before crossing sometimes increases the chances of obtaining a viable hybrid. Colchicine-induced allopolyploids were raised from most of the semi-fertile and completely seed sterile F_1 hybrids that had high pollen fertility and seed set (Pande et al. 1990). Some of the allopolyploids have been used as bridge species in wide crosses.

3.9 Molecular Markers, Genome Mapping and Genomics as an Adjunct to Breeding

3.9.1 Molecular Markers

Molecular markers which are abundant and have high level of polymorphism and can be subjected to high-throughput analysis are desired for applications in genomic studies and crop improvement (Sharma et al. 1995). Isozymes were the first molecular markers used in chickpea genetic studies, but these markers were small in numbers and showed very low level of polymorphism in the cultivated species. Nevertheless, these markers were used in developing the first linkage map of chickpea (Gaur and Slinkard 1990a, b) and establishing phylogenetic relationships among annual *Cicer* species (Kazan and Muehlbauer 1991; Ahmad et al. 1992). The genetic maps developed from isozyme markers were further expanded using restriction fragment length polymorphisms (RFLP) and randomly amplified polymorphic DNA (RAPD) markers (Simon and Muehlbauer 1997). The use of these markers was restricted, because of some limitations associated with them. The extensive use of molecular markers in chickpea genetics and breeding started only after the development of simple sequence repeat (SSR) markers. The multi-allelic and codominant nature of these markers made them ideal for genomic studies and for use in plant breeding. The SSR markers have been developed from sequence information obtained from various sources, including genomic libraries (Hüttel et al. 1999; Winter et al. 1999; Sethy et al. 2006a, b; Nayak et al. 2010), bacterial artificial chromosome (BAC) libraries and BAC-end sequences (Lichtenzveig et al. 2005; Choudhary et al. 2006; Thudi et al. 2011), tentative unique sequences (TUS) (Hiremath et al. 2011) and expressed sequence tags (ESTs) (Coram and Pang 2005; Varshney et al. 2005, 2009a; Gujaria et al. 2011; Hiremath et al. 2011). Over 2,000 SSR markers are now available for chickpea molecular analysis. The recently published draft genome sequence of chickpea identified over 48,000 SSRs suitable for PCR primer design for use as genetic markers (Varshney et al. 2013a).

Diversity arrays technology (DArT), which utilizes the microarray platform to analyse DNA polymorphisms, is a high-throughput genome analysis method enabling a rapid and economical approach for screening a large number of marker loci (Jaccoud et al. 2001). In chickpea, 15,360 DArT markers were generated from 94 diverse genotypes and 5,397 of these were found polymorphic (Thudi et al. 2011). The level of polymorphism observed for DArT markers was comparable to other crops like sorghum and cassava. Single nucleotide polymorphism (SNP) markers are the new class of markers and have become the preferred choice of markers because of their abundance, codominant nature and amenability to high-throughput analysis. Several thousand SNPs have been identified from transcriptomic analysis in chickpea (Coram and Pang 2005; Varshney et al. 2009b; Gujaria et al. 2011; Hiremath et al. 2011). The draft genome sequence of chickpea is now available (Varshney et al. 2013b) and identified 76,084 SNPs in 15,526 genes. Of these, 27,117 SNPs were identified within the cultivated species and 54,178 SNPs between the cultivated species and its progenitor *C. reticulatum.*

3.9.2 Genome Mapping

All the markers developed in chickpea showed low level of polymorphism within the cultivated species as compared to between the cultivated and wild species. For this reason, the initial studies on genome mapping in chickpea used interspecific mapping populations. The first linkage map of chickpea was developed by Gaur and Slinkard (1990a, b) using isozyme markers and interspecific crosses of *C. arietinum* with *C. reticulatum* and *C. echinospermum*. The DNA-based molecular markers, such as RAPDs and RFLPs, were later integrated to this map by Simon and Muehlbauer (1997). These initial studies used F_2 populations for the development of linkage maps. The first mapping population of recombinant inbred lines (RILs) was developed from the interspecific cross *C. arietinum* (ICC 4958) × *C. reticulatum* (PI 489777) and has been considered as the reference mapping population for genome mapping in chickpea. The genotyping of this mapping population gave the first large genetic map of chickpea consisting of 351 markers and covering a total distance of 2,077.9 cM (Winter et al. 2000). This map was further expanded by later studies using this reference mapping population. Nayak et al. (2010) developed a map with 521 markers and spanning 2602.1 cM. Thudi et al. (2011) developed a comprehensive genetic map including 1,291 markers and spanning 845.56 cM. Recently, Hiremath et al. (2011) developed a genetic map comprising 1,328 marker loci for this reference population. Several studies have been conducted on the development of genetic maps of chickpea based on intraspecific mapping populations (Cho et al. 2002; Flandez-Galvez et al. 2003a, b; Cobos et al. 2005; Radhika et al. 2007; Anuradha et al. 2011; Garg Tosh 2012). Because of limited polymorphism in the cultivated chickpea, maps developed from intraspecific mapping populations had fewer markers (<250 markers) and less genome coverage (<800 cM). Consensus genetic maps using both interspecific and intraspecific populations were also developed. A consensus map based on five interspecific (*C. arietinum* × *C. reticulatum*) and five intraspecific populations was developed by Millán et al. (2010). This map had integrated 555 marker loci. BAC and binary bacterial artificial chromosome (BIBAC) libraries have been developed (Rajesh et al. 2004; Lichtenzveig et al. 2005; Zhang et al. 2010) and used in the development of a physical map of chickpea (Zhang et al. 2010). This physical map comprised 1,945 contigs, spanning about 1,088 Mb. Efforts have also been made to assign linkage groups to specific chromosomes using flow cytometry and PCR-based primers that amplify sequence-tagged microsatellite site markers (Tekeoglu et al. 2002; Zatloukalová et al. 2011). The linkage group 8 (LG8) was assigned to chromosome H, LG5 to chromosome A, LG4 to chromosome E and LG3 to chromosome B. In other cases the chromosomes could not be sorted out separately, so LG1 and LG2 were jointly assigned to chromosomes F and G, and LG6 and LG7 were jointly assigned to chromosomes C and D.

3.9.3 Molecular Mapping of Genes/Quantitative Trait Loci Controlling Agronomically Important Traits

Several RIL mapping populations have been developed in chickpea at ICRISAT (Gaur et al. 2014b) and several other institutes. Molecular markers have been identified for the genes/QTLs linked to resistance to several diseases, including *Fusarium* wilt (Sharma et al. 2004, 2005; Cobos et al. 2005; Gowda et al. 2009; Garg Tosh 2012; Sabbavarapu et al. 2013), *Ascochyta* blight (Millán et al. 2003; Rakshit et al. 2003; Collard et al. 2003; Udupa and Baum 2003; Cho et al. 2004; Iruela et al. 2006; Lichitenzveiz et al. 2006; Anbessa et al. 2009; Kottapalli et al. 2009; Aryamanesh et al. 2010; Garg Tosh 2012; Sabbavarapu et al. 2013), *Botrytis* grey mould (Anuradha et al. 2011) and rust (Madrid et al. 2008); salinity tolerance (Vadez et al. 2012); traits related to drought tolerance (Chandra et al. 2004; Molina et al. 2008; Rehman 2009; Rehman et al. 2012; ICRISAT unpublished results), growth habit and podding (Rajesh et al. 2002; Radhika et al. 2007; Gowda et al. 2009; Aryamanesh et al. 2010; Garg Tosh et al. 2012); phenology (Aryamanesh et al. 2010; Rehman 2009; Rehman et al. 2012); and seed characteristics (Gowda et al. 2009). Several of these studies have been summarized in earlier reviews (Varshney et al. 2007; Upadhyaya et al. 2011; Gaur et al. 2012b, 2014b).

3.9.4 Marker-Assisted Breeding

Marker-assisted breeding can greatly improve the precision and efficiency of breeding programmes. Recent advances in the development of molecular markers and identification of molecular markers linked to genes/QTLs controlling traits of breeders' interest have encouraged applications of marker-assisted backcrossing (MABC) in chickpea improvement. A 'QTL-hotspot' containing QTLs for several root and drought tolerance traits was transferred from the drought-tolerant line ICC4958 to a leading *desi* chickpea cultivar JG 11 through MABC (Varshney et al. 2013b). Varshney et al. (2013b) reviewed the status of genomic resources available for chickpea improvement and suggested ways to go for genomic-assisted breeding in chickpea. Multilocation evaluations of introgression lines (ILs) in India, Ethiopia and Kenya led to the identification of lines with significantly higher yield than JG11 at each location and in each growing condition (rainfed/irrigated) (Gaur et al. 2013b). MABC is also being used for introgressing resistance to various diseases in chickpea. ICRISAT is pyramiding resistances to two races of *Fusarium* wilt (*foc1* and *foc3*) from WR315 and 2 QTLs for *Ascochyta* blight resistance from ILC3279 line into C214. Marker-assisted recurrent selection (MARS) has also been initiated in chickpea for the improvement of yield, particularly under moisture stress conditions. Two good-by-good crosses (JG 11×ICCV 04112 and JG 130×ICCV 05107) are being used at ICRISAT to implement MARS (Gaur et al. 2014b). QTLs were identified specific to these crosses by genotyping in F_3 and phenotyping of $F_{3:5}$ progenies. A set of eight lines were selected for each cross using OptiMAS 1.0 to pyramid superior alleles of the favourable QTLs identified. Superior lines will be developed by accumulating favourable alleles through successive intercrossing using genotypic selection. In addition to MARS, genome-wide selection (GWS) or genomic selection (GS) has been proposed as a potential approach for improving complex traits governed by many genes/QTLs. In this approach, both phenotyping and genotyping data are used to predict genomic estimated breeding values (GEBVs) of progenies and superior progenies are selected based on GEBVs.

3.10 Conclusions

The narrow genetic base of cultivated chickpea warrants systematic collection, documentation and evaluation of chickpea germplasm, particularly wild annual *Cicer* species for effective and efficient use in chickpea breeding programmes. Researchers have clearly demonstrated that desirable alien genes conferring resistance against biotic and abiotic stresses can be successfully introgressed from unexploited wild annual *Cicer* species to the cultivated chickpea. The valuable genetic resources present in the primary gene pool comprising progenitor species can be successfully utilized for genetic enhancement. However, most of the wild species possessing high degree of resistance to multiple stresses are present in the secondary and tertiary gene pools, where hybridization with cultivated species is often limited due to reproductive barriers. There are indications that useful traits which are not available in cultigens may be recovered in the segregating generation of crosses involving wild *Cicer* species. Further the molecular markers can be effectively used to monitor alien gene introgression into the elite cultivars. Besides major genes, the transfer of agronomically important traits into elite cultivars has been made easy and practical through marker-assisted selection. Molecular markers such as SSR and SNP are useful for the construction of high-density genetic maps of chickpea which will be very useful to identify genes/QTLs for stress resistance, quality traits and other yield contributing characters.

References

Abbo S, Molina C, Jungma NNR, Grusa KA, Berkovitch Z, Reifen R, Kahl G, Winter P, Reifen R (2005) Quantitative trait loci governing carotenoid concentration and weight in seeds of chickpea (*Cicer arietinum* L.). Theor Appl Genet 111:185–195

Ahmad F, Slinkard AE, Scoles GJ (1987) The cytogenetic relationship between *Cicer judaicum* Boiss and *Cicer chorassanicum*. Genome 29:883–886

Ahmad F, Slinkard AE, Scoles GJ (1988) Investigations into the barrier/s to interspecific hybridization between *Cicer arietinum* L. and eight other annual *Cicer* species. Plant Breed 100:193–198

Ahmad F, Gaur PM, Slinkard AE (1992) Isozyme polymorphism and phylogenetic interpretations in the genus *Cicer* L. Theor Appl Genet 83:620–627

Ahn AS, Hartmann RW (1978) Interspecific hybridization among four species of the genus *Vigna* Savi. In: Report of first international mungbean symposium, August 16–19, 1977, Univ. Philippines, Los Baños, Asian Vegetable Research and Development Center, Tainan, pp 240–246

Al-Yasiri S, Coyne DP (1966) Interspecific hybridization in genus Phaseolus. Crop Sci 6:59–60

Anbessa Y, Taran B, Warkentin TD, Tullu A, Vandenberg A (2009) Genetic analyses and conservation of QTL for Ascochyta blight resistance in chickpea (*Cicer arietinum* L.). Theor Appl Genet 119(4):757–765

Anuradha C, Gaur PM, Pande S, Gali KK, Ganesh M, Kumar J, Varshney RK (2011) Mapping QTL for resistance to botrytis grey mould in chickpea. Euphytica. doi:10.1007/s10681-011-0394-1

Aryamanesh N, Nelson MN, Yan G, Clarke HJ, Siddique KHM (2010) Mapping a major gene for growth habit and QTLs for Ascochyta blight resistance and flowering time in a population between chickpea and *Cicer reticulatum*. Euphytica 173(3):307–319

Badami PS, Mallikarjuna N, Moss JP (1997) Interspecific hybridization between *Cicer arietinum* and *C. pinnatifidum*. Plant Breed 116:393–395

Basu PS, Ali M, Chaturvedi SK (2009) Terminal heat stress adversely affects chickpea productivity in Northern India – strategies to improve thermo-tolerance in the crop under climate change. In: Panigrahy S, Shankar SR, Parihar JS (eds) ISPRS archives XXXVIII-8/W3 workshop proceedings: impact of climate change on agriculture, 23–25 February, ISPRS Archives, Ahmedabad, pp 189–193

Berger JD, Abbo S, Turner NC (2003) Ecogeography of annual wild *Cicer* species: the poor state of the world collection. Crop Sci 43:1076–1090

Berger JD, Kumar S, Nayyar H, Street KA, Sandhu J, Henzell JM, Kaur J, Clarke HC (2012) Temperature-stratified screening of chickpea (*Cicer arietinum* L.) genetic resource collections reveals very limited reproductive chilling tolerance compared to its annual wild relatives. Field Crop Res 126:119–129

Bhardwaj R, Sandhu JS (2009) Pollen viability and pod formation in chickpea (*Cicer arietinum* L.) as a criterion for screening and genetic studies of cold tolerance. Indian J Agric Sci 79:63–65

Bhardwaj R, Sandhu JS, Kaur L, Gupta SK, Gaur PM, Varshney R (2010) Genetics of resistance to Ascochyta blight in chickpea. Euphytica 171:337–343

Biswas MR, Dana S (1976) *Phaseolus aconitifolius* x *P. trilobus* cross. Indian J Genet 36:125–131

Chandra S, Buhariwalla HK, Kashiwagi J, Harikrishna S, Sridevi KR, Krishnamurthy L (2004) Identifying QTL-linked markers in marker-deficient crops. In: 4th international crop science congress, Brisbane, Australia, 26 Sep–1 Oct

Chaturvedi SK, Kumar S, Dua RP (2003) Chickpea breeding. In: Ali M et al (eds) Chickpea research in India. Indian Institute of Pulses Research, Kanpur, pp 69–98

Chaturvedi SK, Mishra DK, Vyas P, Mishra N (2009) Breeding for cold tolerance in chickpea. Trends Biosci 2(2):1–6

Chaturvedi SK, Gaur PM, Sandhu JS, Mishra N (2010) Efforts for improving seed size and yield of *kabuli* chickpea. Prog Res 5(2):5–8

Chen YM, Backman K, Magasanik B (1982) Characterization of a gene, glnL, the product of which is involved in the regulation of nitrogen utilization in *Escherichia coli*. J Bacteriol 150:214–220

Chen NC, Baker LR, Homa S (1983) Interspecific crosses among four species of *Vigna* food legumes. Euphytica 32:925–937

Chen HK, Mol MC, Mok DWS (1990) Somatic embryogenesis and shoot organogenesis from interspecific hybrid embryos of *Vigna glabrescens* and *V. radiate*. Plant Cell Rep 9:77–79

Chen W, Coyne CJ, Peever TL, Muehlbauer FJ (2004) Characterization of chickpea differentials for pathogenicity assay of Ascochyta blight and identification of chickpea accessions resistant to *Didymella rabiei*. Plant Pathol 40:401–407

Cho S, Kumar J, Shultz JL, Anupama K, Tefera F, Muehlbauer FJ (2002) Mapping genes for double podding and other morphological traits in chickpea. Euphytica 125:285–92

Cho S, Chen W, Muehlbauer FJ (2004) Pathotype-specific factors in chickpea (*Cicer arietinum* L.) for quantitative resistance to Ascochyta blight. Theor Appl Genet 109:733–739

Choudhary S, Sethy NK, Shokeen B, Bhatia S (2006) Development of sequence-tagged microsatellites site markers for chickpea (*Cicer arietinum* L.). Mol Ecol Notes 6:93–95

Chowdhury RK, Chowdhury JB (1977) Intergeneric hybridization between *Vigna mungo* and *Phaseolus calcaratus*. Indian J Agric Sci 47:117–121

Chowdhury RK, Chowdhury JB (1978) Studies on crossability between *Phaseolus aureus* and *Phaseolus mungo*. Acta Bot 6:185–187

Clarke HJ, Siddique KHM (2004) Response of chickpea genotypes to low temperature stress during reproductive development. Field Crops Res 90:323–334

Cobos MJ, Fernandez M, Rubio J, Kharrat M, Moreno MT, Gil J, Millan T (2005) A linkage map of chickpea (*Cicer arietinum* L.) based on populations from Kabuli x Desi crosses: location of genes for resistance to *Fusarium* wilt race 0. Theor Appl Genet 110:1347–1353

Cohen D, Ladizinksky G, Ziv M, Muehlbauer FJ (1984) Rescue of interspecific *Lens* hybrids by means of embryo culture. Plant Cell Tissue Organ Cult 3:343–347

Collard BCY, Pang ECK, Ades PK, Taylor PWJ (2003) Preliminary investigation of QTLs associated with seedling resistance to Ascochyta blight from *Cicer echinospermum*, a wild relative of chickpea. Theor Appl Genet 107:719–729

Cooper HD, Spillane C, Hodgkin T (2001) Broadening the genetic base of crop production. CAB International, Wallingford, p 452

Coram TE, Pang ECK (2005) Isolation and analysis of candidate Ascochyta blight defence genes in chickpea. Part I. Generation and analysis of an expressed sequence tag (EST) library. Physiol Mol Plant Pathol 66:192–200

Croser JS, Ahmad CF, Clarke HJ, Siddique KHM (2003) Utilisation of wild *Cicer* in chickpea improvement-progress, constraints, and prospects. Aust J Agric Res 54:429–444

Devasirvatham V, Gaur PM, Mallikarjuna N, Tokachichu RN, Trethowan RM, Tan DKY (2012) Effect of high temperature on the reproductive development of chickpea genotypes under controlled environments. Funct Plant Biol 139:1009–1018

Dias CAR, Yadav TD (1988) Incidence of pulse beetles in different legume seeds. Indian J Entomol 50:457–461

Duke JA (1981) Handbook of legumes of world economic importance. Plenum, New York, pp 52–57

Duvick D (1995) Security and long-term prospects for conservation of genetic resources. Res Domest Int Agribus Manag 11:33–45

Erskine W, Hussain A, Tahir M, Bahksh A, Ellis RH, Summerfield RJ, Roberts EH (1994) Field evaluation of a model of photothermal flowering responses in a world lentil collection. Theor Appl Genet 88:423–428

FAOSTAT (2011) http://faostat.fao.org/

Fitt GP (1989) The ecology of *Heliothis* species in relation to agro-ecosystems. Annu Rev Entomol 34:17–52

Flandez-Galvez H, Ades PK, Ford R, Pang ECK, Taylor PWJ (2003a) QTL analysis for Ascochyta blight resistance in an intraspecific population of chickpea (*Cicer arietinum* L.). Theor Appl Genet 107:1257–1265

Flandez-Galvez H, Ford R, Pang ECK, Taylor PWJ (2003b) An intraspecific linkage map of the chickpea (*Cicer arietinum* L.) genome based on sequence tagged based microsatellite site and resistance gene along markers. Theor Appl Genet 106:1447–1453

Galluzzi G, Eyzaguirre P, Negri V (2010) Home gardens: Neglected hot spots of agrobiodiversity and cultural diversity. Biodivers Conserv 19:3635–3654

Garg Tosh (2012) Mapping of quantitative trait loci for resistance to Ascochyta blight, botrytis grey mould and Fusarium wilt in chickpea. Ph.D. thesis, Punjab Agricultural University, Ludhiana, India

Garg Tosh, Singh S, Singh I, Sandhu JS, Gaur PM, Varshney RK (2012) Mapping of genes for double podding and its association with morphological traits in chickpea. Crop Improv 39 (Special issue): 429–430

Gaur PM, Gour VK (2002) A gene producing one to nine flowers per flowering node in chickpea. Euphytica 128:231–235

Gaur PM, Gour VK (2003) Broad-few-leaflets and outwardly curved wings: two new mutants of chickpea. Plant Breed 122:192–194

Gaur PM, Slinkard AE (1990a) Genetic control and linkage relations of additional isozymes markers in chickpea. Theor Appl Genet 80:648–653

Gaur PM, Slinkard AE (1990b) Inheritance and linkage of isozyme coding genes in chickpea. J Hered 81:455–459

Gaur PM, Gowda CLL, Knights EJ, Warkentin TD, Acikgoz N, Yadav SS, Kumar J (2007) Breeding achievements. In: Yadav SS, Redden B, Chen W, Sharma B (eds) Chickpea breeding and management. CABI, Wallingford, pp 391–416

Gaur PM, Gour VK, Srinivasan S (2008) An induced brachytic mutant of chickpea and its possible use in ideotype breeding. Euphytica 159:35–41

Gaur PM, Mallikarjuna N, Knights T, Beebe S, Debouck D, Mejía A, Malhotra RS, Imtiaz M, Sarker A, Tripathi S, Gowda CLL (2010) Gene introgression in grain legumes. In: Gupta S, Ali M, Singh BB (eds) Grain legumes: genetic improvement, management and trade. Indian Society of Pulses Research and Development, Indian Institute of Pulses Research, Kanpur, pp 1–17.102

Gaur PM, Jukanti AK, Srinivasan S, Gowda CLL (2012a) Chickpea (Cicer arietinum L.). In: Bharadwaj DN (ed) Breeding of field crops. Agrobios (India), Jodhpur, pp 165–194

Gaur PM, Jukanti AK, Varshney RK (2012b) Impact of genomic technologies on chickpea breeding strategies. Agronomy 2:199–221

Gaur PM, Jukanti AK, Samineni S, Chaturvedi SK, Singh S, Tripathi S, Singh I, Singh G, Das TK, Aski M, Mishra N, Nadarajan N, Gowda CLL (2013a) Large genetic variability in chickpea for tolerance to herbicides imazethapyr and metribuzin. Agronomy 3(3):524–536

Gaur PM, Thudi M, Nayak S, Samineni S, Krishnamurthy L, Gangarao NVPR, Fikre A, Jayalakshmi V, Mannur DM, Varshney RK (2013b) Progress in marker-assisted breeding for drought tolerance in chickpea. Abstract No. W339 in Plant and animal genome conference XXI,

January 11–16, 2013, San Diego, CA, USA. https://pag.confex.com/pag/xxi/webprogram/Paper6526.html
Gaur PM, Jukanti AK, Srinivasan S, Chaturvedi SK, Basu PS, Babbar A, Jayalakshmi V, Nayyar H, Devasirvatham V, Mallikarjuna N, Krishnamurthy L, Gowda CLL (2014a) Climate change and heat stress tolerance in chickpea. In: Tuteja N, Gill SS (eds) Climate change and plant abiotic stress tolerance. Wiley-VCH Verlag GmbH & Co. KGaA, Weinheim, pp 839–856. doi:10.1002/9783527675265.ch31
Gaur PM, Thudi M, Semineni S, Varshney RK (2014b) Advances in chickpea genomics. In: Gupta S, Nadarajan N, Sen D (eds) Legumes in the omic era. Springer, New York, pp 73–94. doi:10.1007/978-1-4614-8370-04
Gomathinayagam P, Ram SG, Rathnaswamy R, Ramaswamy NM (1998) Interspecific hybridization between *Vigna unguiculata* and *V. vexillata* through in vitro embryo culture. Euphytica 102:203–209
Gopinathan MC, Babu CR, Shivanna KR (1986) Interspecific hybridization between rice bean (*Vigna umbellata*) and its wild relative *V. minima:* fertility: Sterility relationship. Euphytica 35:1017–1022
Gowda CLL, Lateef SS, Smithson JB, Reed W (1983) Breeding for resistance to *Heliothis armigera* in chickpea. In: Proceedings of the national seminar on breeding crop plants for resistance to pests and diseases, 25–27 May 1983. School of Genetics, Tamil Nadu Agricultural University, Coimbatore, pp 36–39
Gowda SJM, Radhika P, Kadoo NY, Mhase LB, Gupta VS (2009) Molecular mapping of wilt resistance genes in chickpea. Mol Breed 24:177–183
Gujaria N, Kumar A, Dauthal P, Dubey A, Hiremath P, Bhanu Prakash A, Farmer A, Bhide M, Shah T, Gaur PM, Upadhyaya HD, Bhatia S, Cook DR, May GD, Varshney RK (2011) Development and use of genic molecular markers (GMMs) for construction of a transcript map of chickpea (*Cicer arietinum* L.). Theor Appl Genet. doi:10.1007/s00122-011-1556-1
Hadley HH, Opeshaw SJ (1980) Interspecific and intergeneric hybridization. In: Fery WR, Hadley HH (eds) Hybridization in crop plants. Iowa State University, Ames, pp 133–159
Hajjar R, Hodgkin T (2007) The use of wild relatives in crop improvement: A survey of developments over the last 20 years. Euphytica 156:1–13
Harlan JR, de Wet MJ (1971) Towards a rational classification of crop plant. Taxonomy 20:509–517
Haware MP, Nene YL (1982) Races of *Fusarium oxysporum f. sp. ciceri*. Plant Dis 66:809–810
Haware MP, Nene YL, Natarajan M (1996) Survival of *Fusarium oxysporum f. sp. ciceri* in soil in the absence of chickpea. Phytopathol Mediterr 35:9–12
Hawkes JG (1977) The importance of wild germplasm in plant breeding. Euphytica 26:615–621
Hegde VS (2011) Morphology and genetics of a new found determinate genotype in chickpea. Euphytica 182:35–42
Hiremath PJ, Farmer A, Cannon SB, Woodward J, Kudapa H, Tuteja R, Kumar A, Bhanu PA, Mulaosmanovic B, Gujaria N, Krishnamurthy L, Gaur PM, Kavi Kishor PB, Shah T, Srinivasan Lohse RM, Xiao Y, Town CD, Cook DR, May GD, Varshney RK (2011) Large-scale transcriptome analysis in chickpea (*Cicer arietinum* L.), an orphan legume crop of the semi-arid tropics of Asia and Africa. Plant Biotechnol J9:922–931
Hodgkin T, Hajjar R (2008) Using crop wild relatives for crop improvement: trends and perspectives. In: Maxted N, Ford-Lloyd BV, Kell SP, Iriondo JM, Dulloo E, Turok J (eds) Crop wild relatives, conservation and use. CAB International, Wallingford, pp 535–548
Hüttel B, Winter P, Weising K, Choumane W, Weigand F (1999) Sequence-tagged microsatellite site markers for chickpea (*Cicer arietinum* L.). Genome 42:210–217
Iruela M, Rubio J, Barro F, Cubero JI, Millan T, Gil J (2006) Detection of two quantitative trait loci for resistance to Ascochyta blight in an intra-specific cross of chickpea (*Cicer arietinum* L.): development of SCAR markers associated with resistance. Theor Appl Genet 112:278–287
Jaccoud D, Peng K, Feinstein D, Kilian A (2001) Diversity arrays: a solid state technology for sequence information independent genotyping. Nucleic Acids Res 29:E25
Jamil FF, Sarwar N, Sarwar M, Khan JA, Geistlinger J, Kahl G (2000) Genetic and pathogenic diversity within *Ascochyta rabiei* (Pass.) Lab. populations in Pakistan causing blight of chickpea (*Cicer arietinum* L.). Physiol Mol Plant Pathol 57:243–254
Jayakumar P, Gossen BD, Gan YT, Warkentin TD, Banniza S (2005) Ascochyta blight of chickpea: infection and host resistance mechanisms. Can J Plant Pathol 27:499–509
Jimenez-Diaz RM, Trapero-Casas A, Cabrera de la Colina J (1989) Races of *Fusarium oxysporum* f. sp. *ciceris* infecting chickpea in southern Spain. In: Tjamos EC, Beckman CH (eds) Vascular wilt disease of plants, vol H28, NATO ASI Sr. Springer, Berlin, pp 515–520
Karmarkar PG, Dana S (1987) Cytogenetic identification of a *Vigna sublobata* collection. Nucleus 30:47–50
Kaur L, Sandhu JS, Gupta SK (2007) Inter and intra-accession variation for resistance to botrytis gray mold of chickpea in wild *Cicer* species. Indian J Agric Sci 77:786–788
Kaur A, Sandhu JS, Gupta SK, Bhardwaj R, Bansal UK, Saini RG (2010a) Genetic relationships among annual wild *Cicer* species using RAPD analysis. Indian J Agric Sci 80(4):309–311
Kaur H, Gupta AK, Kaur N, Sandhu JS (2010b) *Cicer pinnatifidum*: A biochemically important wild species for improvement of chickpea cultivars. J Food Legum 23:86–88
Kaur L, Sirari A, Kumar D, Sandhu JS, Singh S, Singh I, Kapoor K, Gowda CLL, Pande S, Gaur P, Sharma M, Imtiaz M, Siddique KHM (2013) Harnessing Ascochyta blight and Botrytis grey mould resistance in chickpea through interspecific hybridization. Phytopathol Mediterr 52(1):157–165
Kazan K, Muehlbauer FJ (1991) Allozyme variation and phylogeny in annual species of *Cicer* (Leguminosae). Plant Syst Evol 175:11–21
Khanna-Chopra R, Sinha SK (1987) Chickpea: physiological aspects on growth and yield. In: Saxena MC,

Singh KB (eds) The chickpea. CAB International, Wallingford, pp 163–189

Kottapalli P, Gaur PM, Katiyar SK, Crouch JH, Buhariwalla HK, Pande S, Gali KK (2009) Mapping and validation of QTLs for resistance to an Indian isolate of Ascochyta blight pathogen in chickpea. Euphytica 165:79–88

Krishnamurthy L, Gaur PM, Basu PS, Chaturvedi SK, Tripathi S, Vadez V, Rathore A, Varshney RK, Gowda CLL (2010) Large genetic variation for heat tolerance in the reference collection of chickpea (*Cicer arietinum* L.) germplasm. Plant Genet Resour 9(1):59–69

Kumar J, van Rheenen HA (2000) A major gene for time of flowering in chickpea. J Hered 91:67–68

Kumar S, Gupta S, Chandra S, Singh BB (2004) How wide is the genetic base of pulse crops? In: Ali M, Singh BB, Kumar S, Dhar V (eds) Pulses in new perspectives. Indian Institute of Pulses Research, Kanpur, pp 211–221

Kumar S, Gupta S, Chaturvedi SK (2008) Narrow genetic base of pulses. In: Khrakwal MC (ed) Food legumes for nutritional security and sustainable agriculture, vol 2, Indian Society of Genetics and Plant Breeding. New Delhi, India, pp 131–136

Kumar S, Sandhu JS, Gupta SK, Dikshit HK, Gupta S (2009) Broadening genetic base of pulse crops: present status and future strategies. In: Ali M et al (eds) Legumes for ecological sustainability. IIPR, Kanpur, pp 342–358

Kumar S, Thakur P, Kaushal N, Malik JA, Gaur P, Nayyar H (2013) Effect of varying high temperatures during reproductive growth on reproductive function, oxidative stress and seed yield in chickpea genotypes differing in heat sensitivity. Arch Agron Soil Sci 9(6):823–843

Kurdali F (1996) Nitrogen and phosphorus assimilation, mobilization and partitioning in rainfed chickpea (*Cicer arietinum* L.). Field Crops Res 47:81–92

Ladizinsky G, Adler A (1976) The origin of chickpea *Cicer arietinum* L. Euphytica 25:211–217

Lateef SS (1985) Gram pod borer (*Heliothis armigera* (Hub.) resistance in chickpea. Agric Ecosyst Environ 14:95–102

Lateef SS, Pimbert MP (1990) The search for host plant resistance of *Helicoverpa armigera* in chickpea and pigeonpea at ICRISAT. In: Proceedings of the Consultative group meeting on the host selection behavior of *Helicoverpa armigera*, 5–7 March 1990. ICRISAT, Hyderabad, pp 185–192

Lateef SS, Sachan JN (1990) Host plant resistance to *Helicoverpa armigera* (Hub.) in different agro-economical conditions. In: Chickpea in nineties: proceedings of the second international workshop on chickpea, 4–8 December 1989. ICRISAT, Hyderabad, pp 181–189

Lichitenzveiz J, Bonfil DJ, Zhang HB, Shtienberg D, Abbo S (2006) Mapping quantitative trait loci in chickpea associated with time to flowering and resistance to *Didymella rabiei* the causal agent of Ascochyta blight. Theor Appl Genet 113:1357–1369

Lichtenzveig J, Scheuring C, Dodge J, Abbo S, Zhang HB (2005) Construction of BAC and BIBAC libraries and their applications for generation of SSR markers for genome analysis of chickpea, *Cicer arietinum* L. Theor Appl Genet 110:492–510

Lorenzetti F, Negri V (2009) The European seed legislation on conservation varieties. In: Vetelainon M, Negri V, Maxted N (eds) European landraces: on farm conservation, management and use. Biodiversity technical bulletin no. 15. Bioversity International Publication, Rome, pp 287–295

Machado M, Tai W, Baker LR (1982) Cytogenetic analysis of the interspecific hybrid *Vigna radiata* x *V. umbellata*. J Hered 73:205–208

Madrid E, Rubiales D, Moral A, Moreno MT, Millan T, Gil J, Rubio J (2008) Mechanism and molecular markers associated with rust resistance in a chickpea interspecific cross (*Cicer arietinum* x *Cicer reticulatum*). Eur J Plant Pathol 121:45–53

Malhotra RS, Saxena MC (1993) Screening for cold and heat tolerance in cool season food legumes. In: Singh KB, Saxena MC (eds) Breeding for tolerance in cool season food legumes. Wiley, Chichester, pp 227–244

Malhotra RS, Singh KB (1990) The inheritance of cold tolerance in chickpea. J Genet Breed 44:227–230

Malhotra RS, Singh KB (1991) Gene action for cold tolerance in chickpea. Theor Appl Genet 82:598–601

Malhotra RS, Pundir RPS, Slinkard AE (1987) Genetic resources of chickpea. In: Saxena MC, Singh KB (eds) The chickpea. CAB International Cambrian News Ltd, Aberystwyth, pp 67–81

Malhotra RS, Baum M, Udupa SM, Bayaa B, Kababbeh S, Khalaf G (2003) Ascochyta blight research in chickpea-present status and future prospects. In: Sharma RN, Srivastava GK, Rahore AL, Sharma ML, Khan MA (eds.) Proceedings of the international chickpea conference chickpea research for the millennium, 20–22 January 2003, Raipur, Chhattisgarh, India, pp 108–117

Mallikarjuna N, Sharma HC, Upadhyaya HD (2007) Exploitation of wild relatives of pigeonpea and chickpea for resistance to *Helicoverpa armigera*. EJ SAT Agric Res Crop Improv 3(1):4–7

Maxted N, Kell SP, Ford-Lloyd BV (2008) Crop wild relatives conservation and use: establishing the context. In: Maxted N, Ford-Lloyd BV, Kell SP, Iriondo JM, Dulloo E, Turok J (eds) Crop wild relatives, conservation and use. CAB International, Wallingford, pp 3–30

McDonald GK, Paulsen GM (1997) High temperature effects on photosynthesis and water relations of grain legumes. Plant Soil 196:47–58

Meilleur BA, Hodgkin T (2004) In situ conservation of crop wild relatives. Biodivers Conserv 13:663–684

Millán T, Rubio J, Iruela M, Daly K, Cubero JI, Gil J (2003) Markers associated with Ascochyta blight resistance in chickpea and their potential in marker-assisted selection. Field Crops Res 84:373–384

Millán T, Winter P, Jungling R, Gil J, Rubio J, Cho S, Cobos MJ, Iruela M, Rajesh PN, Tekeoglu M, Kahl G,

Muehlbauer FJ (2010) A consensus genetic map of chickpea (*Cicer arietinum* L.) based on 10 mapping populations. Euphytica 175:175–189

Molina C, Rotter B, Horres R, Udupa S, Besser B, Bellarmino L, Baum M, Matsumura H, Terauchi R, Kahl G, Winter P (2008) SuperSAGE: the drought stress-responsive transcriptome of chickpea roots. BMC Genomics 9(1):553

Mookherjee PB, Jotwani MG, Yadav TD, Sircar P (1970) Studies on incidence and extent of damage due to insect pests in stored seeds. II. Leguminous and vegetable seeds. Indian J Entomol 32:350–355

Muehlbauer FJ, Singh KB (1987) Genetics of chickpea. In: Saxena MC, Singh KB (eds) The chickpea. CAB International, Wallingford, pp 99–126

Muehlbauer FJ, Kaiser WJ, Simon CJ (1994) Potential for wild species in cool season food legume breeding. Euphytica 73:109–114

Murashige T (1977) Current status of plant cell and organ culture. Hortic Sci 12:127–130

Nadarajan N, Chaturvedi SK (2010) Genetic options for enhancing productivity of major pulses- retrospect, issues and strategies. J Food Legum 23(1):1–7

Nayak SN, Zhu H, Varghese N, Datta S, Choi H-K, Horres R, Jüngling R, Singh J, Kavi Kishore PB, Sivaramakrishnan S, Hoisington DA, Kahl G, Winter P, Cook DR, Varshney RK (2010) Integration of novel SSR and gene-based SNP marker loci in the chickpea genetic map and establishment of new anchor points with *Medicago truncatula* genome. Theor Appl Genet 120:1415–1441

Nayyar H, Bains TS, Kumar S (2005a) Chilling stressed chickpea seedlings: effect of cold acclimation, calcium and abscisic acid on cryoprotective solutes and oxidative damage. Environ Exp Bot 54:275–285

Nayyar H, Chander K, Kumar S, Bains TS (2005b) Glycine betaine mitigates cold stress damage in chickpea. Agron Sustain Dev 25:381–388

Negri V (2003) Landraces in central Italy: where and why they are conserved and perspectives for their on-farm conservation. Genet Resour Crop Evol 50(8):871–885

Nene YL, Reddy MV (1987) Chickpea diseases and their control. In: Saxena MC, Singh KB (eds) The chickpea. CAB International, Wallingford, pp 233–270

Nene YL, Reddy MV, Haware MP, Ghaneka AM, Amin KS (1991) Field diagnosis of chickpea diseases and their control. Info Bull No. 28, ICRISAT, Hyderabad, India

Pande K, Raghuvanshi SS, Prakash D (1990) Induced high yielding amphiploid of *Vigna radiata* x *V. mungo*. Cytologia 55:249–253

Patrick JW, Stoddard FL (2010) Physiology of flowering and grain filling in faba bean. Field Crops Res 115:234–242

Plucknett DL, Smith N, Williams JT, Anishetty NM (1987) Gene banks and the world's food. Princeton University Press, Princeton

Prescott-Allen R, Prescott-Allen C (1988) Genes from the wild: using wild genetic resources for food and raw materials. International Institute for Environment and Development, Earthscan Publications, London

Pundir RPS, Mengesha MH (1995) Cross compatibility between chickpea and its wild relative *Cicer echinospermum* Davis. Euphytica 83:241–245

Pundir RPS, Rao NK, van der Maesen LJG (1985) Distribution of qualitative traits in the world germplasm of chickpea (*Cicer arietinum* L.). Euphytica 34:697–703

Radhika P, Gowda SJM, Kadoo NY, Mhase LB, Jamadagni BM, Sainani MN, Chandra S, Gupta VS (2007) Development of an integrated intraspecific map of chickpea (*Cicer arietinum* L.) using two recombinant inbred line populations. Theor Appl Genet 115: 209–216

Raghvan V (1980) Embryo culture. Int Rev Cytol (Suppl) 11:209–240

Rajesh PN, Tullu A, Gil J, Gupta VS, Ranjekar PK, Muehlbauer FJ (2002) Identification of an STMS marker for the double-podding gene in chickpea. Theor Appl Genet 105:604–607

Rajesh PN, Coyne C, Meksem K, Sharma KD, Gupta V, Muehlbauer FJ (2004) Construction of a HindIII bacterial artificial chromosome library and its use in identification of clones associated with disease resistance in chickpea. Theor Appl Genet 108:663–669

Rakshit S, Winter P, Tekeoglu M, Juarez Muñoz J, Pfaff T, Benko-Iseppon AM, Muehlbauer FJ, Kahl G (2003) DAF marker tightly linked to a major locus for Ascochyta blight resistance in chickpea (*Cicer arietinum* L.). Euphytica 132:23–30

Reddy MV, Kabbabeh S (1985) Pathogenic variability in Ascochyta rabiei (Pass.) Lab. in Syria and Lebanon. Phytopathol Mediterr 24:265–266

Rehman AU (2009) Characterization and molecular mapping of drought tolerance in kabuli chickpea (*Cicer arietinum* L.), Ph.D. thesis, University of Saskatchewan, Saskatoon, Canada

Rehman AU, Malhotra RS, Bett K, Tar'an B, Bueckert R, Warkentin TD (2012) Mapping QTL associated with traits affecting grain yield in chickpea (*Cicer arietinum* L.) under terminal drought stress. Crop Sci 51:450–463

Rubio J, Hajj Moussa E, Kharrat M, Moreno MT, Millan T, Gill J (2003) Two genes and linked RAPD markers involved in resistance to *Fusarium oxysporum* f.sp. *ciceris* race 0 in chickpea. Plant Breed 122:188–191

Sabbavarapu MM, Sharma M, Chamarthi SK, Swapna N, Rathore A, Thudi M, Gaur PM, Pande S, Singh S, Kaur L, Varshney RK (2013) Molecular mapping of QTLs for resistance to Fusarium wilt (race 1) and Ascochyta blight in chickpea (*Cicer arietinum* L.). Euphytica. doi:10.1007/s10681-013-0959-2

Sandhu JS, Singh G, Bains TS, Sharma YR, Singh I, Sidhu PS, Singh S (2004) PBG 5: A new multiple resistant *desi* chickpea variety for Punjab (India). Int Chickpea Pigeonpea Newsl 11:18–20

Sandhu JS, Arasakesary SJ, Singh P (2005) Evaluation of chickpea (*Cicer arietinum* L) genotypes for cold tolerance. Indian J Pulses Res 18:171–174

Sandhu JS, Gupta SK, Singh G, Sharma YR, Bains TS, Kaur L, Kaur A (2006) Interspecific hybridization between *Cicer arietinum* L. and *Cicer pinnatifidum*

Jaub. et. Spach for improvement of yield and other traits. In: 4th international food legumes research conference, Oct 18–22, 2005, Indian Society of Genetics and Plant Breeding, New Delhi, p 192

Sandhu JS, Gupta SK, Kaur L (2007) Wide hybridization in chickpea and pigeonpea. In: Pulses at a glance. Punjab Agricultural University, Ludhiana, pp 32–37

Sandhu JS, Gupta SK, Singh I, Gill BS, Bhardwaj R (2010) Genetics of bushy growth habit and its implications in chickpea improvement. Indian J Genet 70:383–385

Santra DK, Tekeoglu M, Ratnaparkhe M, Kaiser WJ, Muehlbauer FJ (2000) Identification and mapping of QTLs conferring resistance to Ascochyta blight in chickpea. Crop Sci 40:1606–1612

Saraf CS, Rupela OP, Hegde DM, Yadav RL, Shivkumar BG, Bhattarai S, Razzaque MA, Sattar MA (1998) Biological nitrogen fixation and residual effects of winter grain legumes in rice and wheat cropping systems of the Indo-Gangetic plain. In: Kumar JVDK, Johansen C, Rego TJ (eds) Residual effects of legumes in rice and wheat cropping systems of the Indo-Gangetic plain. Oxford/IBH Publishing, New Delhi, pp 14–30

Savithri KS, Ganapathy PS, Sinha SK (1980) Sensitivity to low tolerance in pollen germination and fruit set in *Cicer arietinum* L. J Exp Bot 31:475–481

Saxena MC (1990) Problems and potential of chickpea production in nineties. In: Chickpea in the nineties: proceedings of the second international workshop on chickpea improvement, 4–8 Dec 1989, ICRISAT Center, Patancheru, India

Saxena MC, Saxena NP, Mohamed AK (1988) High temperature stress. In: Summerfield RJ (ed) World crops: Cool season food legumes. Springer, Dordrecht, pp 845–856

Sethy NK, Choudhary S, Shokeen B, Bhatia S (2006a) Identification of microsatellite markers from *Cicer reticulatum*; molecular variation and phylogenetic analysis. Theor Appl Genet 112:347–357

Sethy NK, Shokeen B, Edwards KJ, Bhatia S (2006b) Development of microsatellite markers and analysis of intraspecific genetic variability in chickpea (*Cicer arietinum* L.). Theor Appl Genet 112:1416–1428

Sharma HC, Ortiz R (2002) Host plant resistance to insects: an eco-friendly approach for pest management and environment conservation. J Environ Biol 23:11–35

Sharma J, Satija CK (1996) *In vitro* hybridization in incompatible crosses of *Vigna* species. Crop Improv 23:29–32

Sharma PC, Winter P, Bunger T, Huttel B, Weigand F, Weising K, Kahl G (1995) Abundance and polymorphism of di-, tri- and tetra-nucleotide tandem repeats in chickpea (*Cicer arietinum* L.). Theor Appl Genet 90:90–96

Sharma KD, Winter P, Kahl G, Muehlbauer FJ (2004) Molecular mapping of *Fusarium oxysporum* f. sp. *ciceris* race 3 resistance genes in chickpea. Theor Appl Genet 108:1243–1248

Sharma KD, Chen W, Muehlbauer FJ (2005) Genetics of chickpea resistance to five races of Fusarium wilt and a concise set of race differentials for *Fusarium oxysporum* f. sp. *ciceris*. Plant Dis 89:385–390

Simon CJ, Muehlbauer FJ (1997) Construction of a chickpea linkage map and its comparison with maps of pea and lentil. J Hered 88:115–119

Singh KB (1990) Patterns of resistance and susceptibility to races of *Ascochyta rabiei* among germplasm accessions and breeding lines of chickpea. Plant Dis 74:127–129

Singh KB (1997) Chickpea (*Cicer arietinum* L.). Field Crops Res 53:161–170

Singh KB, Hawtin GC (1979) Winter planting. Int Chickpea Newsl 1:4

Singh KB, Ocampo B (1993) Interspecific hybridization in annual *Cicer* species. J Genet Breed 47:199–204

Singh KB, Ocampo B (1997) Exploitation of wild *Cicer* species for yield improvement in chickpea. Theor Appl Genet 95:418–423

Singh KB, Reddy MV (1991) Advances in disease resistance breeding in chickpea. Adv Agron 45:191–222

Singh G, Sharma YR (2002) Fungal diseases of pulses. In: Gupta VK, Paul YS (eds) Diseases of field crops. Indus Publishing, New Delhi, pp 155–192

Singh G, Kapoor S, Singh K (1982a) Screening chickpea for gray mold resistance. Int Chickpea Newsl 7:13–14

Singh G, Singh K, Kapoor S (1982b) Screening for sources of resistance to Ascochyta blight of chickpea. Int Chickpea Newsl 6:15–17

Singh BD, Jaiswal HK, Singh RM, Singh AK (1984a) Isolation of early flowering recombinants from the interspecific cross between *Cicer arietinum* and *C. reticulatum*. Int Chickpea Newsl 11:14–16

Singh G, Verma MM, Gill AS, Sandhu TS, Brar HS, Sra SS, Kapoor S (1984b) Screening of gram varieties against Ascochyta blight. Crop Improv 11:153–154

Singh KB, Malhotra RS, Halila MH, Knights EJ, Verma MM (1994) Current status and future strategy in breeding chickpea for resistance to biotic and abiotic stresses. Euphytica 73:137–149

Singh KB, Ocampo B, Robertson LD (1998) Diversity for abiotic and biotic stress resistance in the wild annual *Cicer* species. Genet Resour Crop Evol 45:9–17

Singh S, Gumber RK, Joshi N, Singh K (2005) Introgression from wild *Cicer reticulatum* to cultivated chickpea for productivity and disease resistance. Plant Breed 124:477–480

Singh TP, Deshmukh PS, Kumar P (2008) Relationship between physiological traits in chickpea (*Cicer arietinum* L.) under rainfed condition. Indian J Plant Physiol 13:411–413

Singh M, Khan Z, Kumar K, Dutta M, Pathania A (2012a) Sources of resistance to *Fusarium* wilt and root knot nematode in indigenous chickpea germplasm. Plant Genet Resour 10(3):258–260

Singh RP, Singh I, Singh S, Sandhu JS (2012b) Assessment of genetic diversity among interspecific derivatives in chickpea. J Food Legum 25:150–152

Singh I, Singh RP, Singh S, Sandhu JS (2012c) Introgression of productivity genes from wild to cultivated *Cicer*. In: Sandhu SK, Sidhu N, Rang A (eds) International conference on sustainable agriculture for food and livelihood security, Crop Improve 39 (Special issue), Crop Improvement Society of India, Ludhiana, pp 155–156

Singh I, Sandhu JS, Gupta SK, Singh S (2013) Introgression of productivity and other desirable traits from rice bean (*Vigna umbellata*) into black gram (*Vigna mungo*). Plant Breed 132:401–406

Singh M, Bisht IS, Dutta M, Kumar K, Basandrai AK, Kaur L, Sirari S, Khan Z, Rizvi A, Sarker A, Bansal KC (2014) Characterization and evaluation of wild annual *Cicer* species for agro-morphological traits and major biotic stresses under Northwestern Indian conditions. Crop Sci 54:229–239

Spillane C, Gepts P (2001) Evolutionary and genetic perspectives on the dynamics of crop genepools. In: Cooper HD, Spillane C, Hodgkin T (eds) Broadening the genetic base of crop production. CAB International, Wallingford, pp 25–70

Srinivasan A, Saxena NP, Johansen C (1999) Cold tolerance during early reproductive growth of chickpea (*Cicer arietinum* L.): Genetic variation in gamete development and function. Field Crops Res 60:209–222

Stalker HT (1980) Utilization of wild species for crop improvement. Adv Agron 33:111–145

Stamigna C, Crino P, Saccardo F (2000) Wild relatives of chickpea: multiple disease resistance and problems to introgression in the cultigen. J Genet Breed 54:213–219

Stebbins GL (1966) Chromosomal variation and evolution. Science 152:1463–1469

Stolton S, Maxted N, Ford-Lloyd BV, Kell SP, Dudley N (2006) Food stores: using protected areas to secure crop genetic diversity. WWF arguments for protection series. WWF World Wide Fund for Nature, Gland

Summerfield RJ, Hadley P, Roberts EH, Minchin FR, Rawsthrone S (1984) Sensitivity of chickpea (*Cicer arietinum* L.) to hot temperatures during the reproductive period. Exp Agric 20:77–93

Tanksley SD, Nelson JC (1996) Advanced backcross QTL analysis: a method for the simultaneous discovery and transfer of valuable QTLs from unadapted germplasm into elite breeding lines. Theor Appl Genet 92:191–203

Tayyar RI, Waines JG (1996) Genetic relationships among annual species of *Cicer* (*Fabaceae*) using isozyme variation. Theor Appl Genet 92:245–254

Tekeoglu M, Rajesh PN, Muehlbauer FJ (2002) Integration of sequence tagged microsatellite sites to chickpea genetic map. Theor Appl Genet 105:847–854

Thudi M, Bohra A, Nayak SN, Varghese N, Shah TM, Penmetsa RV, Thirunavukkarasu N, Gudipati S, Gaur PM, Kulwal PL, Upadhyaya HD, KaviKishor PB, Winter P, Kahl G, Town CD, Kilian A, Cook DR, Varshney RK (2011) Novel SSR markers from BAC-end sequences, DArT arrays and a comprehensive genetic map with 1,291 marker loci for chickpea (*Cicer arietinum* L.). PLoS One 6(11):1–12

Toker C, Canci H (2006) Selection for drought and heat resistance in chickpea under terminal drought conditions. In: Kharkwal MC (ed) Food legumes for nutritional security and sustainable agriculture, 4th international food legumes research conference, Indian Agricultural Research Institute, New Delhi, October, 2006, Indian Society of Genetics and Plant Breeding, New Delhi, pp 18–22

Toker C, Canci H (2009) Evaluation of yield criteria for drought and heat resistance in chickpea (*Cicer arietinum* L.). J Agron Crop Sci 195:47–54

Toker C, Canci H, Yildirim T (2007) Evaluation of perennial wild Cicer species for drought resistance. Genet Resour Crop Evol 54:1781–1786

Udupa SM, Baum M (2003) Genetic dissection of pathotype-specific resistance to Ascochyta blight disease in chickpea (*Cicer arietinum* L.) using microsatellite markers. Theor Appl Genet 106:1196–1202

Udupa SM, Weigand F, Saxena MC, Kahl G (1998) Genotyping with RAPD and microsatellite markers resolves pathotype diversity in the Ascochyta blight pathogen of chickpea. Theor Appl Genet 97:299–307

Upadhyaya HD, Smithson JB, Haware MP, Kumar J (1983) Resistance to wilt in chickpea. II: further evidence for two genes for resistance to race 1. Euphytica 32:749–755

Upadhyaya HD, Thudi M, Dronavalli N, Gujaria N, Singh S, Sharma S, Varshney RK (2011) Genomic tools and germplasm diversity for chickpea improvement. Plant Genet Resour 9(1):45–58

Upadhyaya HD, Kashiwagi J, Varshney RK, Gaur PM, Saxena KB, Krishnamurthy L, Gowda CLL, Pundir RPS, Chaturvedi SK, Basu PS, Singh IP (2012) Phenotyping chickpeas and pigeonpeas for adaptation to drought. Front Physiol. doi:10.3389/fphys. 2012.00179

Vadez V, Krishnamurthy L, Thudi M, Anuradha C, Colmer TD, Turner NC (2012) Assessment of ICCV 2×JG 62 chickpea progenies shows sensitivity of reproduction to salt stress and reveals QTL for seed yield and yield components. Mol Breed 30:9–21

van der Maesen LJG (1987) Origin, history and taxonomy of chickpea. In: Saxena MC, Singh KB (eds) The chickpea. CAB International, Cambridge, pp 11–34

van der Maesen LJG, Maxted N, Javadi F, Coles S, Davies AMR (2007) Taxonomy of the genus *Cicer* revisited. In: Yadav SS, Redden RJ, Chen W, Sharma B (eds) Chickpea breeding and management. CABI, Wallingford, pp 14–47

van Dorrestein B, Baum M, Malhotra RS (1998) Interspecific hybridization between cultivated chickpea (*Cicer arietinum* L.) and the wild annual species *C. judaicum, C. pinnatifidum*. In: Proceedings of third European conference on grain legumes, 14–19 November 1998, Valladolid, Spain. AEP, Paris, pp 362–363

van Rheenen HA (1992) Biotechnology and chickpea breeding. Int Chickpea Newsl 26:14–17

van Rheenen HA, Pundir RPS, Miranda JH (1993) How to accelerate the genetic improvement of a recalcitrant crop species such as chickpea. Curr Sci 65:414–417

Varshney RK, Graner A, Sorrells ME (2005) Genic microsatellite markers in plants: features and applications. Trends Biotechnol 23:48–55

Varshney RK, Hoisington DA, Upadhyaya HD, Gaur PM, Nigam SN, Saxena K (2007) Molecular genetics and breeding of grain legume crops for the semi-arid tropics. In: Varshney RK, Tuberosa R (eds) Genomics-assisted crop improvement, vol II, Genomics applications in crops. Springer, Dordrecht, pp 207–242

Varshney RK, Hiremath PJ, Lekha P, Kashiwagi J, Balaji J, Deokar AA (2009a) A comprehensive resource of drought-and salinity-responsive ESTs for gene discovery and marker development in chickpea (*Cicer arietinum* L). BMC Genomics 10:5

Varshney RK, Hiremath PJ, Lekha P, Kashiwagi J, Balaji J, Deokar AA, Vadez V, Xiao Y, Srinivasan R, Gaur PM, Siddique KHM, Town CD, Hoisington DA (2009b) A comprehensive resource of drought- and salinity- responsive ESTs for gene discovery and marker development in chickpea (*Cicer arietinum* L.). BMC Genomics 10:523

Varshney RK, Song C, Saxena RK, Azam S, Yu S, Sharpe AG et al (2013a) Draft genome sequence of chickpea (*Cicer arietinum*) provides a resource for trait improvement. Nat Biotechnol. Epub January 27. Available from: http://www.nature.com/nbt/journal/vaop/ncurrent/full/nbt.2491.html

Varshney RK, Mohan SM, Gaur PM, Gangarao NVPR, Pandey MK, Bohra A et al (2013b) Achievements and prospects of genomics-assisted breeding in three legume crops of the semi-arid tropics. Biotech Adv 31(8):1120–1134. doi:10.1016/j.biotechadv.2013.01.001

Vavilov NI (1926) The mountainous districts as the home of agriculture. Studies on the origin of cultivated plants. Bull App Bot – Plant Breed

Verma MM, Sandhu JS, Brar HS, Brar JS (1990) Crossability studies in different species of *Cicer* (L.). Crop Improv 17:179–181

Verma MM, Sandhu JS, Ravi (1995) Characterization of the interspecific cross *Cicer arietinum x C. judaicum* (Boiss). Plant Breed 114:259–266

Wahid A, Close TJ (2007) Expression of dehydrins under heat stress and their relationship with water relations of sugarcane leaves. Biol Plant 51:104–109

Wang J, Gan YT, Clarke F, McDonald CL (2006) Response of chickpea yield to high temperature stress during reproductive development. Crop Sci 46:2171–2178

Weigand S, Tahhan O (1990) Chickpea insect pests in the Mediterranean zones and new approaches to their management. In: Chickpea in the nineties: proceedings of the second international workshop on chickpea improvement, 4–8 December 1989. ICRISAT, Hyderabad, pp 169–175

Wery J, Turc O, Lecoeur J (1993) Mechanism of resistance to cold, heat and drought in cool-season legumes, with special reference to chickpea and pea. In: Singh KB, Saxena MC (eds) Breeding for tolerance in cool season food legumes. Wiley, Chichester, pp 271–291

Williams PC, Singh U (1987) Nutritional quality and the evaluation of quality in breeding programmes. In: Saxena MC, Singh KB (eds) The chickpea. CAB International, Wallingford, pp 329–390

Winter P, Pfaff T, Udupa SM, Sharma PC, Sahi S et al (1999) Characterization and mapping of sequence-tagged microsatellite sites in the chickpea (*C. arietinum* L.) genome. Mol Gen Genet 262:90–101

Winter P, Benko-Iseppon AM, Hüttel B, Ratnaparkhe M, Tullu A, Sonnante G, Pfaff T, Tekeoglu M, Santra D, Sant VJ, Rajesh PN, Kahl G, Muehlbauer FJ (2000) A linkage map of the chickpea (*Cicer arietinum* L.) genome based on the recombinant inbred lines from a *C. arietinum* x *C. reticulatum* cross: localization of resistance genes for *Fusarium* races 4 and 5. Theor Appl Genet 101:1155–1163

Zatloukalová P, Hřibová E, Kubaláková M, Suchánková P, Šimková H, Adoración C, Kahl G, Millán T, Doležel J (2011) Integration of genetic and physical maps of the chickpea (Cicer arietinum L.) genome using flow-sorted chromosomes. Chromosom Res 19:729–739

Zhang X, Scheuring CF, Zhang M, Dong JJ, Zhang Y, Huang JJ, Lee MK, Abbo S, Sherman A, Shtienberg D, Chen W, Muehlbauer F, Zhang HB (2010) A BAC/BIBAC-based physical map of chickpea, *Cicer arietinum* L. BMC Genomics 11:501

Zohary D, Hopf M (2000) Pulses. In: Domestication of plants in the old world: the origin and spread of cultivated plants in West Asia, Europe, and the Nile Valley, 3rd edn. Oxford University Press, New York, pp 108–111

Faba Bean

4

R. Redden, J. Paull, X. Zong, O. Sass, T. Yang, and Li Ling

Abstract

Faba beans are partially allogamous with both self- and cross-fertilisation enhanced by various bee species. For breeding, this population structure enables population breeding methods or alternatively the development of inbred 'pure' lines in bee-proof enclosures and selection among lines. Heterosis occurs, notably in crosses between seed-type groups. Important diseases are Aschochyta blight, chocolate spot, rust, Cercospora leaf spot, downey mildew, and a range of root rot and viral diseases. Forms of host plant resistance include both major and quantitative gene resistances for Aschochyta and for chocolate spot, race-specific major genes for rust and a range of resistance expressions to various viruses. Resistance to broomrape is rare, incomplete, quantitative and of low heritability. Faba bean is very sensitive to abiotic stresses in the reproductive phase. Genetic variation in stomatal conductance and other component traits may provide drought tolerance, heat tolerance is being investigated, and there is a genetic variation for winter hardiness and for waterlogging tolerance. Genetic control of abiotic stress responses is being investigated with biotechnology tools, though previously few SSR markers were available. The application of next-generation sequencing has identified many high-quality novel SSR markers which will improve the development of

R. Redden (✉)
Australian Grains Genebank, Department of Environment and Primary Industries, Private Bag 260, Horsham Victoria 3401, Australia
e-mail: Bob.Redden@dpi.vic.gov.au

J. Paull
School of Agriculture, Food and Wine, University of Adelaide, Adelaide, SA 5064, Australia
e-mail: jeffrey.paull@adelaide.edu.au

X. Zong • T. Yang
Institute of Crop Sciences/National Key Facility for Crop Germplasm Resources and Genetic Improvement, Chinese Academy of Agricultural Sciences, Beijing 100081, China
e-mail: zongxx@mail.caas.net.cn; ytytyt2209@163.com

O. Sass
Norddeutsche Pflanzenzucht, Hans-Georg Lembke KG, 24363 Holtsee, Germany
e-mail: O.Sass@npz.dewww.npz.de

L. Ling
Liaoning Institute of Cash Crops, Liaoning Academy of Agricultural Sciences, Liaoyang 111000, China
e-mail: lilingsws@126.com

M. Singh et al. (eds.), *Broadening the Genetic Base of Grain Legumes*,
DOI 10.1007/978-81-322-2023-7_4,

linkage maps, but at present linkage maps have not been resolved to six linkage groups that represent each chromosome. Markers will be useful for identification of marker-locus-trait associations.

4.1 Introduction

Faba bean (*Vicia faba* L.) is grown over a wide geographic range, with significant production zones from 50°N to 40°S and up to an altitude of 3,000 m above sea level. It is grown as an autumn/winter crop in regions with mild winter such as in the Middle East, North Africa, the Mediterranean region, maritime Europe, Southern China and Australia or a spring-sown crop where winters are severe such as in continental Europe, Northern China and Canada. It is grown under both rainfed and irrigated conditions. The crop is harvested as dry seed which is used for food, particularly in the Middle East and North Africa where it is a staple component of the human diet or used as a protein component in animal feed. The average world production of faba bean for 2009–2011 was 4.1 million metric tonnes (mmt), and the major producing countries were China (1.53 mmt), Ethiopia (0.67 mmt), France (0.42 mmt), Australia (0.26 mmt), Egypt (0.24 mmt), the UK (0.16 mmt), Morocco (0.15 mmt) and Sudan (0.13 mmt) (FAOSTAT 2013). Correspondingly, the mean percentages of the crop area were 36 % for China, 20 % for Ethiopia, 7 % for Morocco, 6 % for Australia and below 5 % for the other major producers. About 25 % of world faba bean production is internationally traded, with Egypt being the major importer and France, the UK and Australia the major exporters.

The range in production environments and cultural conditions results in many factors, such as diseases, soil constraints, temperature extremes and moisture deficit, limiting crop production. Some limiting factors are widespread and form the topics of significant international research while others are restricted to specific regions. Understanding the regional significance, genetic control and sources of resistance/tolerance to potential limiting factors will assist in determining the priorities for crop improvement programmes. China is the world's dominant producer of faba beans, and it is useful to understand its production environments. Faba bean production in China is mainly done in the paddy fields of Southern China after harvest of rice, as an irrigated winter legume crop for fresh green seeds and for dry grain. The major biotic limiting factors are chocolate spot, rust, aphids and leaf minors. The major abiotic limiting factors for winter faba bean are frost damage, during both flowering and podding stages, and terminal drought or waterlogging at maturity. Although spring-sown faba bean in Northern China shared only 17 % of the total faba bean production area in China, half of the spring faba bean sowing areas are irrigated and produce high-yielding large-seeded faba beans. The major biotic factors limiting spring sowing areas are aphids, *Epicauta gorhami* (Marseul) and leaf minors. The major abiotic factors limiting spring faba bean are salinity, drought at seedling stage and waterlogging at maturity stage. A shortage of suitable machinery for sowing and harvesting coupled with an increasing shortage of rural labour, both for fresh green seeds and for dry grain, are the major factors that discourage faba bean growers.

4.2 Evolution and Taxonomy

Faba bean has a very wide range in seed size, which is usually regarded as seed-type groups, viz. major (large flattened or broad bean, 100–200 g/100 seed), equina (horse bean, 60–100 g/100 seed), minor (tic bean with ellipsoidal seed, 40–60 g/100 seed) and *paucijuga* (regarded as primitive, 31–40 g/100 seed) (Cubero 1973; Serradilla et al. 1993). Both Muratova (1931) and Hanelt et al. (1972) recognised these four seed types but classified *paucijuga* as a subspecies. *V. faba* var. *paucijuga* has a higher level of dehiscent pods than the other groups; it is self-fertile with short stature and strong tillering and closest to the wild type in traits (Ladizinsky 1998), occurring in the Afghan-north Indian subcontinent region. The small-seeded features are dominant over that of the larger more recent seed types (Cubero and Nadal 2005). Fossil remains of faba bean date back to 8,500 BP in Israel and 9,300 BP in Syria and even 11,200 BP in Iraq (Ladizinsky 1998; Tanno and Willcox 2006), with seed from small, flat to wedge-shaped at the hilum end, distinct from the more globular seed of wild *V. narbonensis* and *V. johannis* which still occur in the vicinity of the Israeli site (Ladizinsky 1998). Medium-sized seed were found in Iberia and central Europe in 5,000 BP, and larger more flattened seed were not known before 1,500 BP (Ladizinsky 1998). There is archaeological evidence that indicates faba bean arrived in China/Japan around 4,000–5,000 BP (Ye et al. 2003), and it has developed into a very distinct gene pool, as discussed below.

4.3 Absence of Primary and Secondary Gene Pools

Faba bean has a basic chromosome number, $n=6$, with twice the DNA content of the nearest wild relatives (*Vicia* species: *narbonensis*, *johannis* and *galilea*) in the subgenus *Vicia* of the family *Viceae*, which have $n=7$ and appear to be only distantly related although morphologically similar (von Hanelt et al. 1972; Ladizinsky 1998). *V. faba* is cross incompatible with these and other *Vicia* relatives and also differs from them in mitochondrial and nuclear DNA (Ven et al. 1993). Thus, no wild relative is known for *V. faba*, and no genetic divergence accompanied its geographic dispersal (Serradilla et al. 1993). Possibly, *V. faba* had multiplicity in the number of lateral chromosome strands which results in disjunct karyotype groups with DNA content per chromosome in 1:2:4 ratios with one chromosome being very large, possibly with segmental duplications (Chooi 1971). There is a high cross compatibility between the different seed groups of *V. faba*, indicating that they belong to one species (Cubero 1973). The greatest morphological diversity in *V. faba* is found in the Fertile Crescent region of SW Asia (Muratova 1931), which may have been the centre of origin for this species. It is not clear whether the *paucijuga* group located in the region from Afghanistan to north India is derived from the main *V. faba* diversity in SW Asia or could have been independently domesticated. Wild *V. faba* was reported in Algeria and Morocco as *V. pliniana* in the nineteenth century, but is apparently extinct today (Muratova 1931).

4.4 Morphological Diversity and Conservation of Germplasm

Faba beans require long days to flower and mature, with temperatures below 20 °C for germination and below 30 °C for optimal growth (Duc 1997; Cubero and Nadal 2005). Although indeterminate in habit, a radiation-induced mutation has provided a gene for a determinate growth habit used in the International Center for Agricultural Research in the Dry Areas (ICARDA) and other breeding programmes for the 'major' seed types (Cubero and Nadal 2005). The seed-type classification as described in the chapter 'History' is widely accepted as a description of the range in seed size and associated end use as food or feed.

At ICARDA one representative plant per landrace was inbred for at least five generations as a bean pure line (BPL) for much of the faba bean collection. Compared to the initial landrace accessions, the pure lines provided the advantages of ease of maintenance; screening could be undertaken on several plants of the same genotype; the possibility of genetic drift due to intercrossing among accessions was reduced; specific genotypes could be maintained even if a destructive screening method was used; and recessive genes that were hidden by heterozygosity were uncovered (Hawtin 1982; Robertson and El-Sherbeeny 1991). A BPL collection of 840 accessions was field-evaluated for 22 discretely scored vegetative, phenologic and seed traits as per IBPGR/ICARDA descriptor lists (Robertson and El-Sherbeeny 1991). All accessions were indeterminate, and over 98 % had stipule spot pigmentation and no secondary branching. Although 72 % were susceptible to lodging, 7.3 % were rated as partially resistant due to stiffer straw and with similar height. Leaf size ranged from small

(38 %) to medium (51 %) to large (11 %). There was little variation in flower colour (99.8 % white with 99.5 % of the wings spotted), with the uniform white colour significant as an indicator of low to zero tannin content which confers higher digestibility for mono-gastrics (Bond 1976). The flower standard was moderately streaked for the majority with variation from no streaks (0.4 %) to intensely streaked (12 %). Pods were non-shattering for 90 % of the BPLs, and 97 % had a matte surface, with mostly (93 %) uniform distribution on the stem but 6 % with basal distribution. For ease of harvest, 43 % of the BPLs had erect pods, in contrast to 17 % pendant types. Over 60 % of the BPLs had dark-coloured mature pods, and almost all had plain testa with black hilum (94 %). There was a variation among BPLs in levels of autofertility. The seed was 28 % major type, with a range through the *paucijuga* type, reflecting the preponderance of origins from the Fertile Crescent, North Africa, Ethiopia and Afghanistan.

Patterns of trait combinations were associated with BPL origins: low stem pigmentation in Sudan and the Nile valley, a high frequency of small leaves in North Europe which also had the least lodging, erect pods in the Indian subcontinent but pendant in south Europe and both dark pods and severe lodging in Ethiopia (Robertson and El-Sherbeeny 1991). Regional adaptation included resistance to 'chocolate spot' and to rust from Ecuador, tolerance to heat stress from Bangladesh, frost tolerance in winter types from Europe and drought tolerance from the Mediterranean region, plus genetic variation for resistances to Ascochyta, nematodes, bean yellow mosaic virus and bean leaf roll virus with two BPLs from Afghanistan resistant to both viruses (Robertson et al. 2000).

Faba bean germplasm accessions (500) were evaluated for morphologic traits, disease resistance and abiotic stress tolerances (Lang Li-jaun et al. 1993). Some accessions from the Chinese provinces Zhejiang, Fujian, Gansu and Shaanxi were salinity tolerant, consistent with another screening of 504 diverse sources of germplasm in which there were 16 tolerant lines especially from Greece and China, including Zhejiang (Enneking et al. personal communication; Duc et al. 2010). In Southern China, faba beans sown prior to winter are known as winter types. They are adapted to short mild winters and tolerate temperatures up to –5 °C, whereas in Northern China, where there is snow cover through a long winter, only spring types which tolerate temperatures of only 3–5 °C at seedling stage are sown (Lang Li-jaun et al. 1993). The differences between these two ecotypes can be characterised by responses to both photoperiod and temperature that operate independently, with a lower base temperature for growth in the winter types which may have a higher optimum temperature for rate of progress towards flowering. Though both ecotypes require 1,000 degree days for flowering at suboptimal temperatures, varietal differences in base temperature and in sensitivity to temperature range characterise the geographic range in adaptation of different Chinese faba bean landraces to their respective local environments.

In Ethiopia faba landrace variation showed accessions from the north to be closely related with small seed and lower yield, plant height and greater susceptibility to chocolate spot disease, whereas in the south the diversity of morphologies and seed types was greatest and included those with large seed, superior yield and disease resistance. Thus, the clustering of landraces in Ethiopia did not correspond to the geographic diversity of their origins. Faba bean germplasm of over 38,000 accessions of landraces and varieties is conserved worldwide in at least 43 national gene banks as well as at ICARDA (Duc et al. 2010). The size of the collections ranges from below 100 to 9,000 at ICARDA, exceeding 1,000 in China, Australia, France, Germany, Italy, Russia, Spain, Poland, Morocco and Ethiopia. Although 52 % of accessions originated in Europe, the range of habitats may be wider in Ethiopia and China, with cultivation up to 3,000 m in altitude.

4.5 Genetic Diversity and Gene Pools

The largest genetic diversity study of faba bean landraces was of 802 landraces and varieties from China, North Africa, Europe and Asia, using ISSR markers (Wang et al. 2012). The Chinese landraces were widely separated from a combined grouping of African, European and Asian (outside of China) germplasm which were closely similar though still distinct. Within China there was a major separation of the spring- and winter-grown landraces which corresponded with northern and southern regions, respectively. Spring landraces from the northern group of Shaanxi, Hebei, Inner Mongolia and Xinjiang provinces/regions were the most diverse, followed by the European group and then spring landraces from Northwest China (Qinghai, Ningxia and Gansu provinces), while a winter group from Central China (Hubei, Hunan, Anhui and Jiangsu provinces) was the least diverse. Two of the Chinese winter groups (Sichuan and Giuzhou provinces) and (Jiangxi and Sheijiang provinces), respectively from medium to low elevation (200–1,000 m) regions in SW China and SE China, were widely separated (a) from the Central China group (mainly low, 0–400 m, elevation including the lower reaches of the Yangtze River) and (b) from the Yunnan group above 1,500 m. These various groupings in China reflect in part selection for adaptation to different crop ecologies.

Faba bean landraces/varieties from outside of China were clustered in groups for North Africa with the Egypt and Sudan accessions as the most related; for Europe, from Turkey to Germany with distinct separation of those from Portugal; and for Asia, from Syria to Afghanistan with distinct separation of those from Japan. Again this is a general ecogeographical separation, indicating a trend for a crop with medium to high levels of outcrossing to evolve a dynamic population in equilibrium with local seasonal conditions provided that sufficient genetic variation is available, thus a degree of local adaptation. An association of diversity across a wide range of geographic origins was also found by Kwon et al. (2010), who evaluated a world collection of 151 faba bean landraces and varieties with 'target region amplified polymorphism' (TRAP) markers. Five major groups were identified with a dendrogram: I with single entries from France and Nepal, II with 13/15 accessions from Afghanistan, III with four accessions from China plus one from Afghanistan, IV with 25 European accessions plus three from Nepal and China and two commercial varieties and V with 101 accessions from China. A wider variation was found in landraces than the advanced breeding line included for comparison in this analysis.

In contrast, in a less geographically diverse study, no clear geographic association was found for 34 Middle Eastern, European and diverse ICARDA accessions of faba beans, assayed with inter-simple sequence repeat (ISSR) markers. Twenty Greek landrace populations were analysed with ISSR by Terzopoulos and Bebeli (2008), with separation of the small-seeded 'minor' types from the medium- to large-seeded Mediterranean types which were divided into two subgroups. Although on morphological traits the 'minor' types were separated also from the Mediterranean types, the morphological and genetic groupings were not significantly correlated. In a limited study of faba bean landraces from Tunisia compared with the variety Aquadulce, one large-seeded 'major' landrace was widely separated from other 'major' and 'minor' landraces and commercial varieties/reselections (Ouji et al. 2012). AFLP diversity analysis of 22 elite faba bean cultivars from Europe revealed the most shared genetic similarity among the spring types with one winter type classified as separate, and the closest relationship was between two advanced lines from the same parents (Zeid et al. 2001). In an assay with transposon-based markers, a diverse collection of 20 faba bean accessions from a gene bank in Spain separated into groups mostly unrelated to geographic origins or to morphologic groupings (Sanz et al. 2007). Within relatively small geographic regions and relatively small number of landraces, associations of genetic diversity with geographic distance of origins may not be found or are weak, with morphology being an unreliable guide to genetic diversity. However associations of genetic diversity with geographic distance tend to be found in more extensive studies including cross-continental sources of landraces and varieties, especially in spring and winter habits of landraces reflecting the agricultural practices in respective diverse environments and geographic origins.

4.6 Assessment of Gene Flow for Crop Improvement

Faba bean is a partially allogamous crop with both self- and cross-fertilisation occurring within a plant and even for individual seeds within a single pod. Populations are therefore a mixture of inbred and hybrid individuals with hybrid individuals being more vigorous. In addition, inbreeding results in lower set seed without manipulation of the flower, for example, by bee visitation; therefore, the inbred plants within a population will contribute less seeds to the following generation than hybrid plants when pollinators are at a low frequency (Drayner 1956). Cross-pollination of faba bean is effected by a number of bee species, including bumblebees (*Bombus* spp.), honeybees (*Apis mellifera*) and solitary bees (*Anthophora* and *Xylocarpa*), with the dominant species varying between regions (Bond and Poulson 1983). In addition to transferring pollen between flowers, foraging insects trip the flowers, or cause the style to be released and rupture the stigmatic surface, which can result in self-pollination. A wide range of cross-pollination of faba bean has been observed, and factors that contribute to variation include environment, germplasm, incidence of insect pollinators and distance over which cross-pollination is measured. Bond and Poulson (1983) reported a range from 4 to 84 %, with an average value in the order of 35 %. Faba bean genotypes differ in autofertility (Siong and Knight 1980; Robertson and El-Sherbeeny 1988), or the ability to set seed in the absence of insect pollinators or without tripping of flowers. The potential for cross-fertilisation has a very significant impact in maintaining germplasm collections, devising breeding strategies, multiplying breeding lines and managing seed for commercial crops of faba bean.

4.7 Gene Flow Constraints

Multiplication of faba bean germplasm, whether lines in a breeding programme or accessions in an ex situ genetic resource collection, can be undertaken either in insect-proof conditions to ensure self-fertilisation or in open-pollinated conditions where there might be contamination of pollen between lines. The method selected will depend on the specific germplasm, objectives and available resources. In a breeding programme, the early generations such as F_1, F_2 and F_3 might be grown inside bee-proof cages to ensure self-fertilisation and a high frequency of homozygosity for genes that control major breeding traits prior to selection and field evaluation. Several methods can be employed for subsequent multiplication to obtain sufficient seed for widespread field evaluation. Individual lines can be grown in spatially isolated plots that are spaced at sufficient distance to ensure there is minimal pollen transfer between plots. Pope and Bond (1975) reported the rate of cross-fertilisation at a range of distances, and there was less than 5 % cross-fertilisation between plots 50 m apart. Only a limited number of lines can be grown in isolated plots due to the required space; therefore this means of multiplication is not suitable when numerous small plots are required. An alternative method of reducing cross-fertilisation is to grow a barrier crop between the faba bean plots. A number of crops, including canola, male sterile and autotetraploid faba bean and narbon bean, have been evaluated as barrier crops and compared with barren ground between faba bean plots (Robertson and Cardona 1986; Suso et al. 2006). There was less inter-plot gene flow when small multiplication plots were surrounded by a barrier crop such as narbon bean compared to barren ground (Suso et al. 2006). However, results were inconsistent with the significant effects of faba bean genotype, location and isolation zone, and there was variation among genotypes in the level of inter-plot gene flow when measured as female (pollen recipient) as compared to male (pollen donor) parent. Further steps can be taken to minimise the impact of gene flow between multiplication plots. For instance, plots can be arranged according to phenotype, such as seed traits or disease resistance, or to pedigree groups so that any cross-pollination is among similar types.

Faba bean accessions and open-pollinated varieties are populations with varying degrees of heterozygosity at the gene level and heterogeneity among plants, which are maintained through insect-mediated cross-fertilisation. This population structure enables population breeding methods, such as recurrent mass selection, to be applied for crop improvement. An alternative breeding method is to develop inbred 'pure' lines through self-fertilisation in bee-proof enclosures and selection among these lines. Pure lines can either be evaluated and released as varieties or be included as components of synthetic varieties. In a comparison of a number of population structures of faba bean, including inbred lines, F_1 hybrids, blends, synthetics and open-pollinated populations, Stelling et al. (1994) identified that yield stability improved with an increase in heterozygosity and heterogeneity. The reproductive system of faba bean is well suited to developing and maintaining synthetic varieties and potentially F_1 hybrids.

Heterosis for yield where F_1 hybrids yield more than both the mid-parent and the better parent means has been reported for faba bean (Schill et al. 1998; Zeid et al. 2004; Ibrahim 2010). Heterosis has also been observed in the F_2 with its yield at approximately the midpoint of the F_1 hybrid and parental mean (Zeid et al. 2004), as theoretically expected. Inter-pool crosses such as Mediterranean × minor or minor × major produce greater heterosis than intra-pool crosses (Schill et al. 1998); therefore, a high level of diversity should be maintained among component lines of a synthetic variety. Diversity can be introduced to the breeding programme by backcrossing between elite breeding lines as recurrent parents and exotic germplasm in a pre-breeding strategy. Selected lines can then be assessed for potential as components of a synthetic variety. F_1 hybrid varieties have been proposed for faba bean, and several cytoplasmic male sterility (CMS) lines have been identified (Bond 1989). The first, CMS477, was spontaneous and discovered by Bond in 1956 (Bond 1957). Additional spontaneous male sterile cytoplasm (CMS350) was reported by Berthelem and Le Guen (1974), while Link et al. (1997) identified cytoplasmic male sterility in the F_2 following wide crosses (CMS199 and CMS297). Cytoplasmic male sterility was also induced by chemical mutagens (Duc et al. 1985). All of the CMS systems were found to revert to a proportion of pollen fertility after a few generations, or in response to the environment. Thus, despite a high level of heterosis being observed in faba bean and effective insect vectors to transfer pollen, F_1 hybrid varieties have not been developed yet for faba bean.

4.8 Production-Related Problems

4.8.1 Pests and Diseases

Faba bean is affected by a range of diseases and pests that vary in regional significance and priority for crop improvement. While there are numerous fungal diseases that infect faba bean, the three major diseases of widespread international significance with respect to both crop damage and breeding objectives are Ascochyta blight (incited by *Ascochyta fabae* Speg.), chocolate spot (*Botrytis fabae* Sard. and *B. cinerea* Pers.) and rust (*Uromyces viciae-fabae* (Pers.) J. Shört.). Fungal diseases with restricted distribution or lesser impact on crop production include Cercospora leaf spot (*Cercospora zonata* (G. Winter)) which has recently been reported to be damaging in Southern Australia (Kimber and Paull 2011) and downy mildew (*Peronospora viciae* (Berk.) de Bary f.sp. *fabae*) which is most frequent and severe in Northwestern Europe. A number of soilborne fungal diseases infect faba bean, including *Fusarium* spp., *Rhizoctonia solani* Kühn, *Pythium* spp., *Phoma* spp. and *Aphanomyces euteiches* Drechs. (recently reviewed by Sillero et al. 2010), but these are not major priorities in breeding programmes. Internationally, major breeding efforts are directed towards improving resistance to fungal diseases Ascochyta blight, chocolate spot and rust, while breeding is undertaken for other fungal diseases at the local level.

Ascochyta blight is a cool-season disease and is therefore particularly important for autumn- and winter-sown crops where major loss in yield and seed quality can occur in susceptible cultivars if disease is not controlled. It is widespread and has been reported in most faba bean-growing regions.

Resistance to Ascochyta blight has been identified by a number of researchers, and resistance appears to be widespread internationally (e.g.

Hanounik and Robertson 1989; Rashid et al. 1991; Maurin and Tivoli 1992; Sillero et al. 2010). Initial screening of germplasm accessions at ICARDA revealed a considerable variation in resistance although no accession was immune (Hanounik and Maliha 1986). A higher level of resistance was identified in inbred lines that were more homozygous and homogeneous than the original accessions (Hanounik and Robertson 1989). Many accessions and cultivars have been identified as heterogeneous in resistance to Ascochyta blight, and selection within accessions through several generations under self-pollinating conditions, as undertaken at ICARDA to develop the BPLs (Hawtin and Omar 1980), has resulted in new sources of resistance for use in breeding programmes (Kimber et al. 2006) and the release of new resistant cultivars (Paull et al. 2006). A number of genes control resistance to Ascochyta blight and the genetic control of resistance differ depending on the source of resistance and combination of parents (Sillero et al. 2010). However, in general resistance is highly heritable and under the control of relatively few major genes. A single dominant gene for resistance was identified in ILB 752 (Kohpina et al. 2000), while seven genes for resistance were identified among 19 inbred lines when inoculated with five individual isolates (Rashid et al. 1991). Six QTLs associated with resistance to Ascochyta blight were identified by Avila et al. (2004). Four of the QTLs were associated with the resistance of both leaves and stems, while resistance associated with the two other QTLs was effective only in leaves or stems. Diaz-Ruiz et al. (2009) identified two QTLs in a different mapping population, one of which mapped to a similar region to one of the QTLs reported by Avila et al. (2004). The relatively large number and wide diversity of sources of resistance to Ascochyta blight enable breeding programmes to use a range of sources of resistance as a precautionary measure against breakdown in resistance of one source.

Chocolate spot is the most devastating disease of faba bean and can result in very significant yield losses. Chocolate spot is a significant problem in many faba bean-producing regions including the Nile Delta, Yangtze valley and areas of Southern China, coastal areas of the Mediterranean and oceanic region of western France and the western UK (Bond et al. 1994) and Australia. Resistance to chocolate spot has been identified among a limited range of germplasm, and the most resistant lines have originated from the mountainous region in Ecuador (Bond et al. 1994). Hanounik and Maliha (1986) and Hanounik and Robertson (1988) undertook two cycles of selection within a large germplasm collection at ICARDA, and 14 accessions, all originating from Colombia, were identified as resistant in the two cycles. While the accessions were classified as originating from Colombia, they were obtained from CIAT and potentially from other countries in the region. BPL 710 was one of the most resistant lines from the early ICARDA screening, and the resistance was demonstrated across environments in international trials (Hanounik and Maliha 1986). The resistance of BPL 710 has proven to be difficult to transfer in full to new cultivars through hybridisation, and this reflects the quantitative inheritance of the trait (Robertson and Saxena 1993). Paull and Kimber (2006) demonstrated that minor genes conferring resistance to chocolate spot could be combined through several cycles of hybridisation to develop breeding lines with greater resistance than any of the parents and equivalent to cv. Icarus, which was selected from BPL 710. Resistance to chocolate spot has been reported in germplasm from other origins. Bouhassan et al. (2004) identified nine genotypes from the Magreb region that were equivalent to or more resistant than BPL 710. Villegas-Fernández et al. (2009) identified several genotypes with greater stability of chocolate spot resistance than BPL 710 in trials across five countries. One of these, BPL 1763, was selected from ILB 1226 which originated from Ethiopia, while several other resistant lines (designated Sel.97 Lat.97) were derived from germplasm collected in Ecuador in 1996. These alternative sources of resistance are important as they enable a greater diversity of disease resistance genes to be used in breeding programmes and also minimise adverse effects that might result from linkage drag when a single source of resistance is used.

Rust is a serious disease of faba bean in the Middle East, North Africa, Europe, China and Australia where it can cause significant loss in yield and reduction in seed size if infection is early. A number of races of rust have been reported (Conner and Bernier 1982; Emeran et al. 2001). Early studies in Canada identified distinct virulence patterns among 17 genotypes and two rust isolates (Conner and Bernier 1982), while 15 physiological races were identified among 27 Spanish and Portuguese isolates (Rojas-Molina et al. 2006). Rust resistance has been identified among geographically diverse materials, and several cycles of inbreeding and selection within accessions increase the success of identifying resistance. Many sources of rust resistance are summarised by Rashid and Bernier (1984), Bond et al. (1994) and Sillero et al. (2010). The race specificity of resistance is indicated by the fact that only one resistant line out of 65 reported by Rashid and Bernier (1986) was resistant at ICARDA (Bond et al. 1994). Resistance to rust in the study by Conner and Bernier (1982) was controlled by major genes, with monogenic segregation for several combinations of resistant × susceptible parents, and transgressive segregation including susceptible progeny was observed in one resistant × resistant combination. Incomplete (Rashid and Bernier 1984) and hypersensitive resistance to rust has been identified, with the hypersensitive resistance controlled by major genes (Avila et al. 2003). The highly variable nature of faba bean rust indicates that a single gene resistance might not lead to long-term disease control, and breeding programmes should include a range of sources of resistance to rust among parents. Genetic- and race-specific studies to better understand the relationship between available sources of rust resistance are required to enable the efficient use of alternative genes. The development of linked markers would facilitate pyramiding rust-resistant genes.

Many viruses damage faba beans, including bean yellow mosaic virus (BYMV), pea enation mosaic virus (PEMV), bean leaf roll virus (BLRV), faba bean necrotic yellows virus (FBNYV), true broad bean mosaic virus (TBBMV), broad bean mottle virus (BBMV) and broad bean stain virus (BBSV), as reviewed by Sillero et al. (2010). Viruses differ in importance, both on a regional and on a yield loss basis, and the degree of resistance to individual viruses also varies. Selection for virus resistance can be undertaken in fields with either natural infection or introduced inoculation or in screen houses with inoculation where self-pollination of resistant plants is assured. Progeny testing of resistant plants can result in improved resistance through identification of homozygous resistant plants (Makkouk and Kumari 1995). Resistance to viruses can be assessed on the basis of symptom expression, concentration of virus in plant tissue based on tissue blot immunoassay reactions and seed yield in the presence of virus. Resistance to several viruses that affect faba bean has been identified, including bean yellow mosaic virus (Robertson and Saxena 1993; Makkouk and Kumari 1995), pea enation mosaic virus-1 (Schmidt et al. 1988) and bean leaf roll virus (Robertson and Saxena 1993; Makkouk et al. 2002). Makkouk et al. (2002) reported 15 lines from 10 countries with a high level of resistance to BLRV following several cycles of screening and selection.

Broomrapes (*Orobanche* spp.) are parasitic weeds of faba bean and a range of other crops and can lead to a very significant reduction in yield. *O. crenata* is widespread throughout the Mediterranean and Middle East regions where it is one of the major constraints to production, whereas *O. foetida* is a significant weed only in Tunisia although it does occur more widely (Sillero et al. 2010). Management of broomrape requires an integrated approach incorporating a range of agronomic practices, the use of herbicides that are selective for broomrape but not the host plant, biological control, crop hygiene measures to prevent spread of broomrape seeds and resistance of the host (Pérez-de-Luque et al. 2010). Resistance of faba bean to broomrape is rare, incomplete, quantitative and of low heritability; consequently breeding for resistance has been slow. The first *O. crenata*-resistant line to be developed was F 402 in Egypt, subsequently released as Giza 402 (Nassib et al. 1982). This source of resistance has been used extensively and has been incorporated in several varieties, for instance, Baraca (Nadal et al. 2004). Additional sources of resistance to broomrape, including *O. foetida*, have been summarised by Sillero et al. (2010). Selection for resistance to broomrape is conducted in field plots naturally infested with *Orobanche* and is assessed on the basis of weight and/or number of *Orobanche* shoots per faba bean plant or yield of faba bean lines. The inherent variability in field plots with heterogeneity of distribution of *Orobanche* seeds and other environmental factors results in low heritability of resistance. Alternative selection methods, such as the use of molecular markers, could increase efficiency of selection. Three QTLs controlling broomrape resistance were identified in an F_2 population (Román et al. 2002), and two of these QTLs were confirmed in RILs across environments (Diaz et al. 2005).

4.8.2 Abiotic Stresses

Faba bean has the highest yield potential among cool-season pulse crops, but yield stability is low compared to other winter crops. Much of this can be attributed to abiotic constraints such as cold, heat and drought at any stage of plant development, but the flowering stage is particularly vulnerable.

4.8.2.1 Heat Stress

Faba beans produce an abundance of flowers but the number that is retained and developed into pods is influenced by a number of environmental factors. These include partitioning of assimilates between the shoot tip and flowers, with cold and overcast conditions at the time of flowering resulting in reduced photosynthesis and reduced pod set (Stoddard 1993), and moisture stress (El Nadi 1969). Faba bean is sensitive to high temperature during the reproductive stage (McDonald and Paulsen 1997; Saxena et al. 1988), and temperature above 23 °C can lead to the termination of flowering (Evans 1959). Faba bean is relatively intolerant of drought, as also discussed in Sect. 4.9 of this chapter, and temporary moisture stress can occur at any stage of the development, while terminal drought during the reproductive stage is particularly important. Physiological responses to moisture stress include stomatal closure which enables the plant water status to be maintained. Genetic variation in stomatal conductance among eight faba bean genotypes was reported by Khan et al. (2007), with drought-avoiding accessions having lower $\Delta\ ^{13}C$ than a drought-sensitive line. Leaf temperature, as measured by infrared thermometer, was positively correlated with transpiration efficiency and negatively correlated with stomatal conductance and $\Delta\ ^{13}C$, indicating that leaf temperature might be a simple parameter to measure as an indicator of water use (Khan et al. 2007). The impact of terminal drought can be minimised via drought escape by developing early-flowering faba bean varieties with the reproductive stage completed prior to the onset of drought. However, drought escape is not suitable for managing transient drought during earlier stages of development (Khan et al. 2010).

Faba bean is sown in winter in the areas of Europe where winters are mild and young plants are able to survive the winter. Crops sown at this time utilise moisture during winter; flower and mature earlier than spring-sown crops, thereby reducing exposure to summer drought; and have higher yield potential. An increase in the level of winter hardiness to reduce the likelihood of winterkill would increase the potential extent of this type of variety (Link et al. 2010). At present there are limited sources of tolerance to winter hardiness, with the old variety Côte d'Or being the most tolerant line identified, while several other genotypes that can withstand temperatures up to −14 °C or lower have been reported recently by Olszewski (cited in Link et al. 2010). Several physiological processes are involved in determining winter hardiness, including (1) vernalisation to ensure plants do not flower until they have been exposed to a sufficiently long period of cold and thereby avoid flowering during winter and (2) hardening which occurs as plants are exposed to a prolonged period of low but non-freezing temperatures prior to extreme cold. Hardening involves changes in the plant cell membrane such as changes in lipid composition and accumulation of proline (Arbaoui et al. 2008a, b) that protects the cell against freezing. Breeding for winter hardiness requires combining of both processes and also tolerance to excess moisture that might occur during winter and fungal diseases such as Ascochyta blight that infects winter beans to a greater extent than it does with spring beans. Faba bean is adapted to heavy texture, often poorly drained soils, sometimes with irrigation, and these situations can result in waterlogging or soil oxygen deficiency. While management of water application and drainage can reduce waterlogging in irrigated situations, identification of varieties tolerant to waterlogging would have wide application. In a comparison of seven grain legume species, Solaiman et al. (2007) identified faba bean to be more tolerant to waterlogging than six other species on the basis of root and shoot growth in comparison to a control. Variation was also identified among six faba bean genotypes, and Acc794 (origin Afghanistan) was significantly more tolerant than the other genotypes and was the only genotype capable of maintaining root growth after the waterlogged treatment.

4.9 Traits of Importance for Base Broadening

4.9.1 The Challenges of Climate Change

Climate change is now unequivocal, particularly in terms of increasing temperature, as is evident from observations of increases in global average air and ocean temperatures, widespread melting of snow and ice and rising global average sea level (IPCC 2007). Temperatures across the world could rise 1 °C by 2050 (Lobell et al. 2011), associated with increasing CO_2 concentration. Consequent adverse effects expected are likely increased frequencies of abiotic stresses such as heat and drought and increased frequency of biotic stresses (pests and diseases). Especially for drought, an interval of water deficit leading to a significant reduction in yield is widely considered to be the most important environmental constraint to crop productivity (Fischer and Turner 1978; Borlaug and Dowswell 2005), and faba bean is reputed to be more sensitive to water deficits than some other grain legumes (McDonald and Paulsen 1997; Amede and Schubert 2003). Climate change models predict greater variability in rainfall patterns, and increased periods of summer drought will affect many regions. Population growth will also require more of the available water to be used for domestic and industrial use, rather than for irrigating crops, giving a double benefit for productive crops exhibiting enhanced drought resistance and water use efficiency. In addition, climate change is also expected to cause losses of biodiversity, mainly in more marginal environments (Ceccarelli et al. 2010). Climate change will threaten food security for even more people in the decades to come and add new challenges for agriculture and food production including more frequent, erratic and severe droughts, increased frequency of floods and higher pressure from insects and diseases.

4.9.2 Breeding to Tackle Climate Change

Genetic diversity and plant breeding are key elements in tackling climate change. The genetic diversity of crop plants is the foundation for the sustainable development of new varieties for present and future challenges. Resource-poor farmers have been using genetic diversity intelligently over centuries to develop varieties adapted to their own environmental stress conditions. Assessment research of germplasm resources with tolerance to abiotic stresses is being launched. Two thousand accessions of pea and 300 accessions of faba bean germplasm from the Western Regional Plant Introduction Station, Pullman, USA, are being evaluated by scientists from America and China in a US Department of Agriculture project. For drought resistance and heat tolerance, screening will be conducted in Yunnan province. For salinity tolerance and winter hardiness, screening will be conducted during winter in Qingdao and Beijing. Efficient methods to explore plant agro-biodiversity for climate change-adaptive traits are also researched. The Focused Identification of Germplasm Strategy (FIGS) approach could enhance the discovery of new genes for abiotic stress adaptation. Khazaei et al. (2013) evaluated the effectiveness of FIGS to search a large faba bean collection for traits related to drought adaptation. Two sets of faba bean accessions were created, one from moisture-limited environments and the other from wetter sites. The two sets were grown under well-watered conditions, and leaf morphophysiological traits related to plant water use were measured. Machine-learning algorithms split the accessions into two groups based on the evaluation data, and the groups created by this process were compared to the original climate-based FIGS sets. The sets defined by trait data were in almost perfect agreement to the FIGS sets, demonstrating that ecotype differentiation driven by moisture availability has occurred within the faba bean gene pool. Leaflet and canopy temperature as well as relative water content contributed more than other traits to the discrimination between sets, indicating their utility as drought-tolerance selection criteria for faba bean germplasm. This study supports the assertion that FIGS could be an effective tool to enhance the discovery of new genes for abiotic stress adaptation.

Plant breeding can develop varieties to cope with climate change through many different techniques ranging from simply selecting plants with desirable traits to more complex classical or molecular techniques. With the discovery of genetics, plant breeding developed reliable procedures to improve varieties. Now, with the application of modern biotechnology in plant breeding, the development of new adapted varieties has become a more precise and rapid process (Singh et al. 2012). In the context of the Grain Legumes Integrated Project (GLIP), a European project, abiotic stress tolerance has been focused on species such as *Medicago truncatula*, pea and chickpea leading to the identification of factors potentially involved in abiotic stress adaptation and tolerance (Singh et al. 2012). The involvement of some genes in abiotic stress responses has been already analysed in different legumes. For instance, alfalfa over-expression of the gene for chloroplastic MnSOD showed lower cold-induced membrane injuries, although these transgenic lines did not present better tolerance to drought stress (Singh et al. 2012). The transcriptional regulator, Alfin1, over-expressed in alfalfa was shown to regulate endogenous NaCl-inducible gene expression, resulting in salinity tolerance. Similarly, a drought-responsive AP2-type TF induced several wax-related genes resulting in increased drought tolerance when over-expressed in alfalfa (*M. sativa*). These approaches are relevant to the improvement of minor legume crops, such as faba bean. Several in vitro techniques would be very useful for faba bean. Based on the forgoing discussion, emphasis should be given on the following aspects for further improvement in faba bean:

1. Characterisation of phenological, morphological, physiological and biochemical traits of faba bean that will contribute to adaptation in target environments
2. Characterisation of the effect of gene and QTLs for selecting suitable genotypes with resistance to biotic and abiotic stresses
3. Defining precisely the breeding targets in abiotic stress-prone environments with delineation of the predominant type of stress, and identification of the faba bean varieties preferred by farmers
4. Using simulation modelling and system analysis to evaluate crop response to major abiotic and biotic stress patterns (Singh et al. 2012)
5. Collaborative international evaluation, exchange of elite BPLs and open-pollinated varieties

4.10 Molecular Marker Analysis

Faba bean suffers from several major biotic and abiotic factors that constrain productivity. Although significant genetic gain to overcome these constraints has been made through conventional selection and breeding efforts (Rispail et al. 2010), progress through the use of genomics and associated biotechnologies is limited. This is due mainly to the large genome size (13 GB) (Johnston et al. 1999), which is approximately 25 times larger than that of the model legume,

M. truncatula, and 2.5 times larger than *Pisum sativum* (Rispail et al. 2010), together with a lack of financial investment in this crop species.

4.10.1 SSR Marker Development

Initially, Požárková et al. (2002) developed 25 SSR primer pairs detected in chromosome 1 DNA libraries. Subsequently, Zeid et al. (2009) developed 54 SSR primer pairs, and Gong et al. (2010) developed 11 EST-SSR loci primers. Recently, EST sequences within the public domain databases were screened, and an additional 21 novel SSR loci were characterised and validated among 32 faba bean accessions (Ma et al. 2011). Most recently, a library with 125,559 putative SSR sequences was constructed and characterised for repeat type and length from a mixed genome of 247 spring- and winter-sown faba bean genotypes using the Roche 454 GS FLX Titanium Platform sequencing (Yang et al. 2012). A suite of 28,503 available primer pair sequences were designed, and 150 were randomly selected for validation. Of these, 94 produced reproducible amplicons that were polymorphic among 32 faba bean genotypes selected from diverse geographic locations. The validation by UPGMA cluster analysis of 32 genotypes based on Nei's genetic distance showed a high quality and effectiveness of those novel SSR markers developed via next-generation sequencing technology. Large-scale SSR marker development was successfully achieved using next-generation sequencing of the *V. faba* genome. These novel markers are valuable for constructing genetic linkage maps, future QTL mapping and marker-assisted trait selection in faba bean breeding efforts.

4.10.2 Genetic Linkage Map Construction Based on SSR Markers

Faba bean genetic linkage maps have been developed with different types (morphological, isozyme, seed-protein gene, RAPD and RFLP) of markers (Ven et al. 1991; Torres et al. 1993; Satovic et al. 1996; Patto et al. 1999; Román et al. 2002; Avila et al. 2004; Arbaoui et al. 2008b; Ellwood et al. 2008); however, few SSR markers have been developed for constructing faba bean linkage maps. Until recently, only ten SSR markers were mapped in faba bean linkage maps (Román et al. 2004; Dõaz et al. 2010). Genetic studies about faba bean are behind other legume crops because of its large genome size (Johnston et al. 1999; Kaur et al. 2012). The lack of a reliable faba bean linkage map affected marker-assisted breeding and discovery of candidate genes of important traits. Simple sequence repeat (SSR) marker is a powerful tool for the construction of genetic linkage maps which can be applied for quantitative trait loci (QTLs) and marker-assisted selection (MAS). A genetic map of faba bean was constructed with SSR markers using a population of 129 F_2 individuals derived from the cross of Chinese native variety 91825 (large seed) and K1563 (small seed). By screening 11,551 SSR primers between two parents, 149 primer pairs were detected polymorphic and used for F_2 population analysis. This SSR-based genetic linkage map consisted of 15 linkage groups with 128 SSR (Fig. 4.1). The map encompassed 1,587 cM with an average genetic distance of 12.4 cM. This SSR-based genetic map will be beneficial for genetic studies of faba bean for identification of marker-locus-trait associations as well as marker-assisted selection in faba bean breeding and comparative mapping among faba bean, pea and grass pea (Ma et al. 2013).

4.11 Conclusions

Faba bean is among the highest in yield potential of the cool-season food legumes and is adapted to a wide range of environments, both in terms of latitude and altitude; however, the area of production of faba bean has decreased in recent decades despite an increase in the total area of cropping in the world. The reduction in production has been attributed to instability in yield resulting from both biotic and abiotic stresses, and it is also likely that limited markets and competition from higher value

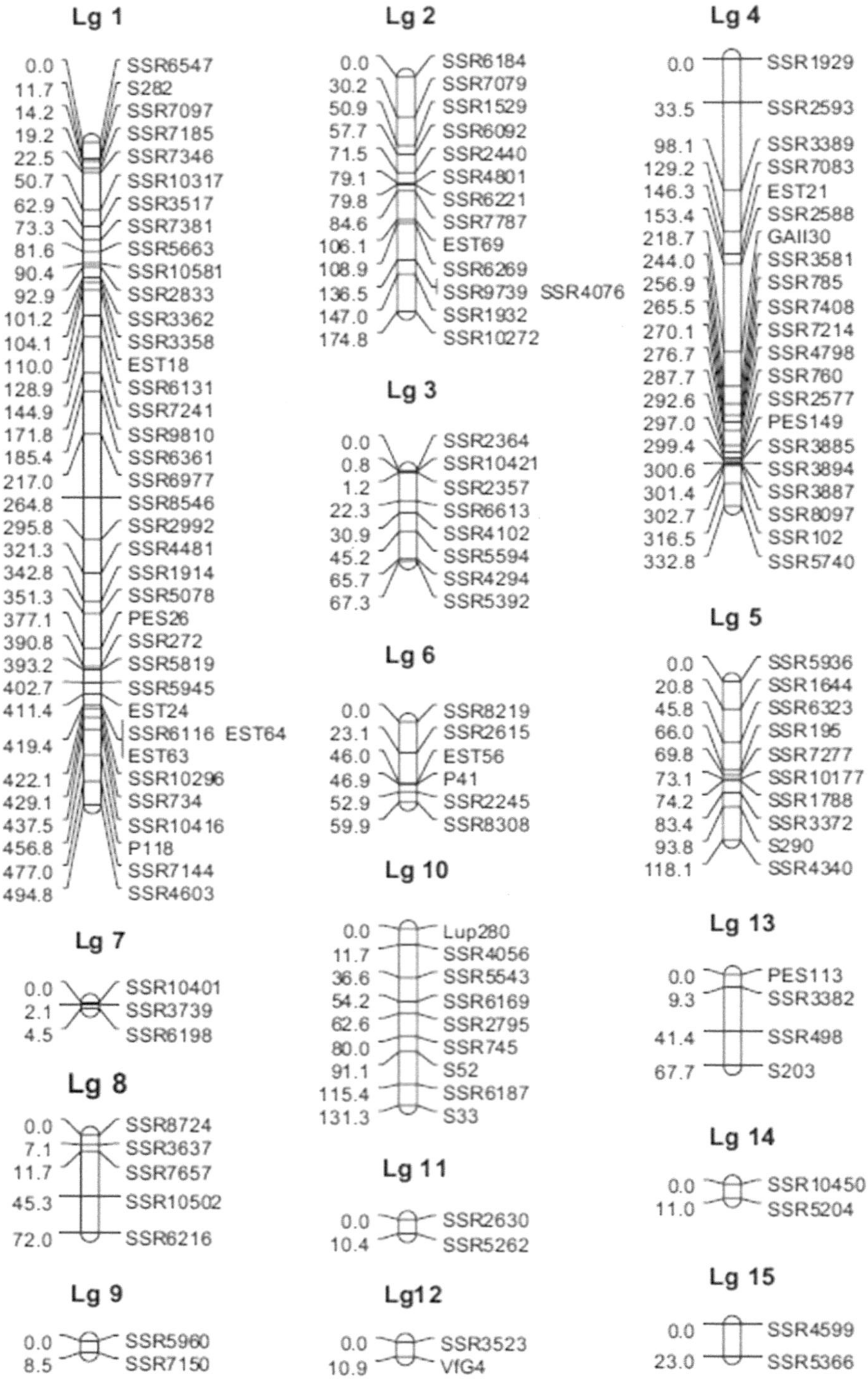

Fig. 4.1 Genetic linkage map based on a population of 129 F2 individuals of faba bean (*Vicia faba* L.). Simple sequence repeat (SSR) markers are listed to the *right* and the interval recombination distance (cM) to the *left* of each linkage group (Ma et al. 2013)

crops have impacted on cultivation of faba bean. Faba bean is a minor crop, and the total international input to research and breeding of faba bean is much less than for the major crops. As a consequence the basic knowledge of faba bean and improvements in yield through breeding and advances in crop management are restricted. Nevertheless, this review has identified a number of areas where the use of genetic resources to broaden the genetic base of faba bean breeding programmes can have a very significant positive impact on outcomes.

Faba bean is reproductively isolated; therefore, genetic variation for crop improvement must be identified within the species, induced by mutagenesis (e.g. the studies of Sjodin (1971)) or introduced by transformation. Significant variation at the molecular level has been demonstrated among faba bean germplasm in a number of studies, with distinct regional and ecotype groupings identified. Variation has been identified between European and Mediterranean region germplasm (Link et al. 1995), between winter- and spring-sown types, between regions within China and between China and other regions (Zong et al. 2009, 2010; Wang et al. 2012). Genetic diversity within faba bean breeding programmes is important, not only for the identification of novel traits but also because the outcrossing nature of faba bean that ensures a level of heterozygosity is maintained within a crop. Heterosis for yield is associated with heterozygosity, and a higher level of heterosis is observed for crosses between gene pools (Schill et al. 1998). While heterosis has not been utilised in the development of F_1 hybrid faba bean cultivars, it is expressed in hybrids between inbred lines, including mutations for a terminal inflorescence and reduced height, and the hybrids which are generally higher yielding and more stable than their component lines (Stelling et al. 1994). Therefore, maximising the diversity within a breeding programme should contribute to greater yield and stability of yield when synthetic varieties are developed.

To date, there has been very little interchange and utilisation of germplasm between China and other regions, and the diversity that has been identified between Chinese and other germplasm pools offers potential for new improvements in yield. The partially allogamous reproductive system of faba bean results in a high level of heterozygosity and heterogeneity within germplasm accessions. Many reports of screening germplasm collections for resistance or tolerance to a particular trait have identified no or very limited resistance/tolerance, particularly when assessing the overall performance of an accession. However, when accessions are inbred for several generations to increase the level of homozygosity, and assessed on the basis of single plants or progeny derived from single plants, useful sources of variation have been identified, with many examples in the area of disease resistance (e.g. Ascochyta blight, (Hanounik and Robertson 1989), rust (Conner and Bernier 1982) and bean yellow mosaic potyvirus (Makkouk and Kumari 1995)). The inherent variation within faba bean accessions greatly increases the potential sources of genes for a particular trait above the average performance of each accession within a gene bank. This review has described a number of molecular marker technologies that have been applied to understand the genetic variation present within faba bean germplasm and has enabled distinct gene pools to be identified. These technologies have also been applied to the development of linkage maps, and a number of maps are currently available. The challenge now is to identify linkage to a significant number of genes controlling the major traits of interest. This would enable better characterisation of the various sources of resistance to diseases that have been described and also enable marker-based selection to be adopted in breeding programmes to increase the efficiency of selection and reduce the confounding effect of environmental variation that reduces heritability and slows genetic gain.

References

Amede T, Schubert S (2003) Mechanisms of drought resistance in grain legumes. I. Osmotic adjustment. Ethiop J Sci 26:37–46

Arbaoui M, Balko C, Link W (2008a) Study of faba bean (*Vicia faba* L.) winter-hardiness and development of screening methods. Field Crop Res 106:60–67

Arbaoui M, Link W, Satovic Z, Torres A (2008b) Quantitative trait loci of frost tolerance and physiologically related trait in faba bean (*Vicia faba* L.). Euphytica 164:93–104

Avila CM, Sillero JC, Rubiales D, Moreno MT, Torres AM (2003) Identification of RAPD markers linked to *Uvf-1* gene conferring hypersensitive resistance against rust (*Uromyces viciae-fabae*) in *Vicia faba* L. Theor Appl Genet 107:353–358

Avila CM, Satovic Z, Sillero JC, Rubiales D, Moreno MT, Torres AM (2004) Isolate and organ-specific QTLs for ascochyta blight resistance in faba bean. Theor Appl Genet 108:1071–1078

Berthelem P, Le Guen J (1974) Rapport d'Activité de Station d'Amélioration des Plantes. Le Rheu, Rennes, pp 1971–1974

Bond DA (1957) Investigations of the breeding system of *Vicia faba*. Ph.D. thesis, University of Durham

Bond DA (1976) In vitro digestibility of the tests in tannin-free field beans (*Vicia faba* L.). J Agric Sci (Camb) 86:561–566

Bond DA (1989) A short review of research on male sterility and prospects for F_1 hybrid varieties in field beans (*Vicia faba* L.). Euphytica 41:87–90

Bond DA, Poulson MH (1983) Pollination. In: Hebblethwaite P (ed) The faba bean (*Vicia faba* L.). Butterworth, London

Bond DA, Jellis GJ, Rowland GG, le Guen J, Robertson LD, Khalil SA, Li-Juan L (1994) Present status and future strategy in breeding faba beans (*Vicia faba* L.) for resistance to biotic and abiotic stresses. Euphytica 73:151–166

Borlaug NE, Dowswell CR (2005) Feeding a world of ten billion people: a 21st century challenge. In Tuberosa R, Phillips RL, Gale M (eds) Proceedings of the international congress in the wake of the Double Helix: from the green revolution to the gene revolution, 27–31 May 2003, Bologna, Italy. Avenue Media, Bologna, pp 3–23

Bouhassan A, Sadiki M, Tivoli B (2004) Evaluation of a collection of faba bean (*Vicia faba* L.) genotypes originating from the Maghreb for resistance to chocolate spot (*Botrytis fabae*) by assessment in the field and laboratory. Euphytica 135:55–62

Ceccarelli S, Grando S, Maatougui M, Michael M, Slash M, Haghparast R, Rahmanian M, Taheri A, Al-Yassin A, Benbelkacem A, Labdi M, Mimoun H, Nachit M (2010) Plant breeding and climate changes. J Agric Sci 14:627–637

Chooi WY (1971) Variation in nuclear content in the genus Vicia. Genetics 68:195–211

Conner RL, Bernier CC (1982) Inheritance of rust resistance in inbred lines of *Vicia faba*. Phytopathology 72:1555–1557

Cubero JI (1973) Evolutionary trends in *Vicia faba* L. Theor Appl Genet 43:59–65

Cubero JI, Nadal S (2005) Faba bean (*Vicia faba* L.). In: Singh RJ, Jauhar PP (eds) Genetic resources, chromosome engineering and crop improvement, vol 1, Grain legumes. Taylor & Francis, Boca Raton, Florida, pp 163–186

Diaz R, Satovic Z, Roman B, Rubiales D, Cubero JI, Torres AM (2005) QTL analysis of broomrape resistance in faba bean (*Vicia faba* L.). In: Kovacevic V, Jovanovac S (eds) XL Croatian symposium on agriculture, Opatija, Croatia, pp 181–182

Diaz-Ruiz R, Satovic Z, Avila CM, Alfaro CM, Gutierrez MV, Torres AM, Roman B (2009) Confirmation of QTLs controlling *Ascochyta fabae* resistance in different generations of faba bean (*Vicia faba* L.). Crop Pasture Sci 60:353–361

Dõaz R, Torres A, Satovic Z, Gutierrez M, Cubero J, Roman B (2010) Validation of QTLs for *Orobanche crenata* resistance in faba bean (*Vicia faba* L.) across environments and generations. Theor Appl Genet 120:909–919

Drayner JM (1956) Regulation of outbreeding in field beans (*Vicia faba*). Nature 177:489–490

Duc G (1997) Faba bean (*Vicia faba* L.). Field Crop Res 53:99–109

Duc G, Picard J, Le Guen J, Berthelem P (1985) Note on the appearance of a new nucleocytoplasmic male sterility in Vicia faba after mutagenesis. Agronomy 5:851–854

Duc G, Bao SY, Baum M, Redden B, Sadiki M, Suso MJ, Vishniakova M, Zong XX (2010) Diversity maintenance and use of *Vicia faba* L. genetic resources. Field Crop Res 115:270–278

El Nadi AH (1969) Water relations of beans. I. Effects of water stress on growth and flowering. Exp Agric 5:195–207

Ellwood S, Phan H, Jordan M, Hane J, Torres A, Avila C, Cruzizquierdo S, Oliver R (2008) Construction of a comparative genetic map in faba bean (*Vicia faba* L.); conservation of genome structure with *Lens culinaris*. BMC Genomics 9:380

Emeran AA, Sillero JC, Rubiales D (2001) Physiological specialisation of *Uromyces viciae-fabae*. In: AEP (ed) European Association for Grain Legume Research, towards the sustainable production of healthy food, feed and novel products. Proceedings of the 4th European conference on grain legumes, Cracow, p 263

Evans LT (1959) Environmental control of flowering in *Vicia faba* L. Ann Bot 23:521–546

FAOSTAT (2013) Food and agriculture organisation of the United Nations, Rome

Fischer RA, Turner NC (1978) Plant productivity in the arid and semiarid zone. Annu Rev Plant Physiol 29:277–317

Gong Y, Xu S, Mao W, Hu Q, Zhang G, Ding J, Li Z (2010) Generation and characterization of 11 novel EST derived microsatellites from *Vicia faba* (Fabaceae). Am J Bot 97:e69–e71

Hanelt P, Schäfer H, Schultze-Motel J (1972) Die stellung von I L. in der gattung Vicia L. und betrachtungen zur Enstehung dieser kulturart. Kulturpflanze 20:263–275

Hanounik SA, Maliha N (1986) Horizontal and vertical resistance in *Vicia faba* to chocolate spot caused by *Botrytis fabae*. Plant Dis 70:770–773

Hanounik SB, Robertson LD (1988) New sources of resistance in *Vicia faba* to chocolate spot caused by *Botrytis fabae*. Plant Dis 72:696–698

Hanounik SB, Robertson LD (1989) Resistance in *Vicia faba* germplasm to blight caused by *Ascochyta blight*. Plant Dis 73:202–205

Hawtin GC (1982) The genetic improvement of faba bean. In: Hawtin G, Webb C (eds) Faba bean improvement. Martinus Nijhoff, The Hague, pp 15–32

Hawtin GC, Omar M (1980) International faba bean germplasm collection at ICARDA. FABIS 2:20–22

Ibrahim HM (2010) Heterosis, combining ability and components of genetic variance in faba bean (Vicia faba L.). J King Abdulaziz Univ Meteorol Environ Arid Land Agric Sci 21:35–50

IPCC (Intergovernmental Panel for Climate Change) (2007) Fourth assessment report: climate change. IPCC, Geneva, Switzerland

Johnston J, Bennett M, Rayburn A, Galbraith D, Price H (1999) Reference standards for determination of DNA content of plant nuclei. Am J Bot 86:609

Kaur S, Pembleton L, Cogan N, Savin K, Leonforte T, Paull J, Materne M, Forster J (2012) Transcriptome sequencing of field pea and faba bean for discovery and validation of SSR genetic markers. BMC Genomics 13:104

Khan HR, Link W, Hocking TJH, Stoddard FL (2007) Evaluation of physiological traits for improving drought tolerance in faba bean (*Vicia faba* L.). Plant Soil 292:205–217

Khan HR, Paull JG, Siddique KHM, Stoddard FL (2010) Faba bean breeding for drought-affected environments: a physiological and agronomic perspective. Field Crop Res 115:279–286

Khazaei H, Street K, Bari A, Mackay M, Stoddard FL (2013) The FIGS (Focused Identification of Germplasm Strategy) approach identifies traits related to drought adaptation in *Vicia faba* genetic resources. PLoS ONE 8(5):e63107. doi:10.1371/journal.pone.0063107

Kimber RBE, Paull JG (2011) Identification and genetics of resistance to Cercospora leaf spot (*Cercospora zonata*) in faba bean (*Vicia faba*). Euphytica 177:419–429

Kimber RBE, Davidson JA, Paull JG (2006) Using genetic diversity within faba bean germplasm to develop resistance to Ascochyta blight. In: First international Ascochyta workshop, Le Tronchet, p B17

Kohpina S, Knight R, Stoddard FL (2000) Genetics of resistance to Ascochyta blight in two populations of faba bean. Euphytica 112:101–107

Kwon SJ, Hu J, Coyne CJ (2010) Genetic diversity and relationship among 151 faba bean (*Vicia faba* L.) germplasm entries as revealed by TRAP markers. Plant Genet Resour 8:204–213

Ladizinsky G (1998) Plant evolution under domestication. Kluwer Academic, Dordrecht

Lang Li-jaun, Yu Z, Xu M, Ying H (1993) Faba bean in China: state-of-the-art review. ICARDA, Aleppo, Syria, pp 1–144

Link W, Dixkens C, Singh M, Schwall M, Melchinger AE (1995) Genetic diversity in European and Mediterranean faba bean germ plasm revealed by RAPD markers. Theor Appl Genet 90:27–32

Link W, Ederer W, Gumber RK, Melchinger AE (1997) Detection and characterization of two new CMS systems in faba bean (*Vicia faba*). Plant Breed 116:158–162

Link W, Balko C, Stoddard FL (2010) Winter hardiness in faba bean: physiology and breeding. Field Crop Res 115:287–296

Lobell DB, Schlenker W, Costa-Roberts J (2011) Climate trends and global food production since 1980. Science 333:616–620

Ma Y, Yang T, Guan J, Wang S, Wang H, Sun X, Zong X (2011) Development and characterization of 21 EST-derived microsatellite markers in *Vicia faba* (fava bean). Am J Bot 98:e22–e24

Ma Y, Bao S, Yang T, Hu J, Guan J, He Y, Wang X, Wan Y, Sun X, Jiang J, Gong C, Zong X (2013) Genetic linkage map of Chinese native variety faba bean (*Vicia faba* L.) based on simple sequence repeat markers. Plant Breed. Online publication. doi:10.1111/pbr.12074

Makkouk KM, Kumari SG (1995) Screening and selection of faba bean (*Vicia faba*) germplasm for resistance to bean yellow mosaic potyvirus. J Plant Dis Prot 100:461–466

Makkouk KM, Kumari SG, van Leur JAG (2002) Screening and selection of faba bean (*Vicia faba* L.) germplasm resistant to bean leafroll virus. Aust J Agric Res 53:1077–1082

Maurin N, Tivoli B (1992) Variation in resistance of *Vicia faba* to *Ascochyta fabae* in relation to disease development in field trials. Plant Pathol 41:737–744

McDonald GK, Paulsen GM (1997) High temperature effects on photosynthesis and water relations of grain legumes. Plant Soil 196:47–58

Muratova V (1931) Common beans – Vicia faba L. Trudy po prikladnoi botanike, genetike i selektsii: prilozhenie 50:248–285

Nadal S, Moreno MT, Cubero JI (2004) Registration of 'Baraca' faba bean. Crop Sci 44:1864–1865

Nassib AM, Ibrahim AA, Khalil SA (1982) Breeding for resistance to *Orobanche*. In: Hawtin G, Webb C (eds) Faba bean improvement. Martinus Nijhoff, The Hague, pp 199–206

Ouji A, ElBok S, Syed NH, Abdellaoui R, Rouaissi M, Flavell AJ, Elgazzah M (2012) Genetic diversity of faba bean (*Vicia faba* L.) populations revealed by sequence specific amplified polymorphism (SSAP) markers. Afr J Biotechnol 11:2162–2168

Patto M, Torres A, Koblizkova A, Macas J, Cubero J (1999) Development of a genetic composite map of *Vicia faba* using F_2 populations derived from trisomic plants. Theor Appl Genet 98:736–743

Paull JG, Kimber RBE (2006) Breeding faba beans for resistance to chocolate spot. In: Avila CM, Cubero JI, Moreno MT, Suso MJ, Torres AM (eds) International workshop on faba bean breeding and agronomy, Cordoba, pp 117–120

Paull JG, Rose IA, van Leur JAG, Kimber RBE, Seymour M (2006) Breeding faba beans for the Australian environment. In: Avila CM, Cubero JI, Moreno MT, Suso MJ, Torres AM (eds) International workshop on faba bean breeding and agronomy, Cordoba, pp 66–69

Pérez-de-Luque A, Eizenberg H, Grenz JH, Sillero JC, Avila CM, Sauerborn J, Rubiales D (2010) Broomrape management in faba bean. Field Crop Res 115:319–328

Pope M, Bond DA (1975) Influence of isolation distance on genetic contamination of field bean (*Vicia faba* L.) seed produced in small plots. J Agric Sci (Camb) 85:509–513

Požárková D, Koblížková A, Román B, Torres A, Lucretti S, Lysák M, Doležel J, Macas J (2002) Development and characterization of microsatellite markers from chromosome 1-specific DNA libraries of *Vicia faba*. Biol Plant 45:337–345

Rashid KY, Bernier CC (1984) Evaluation of resistance in *Vicia faba* to two isolates of the rust fungus *Uromyces viciae-fabae* from Manitoba. Plant Dis 68:16–18

Rashid KY, Bernier CC (1986) Selection for slow rusting in faba bean (*Vicia faba* L.) to *Uromyces viciae-fabae*. Crop Prot 5:218–224

Rashid KY, Bernier CC, Connor RL (1991) Genetics of resistance in faba bean inbred lines to five isolates of *Ascochyta fabae*. Can J Plant Pathol 13:218–225

Rispail N, Kal P, Kiss G, Ellis T, Gallardo K, Thompson R, Prats E, Larrainzar E, Ladrera R, Gonzalez E, Arreselgor C, Ferguson B, Gresshoff P, Rubiales D (2010) Model legumes contribute to faba bean breeding. Field Crop Res 115:253–269

Robertson LD, Cardona C (1986) Studies on bee activity and outcrossing in increase plots of *Vicia faba* L. Field Crop Res 15:157–164

Robertson LD, El-Sherbeeny M (1988) Faba bean germplasm catalogue: pure line collection. ICARDA, Aleppo, p 150

Robertson LD, El-Sherbeeny M (1991) Distribution of discretely scored descriptors in a pure line faba bean (*Vicia faba* L.) germplasm collection. Euphytica 57:83–92

Robertson LD, Saxena MC (1993) Problems and prospects of stress resistance breeding in faba bean. In: Singh KB, Saxena MC (eds) Breeding for stress tolerance in cool-season food legumes. Wiley, Chichester, pp 37–50

Robertson LD, Sadiki M, Matic R, Lang L (2000) Vicia spp: conserved resources, priorities for collection and future prospects. In: Knight R (ed) Linking research and marketing opportunities for pulses in the 21st century. Kluwer Academic, Dordrecht, The Netherlands, pp 623–633

Rojas-Molina MM, Rubiales D, Sillero JC (2006) Pathogenic specialization of *Uromyces viciae-fabae* in Spain and Portugal. In: Avila CM, Cubero JI, Moreno MT, Suso MJ, Torres AM (eds) International workshop on faba bean breeding and agronomy, Cordoba, pp 154–156

Román MB, Torres AM, Rubiales D, Cubero JI, Satovic Z (2002) Mapping of quantitative trait loci controlling broomrape (*Orobanche crenata*) resistance in faba bean. Genome 45:1057–1063

Román B, Satovic Z, Pozarkova D, Macas J, Dolezel J, Cubero J, Torres A (2004) Development of a composite map in *Vicia faba*, breeding applications and future prospects. Theor Appl Genet 108:1079–1088

Sanz AM, Gonzalez SG, Syed NH, Suso MJ, Saldaña CC, Flavell AJ (2007) Genetic diversity analysis in Vicia species using retrotransposon-based SSAP markers. Mol Genet Genomics 278:433–441

Satovic Z, Torres A, Cubero J (1996) Genetic mapping of new morphological, isozyme and RAPD markers in *Vicia faba* L. using trisomics. Theor Appl Genet 93:1130–1138

Saxena MC, Saxena NP, Mohamed AK (1988) High temperature stress. In: Summerfield RJ (ed) World crops: cool season food legumes. Kluwer Academic, Dordrecht, The Netherlands, pp 845–856

Schill B, Melchinger AE, Gumber RK, Link W (1998) Comparison of intra- and inter-pool crosses in faba beans (*Vicia faba* L.). II Genetic effects estimated from generation means in Mediterranean and German environments. Plant Breed 117:351–359

Schmidt HE, Carl E, Meyer U (1988) Resistance of field bean (*Vicia faba* L. spp. *Minor* (Peterm, em. Harz) Rothm.) to pea enation mosaic virus. Arch Phytopathol Pflanzenschutz 24:77–79

Serradilla JM, De Mora T, Moreno MT (1993) Geographic dispersion and varietal diversity in *Vicia faba* L. Genet Resour Crop Evol 40:143–151

Sillero JC, Villegas-Fernández AM, Thomas J, Rojas-Molina MM, Emeran AA, Fernández-Aparicio M, Rubiales D (2010) Faba bean breeding for disease resistance. Field Crop Res 115:297–307

Singh AK, Bhatt BP, Upadhyaya A, Kumar S, Sundaram PK, Singh BK, Chandr N, Bharati RC (2012) Improvement of faba bean (*Vicia faba* L.) yield and quality through biotechnological approach: a review. Afr J Biotechnol 11:15264–15271

Siong LE, Knight R (1980) The effect of inbreeding and hybridization on the seed set ability of faba beans (*Vicia faba* L.). Sabrao J 12:99–108

Sjodin J (1971) Induced morphological variation in Vicia faba L. Hereditas 67:155–180

Solaiman Z, Colmer TD, Loss SP, Thompson BD, Siddique KHM (2007) Growth responses of cool-season grain legumes to transient waterlogging. Aust J Agric Res 58:406–412

Stelling D, Ebmeyer E, Link W (1994) Yield stability in faba bean, *Vicia faba* L. 2. Effects of heterozygosity and heterogeneity. Plant Breed 112:30–39

Stoddard FL (1993) Limits to retention of fertilized flowers in faba beans (*Vicia faba*). J Agron Crop Sci 171:251–259

Suso MJ, Gilsanz S, Duc G, Marget P, Moreno MT (2006) Germplasm management of faba bean (*Vicia faba* L.): monitoring intercrossing between accessions with inter-plot barriers. Genet Resour Crop Evol 53:1427–1437

Tanno K, Willcox G (2006) The origins of cultivation of Cicer arietinum L. and Vicia faba L.: early finds from Tell el-Kerkh, north-west Syria, late 10th millennium B.P. Veg Hist Archaeobot 15:197–204. doi:10.1007/s00334-005-0027-5

Terzopoulos PJ, Bebeli PJ (2008) Genetic diversity analysis of Mediterranean faba bean (*Vicia faba* L.) with ISSR markers. Field Crop Res 108:39–44

Torres A, Weeden N, Martõn A (1993) Linkage among isozyme, RFLP and RAPD markers in *Vicia faba*. Theor Appl Genet 85:937–945

Van de Ven WTG, Duncan N, Ramsay G, Phillips M, Powell W, Waugh R (1993) Taxonomic relationships between V. faba and its relatives based on nuclear and mitochondrial RFLPs and PCR analysis. Theor Appl Genet 86:71–80

Ven W, Waugh R, Duncan N, Ramsay G, Dow N, Powell W (1991) Development of a genetic linkage map in *Vicia faba* using molecular and biochemical techniques. Ann Appl Biol 27:49–54

Villegas-Fernández AM, Sillero JC, Emeran AA, Winkler J, Raffict B, Tay J, Flores F, Rubiales D (2009) Identification and multi-environment validation of resistance to *Botrytis fabae* in *Vicia faba*. Field Crop Res 114:84–90

Wang H, Xuxiao Zong X, Jianping Guan J, Yang T, Sun X, Yu Y, Redden R (2012) Genetic diversity and relationship of global faba bean (I L.) germplasm revealed by ISSR markers. Theor Appl Genet 124:789–797

Yang T, Bao S, Ford R, Jia T, Guan J, He Y, Sun X, Jiang J, Hao J, Zhang X, Zong X (2012) High-throughput novel microsatellite marker of faba bean via next generation sequencing. BMC Genomics 13:602

Ye L, Lang L, Xia M, Tu J (2003) Faba beans in China. China Agricultural Press, Beijing. (in Chinese)

Zeid M, Schön CC, Link W (2001) Genetic diversity in recent elite faba bean lines. Czech J Genet Plant Breed 37:34–40

Zeid M, Schön C-C, Link W (2004) Hybrid performance and AFLP-based genetic similarity in faba bean. Euphytica 139:207–216

Zeid M, Mitchell S, Link W, Carter M, Nawar A, Fulton T, Kresovich S (2009) Simple sequence repeats (SSRs) in faba bean: new loci from Orobanche resistant cultivar 'Giza 402'. Plant Breed 128:149–155

Zong X, Liu X, Guan J, Wang S, Liu Q, Paull JG, Redden R (2009) Molecular variation among Chinese and global winter faba bean germplasm. Theor Appl Genet 118:971–978

Zong X, Ren J, Guan J, Wang S, Liu Q, Paull JG, Redden R (2010) Molecular variation among Chinese and global germplasm in spring faba bean areas. Plant Breed 129:508–513

Cowpea

5

François Abaye Badiane, Made Diouf, and Diaga Diouf

Abstract

Cowpea (*Vigna unguiculata* (L.) Walp.) is an important warm-season legume grown mostly by the poor farmers in the semiarid tropics for human consumption and animal feeding. The crop was originated from Africa where it was domesticated from its wild progenitor *V. unguiculata subsp. unguiculata var spontanea*. In addition, single nucleotide polymorphism markers (SNP) analysis suggested different domestication events from East to West Africa or single domestication process in the first region followed by transportation in the second. On the basis of molecular analyses, the genome organization of the crop was intensively studied, leading to the identification of two gene pools and gene flow between cultivated and wild forms or crop to crop can be a threat to the breeding programs. A wide range of biotic (virus, bacteria, fungi, insects, nematodes, and plants) and abiotic (like low phosphorus availability, soil acidity or salinity, drought, and high temperature at night) factors are limiting cowpea production in different parts of the world. To overcome these constraints, diverse programs were implemented for base broadening using interspecific hybridization between cowpea and other members of its genus with limited success because of pre-zygotic and post-zygotic barriers. These failures led the investigators to implement protocols to introduce foreign genes into cowpea. Currently, several genes of interest such as *herbicide imazapyr*, *α-amylase inhibitor 1* (against bruchids), and *Cry1Ab* and *Cry1Ac* (against *Maruca*) have been introduced successfully into commercially

F.A. Badiane
Département de Biologie Végétale, Faculté des Sciences et Techniques, Laboratoire de Biotechnologies Végétales, Université Cheikh Anta Diop, BP 5005, Dakar-Fann, Dakar, Sénégal

Faculté des Sciences et Technologies de l'Education et de la Formation (FASTEF), Dakar, Sénégal
e-mail: abayebadiane@gmail.com

M. Diouf • D. Diouf (✉)
Département de Biologie Végétale, Faculté des Sciences et Techniques, Laboratoire de Biotechnologies Végétales, Université Cheikh Anta Diop, BP 5005, Dakar-Fann, Dakar, Sénégal
e-mail: dioufmade@yahoo.fr; diaga.diouf@ucad.edu.sn

M. Singh et al. (eds.), *Broadening the Genetic Base of Grain Legumes*,
DOI 10.1007/978-81-322-2023-7_5, © Springer India 2014

important cultivars, and these genes are transmitted in Mendelian fashion. In addition, significant genomic resources and a consensus genetic map where agronomic, growth habit, disease, pest resistance, and other trait loci have been placed and are usable in breeding programs.

5.1 Introduction

Cowpea (*Vigna unguiculata* (L.) Walp.) is one of the most important grain legume crops in the semi-arid regions of Africa, Asia, Southern Europe, Southern United States, and Central and South America (Singh 2005; Timko et al. 2007; Diouf 2011). The crop plays a crucial role in human nutrition because of its grain protein quality contents (rich in lysine, tryptophan) and high nutritional value (23–32 %). Its grain contains a substantial amount of mineral and vitamins (e.g., folic acid and vitamin B) and the hay is used for feeding animals during the dry season in many parts of West Africa. In the poorer areas, cowpea is a valuable source of protein cheaper than fish, meat, or poultry helping to fight malnutrition for the low-income farmers. Mature or immature pods are consumed in many parts of Africa particularly during the "hungry period" between August and September. Cowpea is processed into various foods such as "*Akara*", "*Moin-moin*", "*Koki*", "*Couscous*", "*Red-Red Stew*", "*Ndambe*", "*Thiebou Kathiakh*", *Cake*, *Bread*, and *Cookie* and used in infantile aliments. According to Hall (2012), some varieties of cowpea are used to induce suicidal germination of *Striga hermonthica* or suppress the populations of the nematode *Scutellonema cavenessi*, the first parasitizing pearl millet, sorghum, and maize and the second affecting pearl millet, sorghum, and peanut. The plant is an important source of hay for cows in the Southeastern United States and other parts of the world which probably explains the name of cowpea given to the crop. More than 37 names were given to the crops across all of Africa but in West Africa "niébé" is the most popular name. Other local names used worldwide are "wake" and "ewa" in Nigeria, "niao" in Senegal, "luba hilu" in Sudan, "Kon or Kondi" in Cameroon, "Soso" in Mali and in The Gambia, "caupi" in Brazil, and "lobia" and "Chawli" in India. In the United States, it is variously named as "southern peas," "black-eyed peas," "field peas," "pinkeyes," and "crowders". Cowpea is cultivated in 14.5 million hectares with an annual production of 5.5 million tons worldwide. Presently, Nigeria is the largest producer of cowpea followed by Niger, Burkina Faso, Myanmar, Cameroon, and Mali (http://www.cgiar.org/our-research/crop-factsheets/cowpea/; Simon et al. 2007).

Cowpea grows in a wide range of temperature from 18 to 28 °C through all stages of development but with an optimum of 28 °C (Craufurd et al. 1997; Hall et al. 2002). The crop is drought tolerant compared with other crop species and can grow in low-fertility conditions due to its ability to establish nitrogen fixation with the soil bacterium *Bradyrhizobium*. Depending on the genotype, the growth habits are erect, semierect, prostrate (trailing) or climbing, sensitive to the length of the day light (photoperiod), short-day plant, or day-neutral (Craufurd et al. 1997). In West Africa, the photosensitive varieties are sown between the end of September and March resulting in poor survival of the genotypes as this period is the end of the rainy season and moisture availability is the main constraint. The crop is autogamous but around 5 % outcrossing was reported in the cultivated varieties probably due to insect activities (Fery 1985). The full life cycle from germination to dry grain production requires 60 days to more than 150 days depending on the maturity of the genotypes. Its cylindrical pods, which can be curved or straight, contain between 8 and 18 seeds. The seed coat can be white, cream, green, buff, red, brown, or black; the texture is smooth, rough, or wrinkled; and its uniformity ranges from solid, speckled, to patterned. This paper highlights the recent advances carried out on cowpea genome organization, the impact of gene transfer occurring naturally or by conventional and modern approaches. Discussion on other aspects for which a lot of attention was paid in the recent years, such as interspecific hybridization and the use of omics data for base broadening, is also presented.

5.2 Crop Gene Pool, Evolutionary Relationships, and Systematics

Cowpea (*Vigna unguiculata* (L.) Walp.) is a diploid species ($2n=2x=22$) that has been identified since antiquity by Dioscorides but was described by Linné (1760) from a cultivated form collected from the Caribbean, known as *Dolichos unguiculatus* (Pasquet and Baudoin 1997). It is native to Africa, where it was domesticated since the Neolithic age, as wild *Vigna unguiculata* ssp. *unguiculata* var. *spontanea* is endemic in the continent and produces hybrids with the cultivated form (Vanderborght and Baudoin 2001). However, the center of domestication has been the subject of many controversies and two regions were proposed: West Africa and Northeastern Africa.

The theory of cowpea domestication in West Africa was based on the morphological variability of cultivated varieties, the existence of wild hybrids resulting from crossing between the wild and the cultivated form, the presence of cowpea in archaeological excavations carried out in Ghana, and the molecular similarities observed in the chloroplast DNA between the wild species and the cultivated forms from Nigeria (Ng and Marechal 1985; Vaillancourt and Weeden 1992; Ng 1995). However, this theory has been invalidated by isoenzymatic and ethnobotanical studies, which revealed the absence of diversity in the wild form in West Africa (Vaillancourt et al. 1993; Pasquet 1996b).

The theory of cowpea domestication in Northeastern Africa was based on the absence of real varieties ecologically wild in West Africa and the great morphological diversity of the wild form in the region ranging from Ethiopia to Southern Africa (Baudoin and Maréchal 1985). Ethnobotanical studies conducted by Pasquet and Fotso (1994) and linguistic evidences recorded in Cameroon showed that cowpea is linked to the populations of Chadic languages than the ones of Adamawa and Benue-Congo populations. Similarly, the cultigroup *Textilis* would be linked to Nilo-Saharan languages. Both linguistic groups suggest domestication in Northeastern Africa. This hypothesis is validated through the large variability of Ethiopian provenances and genetic characterization with AFLP technique, highlighting cowpea domestication in Northeastern Africa (Pasquet 2000; Coulibaly et al. 2002).

The spread of cowpea in Asia occurred in the third millennium BC and then in America between the sixteenth and seventeenth centuries (Padulosi and Ng 1997). It was introduced in Europe around 300 BC, where it remains a minor crop in the southern part. The Spanish and Portuguese exported cowpea in the seventeenth century to the new world, while other crop cultivars were transported directly from Africa to Latin America, through the slave trade (Fang et al. 2007). However, its introduction in the Southern United States occurred at the beginning of the nineteenth century, explaining its presence in an area ranging from 35° N to 30° S. The crop also grows in the hot areas of Asia and Oceania (Singh and N'Tar 1985; Watt et al. 1985).

The taxonomy of cowpea was based on the major variability parameters including the shape of the leaves, the anthocyanin coloration of internodes, the length of floral stalk, photosensitivity, and the morphology of seeds and pods (Pasquet 1996b). The characteristics of seeds and pods are very diverse in cultivated forms and are widely used to describe and identify cultivars and cultigroups (Pasquet and Fotso 1994; Pasquet 1996b). Four cultigroups were identified in *V. unguiculata* ssp. *unguiculata*, namely, *unguiculata*, *biflora* (or *cylindrica*), *sesquipedalis*, and *textilis* (Ng and Marechal 1985). Thus, the cowpea has been classified as a herbaceous legume belonging to the group of dicotyledons, the order Fabales, family Fabaceae, subfamily Faboideae, tribe Phaseoleae, subtribe Phaseolinae, genus *Vigna*, species *unguiculata*, and subspecies *unguiculata* (Maréchal et al. 1978). The genus *Vigna* includes more than 80 species. Based on morphological characteristics, geographical distribution, and genetic hybridization/reproductive isolation, the genus *Vigna* was subdivided into six subgenera: *Vigna*, *Ceratotropis*, *Plectotropis*, *Sigmoidotropis*, *Lasiosporon*, and *Haydonia*. Recently, the taxonomy within the genus *Vigna* was revisited by Thulin et al. (2004) using chloroplast *trnK*, *ribosomal ITS (Internal Transcribed Spacer)* sequences, and morphological data leading to the transfer of subgenus *Macrorynchus* to genus

Wajira. The genus *Vigna* usually named African *Vigna* contains 39 species encompassing *Vigna unguiculata* (cowpea) and *Vigna subterranea* (bambara groundnut) which are agriculturally important (Maxted et al. 2004). The genus *Vigna* is divided into six sections, namely, *Vigna*, *Comosae*, *Macrodontae*, *Reticulatae*, *Liebrechtsia*, and *Catiang* (Maxted et al. 2004). *V. unguiculata* includes 11 subspecies which comprise the cultivated form var. *unguiculata* and the wild var. *spontanea*. Ten wild perennial subspecies divided into two groups according to their reproduction pattern were identified (Pasquet 1999; Maxted et al. 2004). The allogamous group is formed by ssp. *baoulensis*, ssp. *burundiensis*, ssp. *letouzeyi*, ssp. *aduensis*, and ssp. *pawekiae*; in contrast, the autogamous group contains ssp. *dekindtiana* (var. *spontanea*), ssp. *stenophylla*, ssp. *tenuis*, ssp. *alba*, and ssp. *pubescens*. The cultivated form was divided into five groups: *unguiculata*, *biflora*, *sesquipedalis*, *textilis*, and *melanophthalmus* (Pasquet 2000; Maxted et al. 2004).

In recent years, intensive studies based on morphology (Padulosi 1993), biochemistry (Fotso et al. 1994; Panella and Gepts 1992; Vaillancourt et al. 1993; Pasquet 1999), and DNA markers (chloroplast DNA, Vaillancourt and Weeden 1992; amplified fragment length polymorphism [AFLP], Coulibaly et al. 2002; random amplified polymorphic DNA [RAPD], Sylla-Ba et al. 2004; Diouf and Hilu 2005; microsatellites, Badiane et al. 2012; Asare et al. 2010; Diouf and Hilu 2005) were carried out to assess the genetic diversity of cowpea.

The gene pool organization of cowpea has been widely studied in recent years, resulting in the development of two hypotheses. The first is based on limited crosses between the cultivated and the wild form showing a lack or weak incompatibility barriers between them (Rawal et al. 1976; Baudoin and Maréchal 1985; Sakupwanya et al. 1990; Padulosi 1993). The second, based on hybridization and enzymatic analyses, suggested the existence of significant reproductive barriers between the cultivated and the wild forms and a secondary gene pool in *V. unguiculata* (Fatokun 1991; Pasquet and Baudoin 1997; Echikh 2000). On the basis of these findings, Pasquet and Baudoin (1997) and Echikh (2000) revisited the gene pool organization of *V. unguiculata*. They proposed the existence of a primary gene pool including the cultivated and the autogamous spontaneous forms and the wild allo-autogamous forms and a secondary gene pool encompassing the allogamous forms. Similar classification was made by Kouadio et al. (2007).

Using single-nucleotide polymorphism (SNP) technology, Huynh et al. (2013) analyzed the genome organization of 422 cowpea landraces collected worldwide encompassing 323 from North, East, Central, West, Southern, and Southeastern Africa and 99 distributed in the rest of the world. In addition, 46 accessions of wild cowpea collected from West, East, Southeast, and Southern Africa were included in these studies. Their results showed two major gene pools among the cultivated cowpea throughout Africa. Gene pool 1 encompassed cowpea distributed in West Africa, while gene pool 2 included the Eastern ones, each closely related to the wild cowpea in the same geographical region, suggesting divergent domestication process. The gene flow observed previously between the cultivated and the wild cowpea could also explain their close relationship (Sakupwanya et al. 1990; Vaillancourt and Weeden 1992; Coulibaly et al. 2002). The wider genetic diversity observed in gene pool 2 led Huynh et al. (2013) to conclude on a possible single domestication event which occurred in Eastern Africa followed by transportation with human migration in West Africa. In addition, they hypnotized that given the organization of the cultivated and wild cowpea in two gene pools divergent domestication events might occur in East and West Africa. On the other hand, the close relationship between Asiatic cowpea and those from gene pools 1 and 2 could be explained by the transportation of its seeds during the intense trade (mineral, agricultural products, and slaves) between these two continents as was reported by Collins (2006).

5.3 Assessment of Gene Flow for Crop Improvement

Gene flow is the transfer of gene(s) from one population to another which happens by cross-fertilization between cultivated crop and wild relatives and landraces and can result in the acquisition of new character. For this purpose, a lot of attention is paid on the impact of gene flow on plant evolution. According to the neo-Darwinist theories in the mid-1900s, gene flow was involved on the new adaptive variation observed in plant species but in around 1970, the force of natural selection on species evolution was emphasized (Anderson 1949; Levin and Kerster 1974). Around 1980, the impact of gene flow was specially reevaluated in plants and it was suggested that it was one of the driving forces of evolution and speciation in flowering plants (Ellstrand 2003). Nowadays, with the influx of genetically modified plants on the field, there is a renewed interest on the impact of gene flow on the crops and their wild relatives. Gene flow between cowpea (*Vigna unguiculata* ssp. *unguiculata* var. *unguiculata* (L.) Walp.) and its wild relative (*V. unguiculata* ssp. *unguiculata* var. *spontanea* (Schweinf.) Pasquet) was first reported in 1989 by Sakupwanya et al. (1990) and confirmed later by Vaillancourt and Weeden (1992) who suggested a very limited chloroplast gene exchange between the cultivated and the wild. This view was supported by the findings of Kouam et al. (2012) who reported that the outcrossing rate ranged from 1 to 9.5 % (mean 3.4 %) but Feleke et al. (2006) showed that gene flow can occur in both directions between the cultivated and the wild. Similarly, Coulibaly et al. (2002) concluded that the gene exchange makes it difficult to distinguish between the ancestral state and post-domestication evolution and that the large number of weedy forms observed within *V. unguiculata* resulted from the impact of gene flow on its genetic organization.

The studies carried out on cowpea varieties collected from several parts of Malawi showed a higher genetic variability within than among the populations. These results were not in agreement with previous studies showing a low genetic variability within population of cowpea (Diouf and Hilu 2005; Badiane et al. 2012). These findings suggested an uncontrolled gene flow between populations despite their unchanged morphological characters (seed color and size) due to pollen flow and seed exchange between farmers (Nkongolo 2003). Seed exchange can occur between farmers by way of gift or buying in the market or adoption of improved varieties from the research institutes. According to Langyintuo et al. (2003) a strong trade network exists in Africa between Benin, Burkina Faso, Ghana, Gabon, Niger, Nigeria, and Togo and from West to Central Africa. Planted in closest fields, these varieties from diverse regions pave the way for gene flow between populations.

To anticipate the release of transgenic cowpea in Africa where the cultivated and the wild forms exist, studies were performed in order to assess the gene flow between populations. It is well documented that carpenter bee (*Xylocopa flavorufa*) is one of the most important insect pollinators of cowpea in Coastal Province of Kenya. Using radio-tracking method, Pasquet et al. (2008) reported that *X. flavorufa* can forage up to 6 km from their nest but 64 % of the flight distances observed ranged between 200 m and 1 km. They observed a high level of gene flow within wild populations but a poor gene exchange between populations and concluded that *X. flavorufa* has the potential to move pollen from wild to cultivated cowpea over several kilometers.

5.4 Gene Flow Constraints

According to Hancock (2005), the cultivated crop diverged to their wild progenitor only a 1,000 years ago and reproductive isolation has not occurred yet facilitating gene flow. This gene exchange which occurs when the wild is in the vicinity of the crop could be a concern for breeders, seed producers, and the genebank curators who wish to keep their varieties genetically pure and true identity of their accessions. In addition, the hybrids resulting from a cross between the wild and cultivated crop are more similar to the wild which increases their chance of being not selected while the genes associated with domestication prevent them to spread in a long distance (Doebley et al. 1990; Lelou et al. 2011).

Developing new cowpea lines resistant to the most devastating insects such as pod borer or bruchid was the main objectives of many breeding programs. Unfortunately, due to the low to moderate level of resistance to pod borer or bruchid reported on some cowpea genotypes, the transferability of these genes by classical breeding is problematic. To overcome this constraint, recent investigations were based on contemporary approach for gene transfer to develop cowpea lines genetically modified harboring the insect-resistant gene. Currently, significant progress has been made on cowpea genetic transformation which may become shortly available for the African farmers. The genes used are the *Cry1Ab* expressing the delta endotoxin of *Bacillus thuringiensis* (Bt) ssp. *kurstaki* and the α-amylase inhibitor 1 (αAI-1) to target, respectively, the pod borer (*Maruca vitrata Fabricius*; *Lepidoptera*: *Pyraloidea*: *Crambidae*) and *Callosobruchus maculatus* and *C. chinensis* (Abrol 1999; Popelka et al. 2006; Tarver et al. 2007; Adesoye et al. 2008; Huesing et al. 2011). Due to the outcrossing observed between crops and crop to wild, the introduction of transgenic cowpea harboring insect-resistant gene in African agriculture would be a threat for the non-GM crop and their wild relatives (Williams and Chambliss 1980; Asiwe 2009). The introduction of the gene encoding insecticidal proteins into the wild due to gene flow might enhance the fitness of the wild which could be most aggressive particularly in Africa which is the center of origin and domestication of cowpea. Another threat could be the transfer of the Bt gene into the crop that might reduce the genetic diversity of the landraces and affecting thereby the effectiveness of the breeding programs. For these reasons, appropriate biosafety regulations need to be implemented on priority.

5.5 Level of Diversity in Crop Germplasm

In the last decades, several approaches have been carried out to enhance our knowledge on the nature and the variability of cowpea accessions stored in different germplasm across the world. These approaches are a prerequisite to select the best parent for breeding program or to determine the level of representation of each accession and where particular gene is present. For better understanding of the genetic basis of the accessions stored in the germplasm, genetic diversity studies were performed using morphological or physiological traits and biochemical or DNA markers (chloroplast or nuclear). Using morphological traits such as the leaf, stem, stipule size, calyx lobes, stigma, peduncle characters, and growth habit, Padulosi and Ng (1997) accomplished an interesting work leading to new species description. Combining 35 morphological and 20 isoenzyme markers to study 150 wild accessions of *V. unguiculata*, Pasquet (1996a) revealed their genetic relationship and suggested that *Vigna unguiculata* var. *spontanea*, the wild progenitor of the cultivated form, was more variable in Eastern Africa than in Western part of the continent. In contrast, using 21 enzymes in 271 accessions representing the five cultivar groups (*textilis*, *biflora*, *melanophthalmus*, *unguiculata*, *sesquipedalis*), Pasquet (2000) found a narrow genetic variation within accessions. In the past decades, the intensive use of DNA markers in genetic variation studies of cowpea has provided a better understanding of the genome organization of this crop and the structure of the germplasm. Vaillancourt and Weeden (1992) used chloroplast DNA to describe only two plastome types in the cultivated cowpea accessions suggesting a low genetic variation. These findings were confirmed by Feleke et al. (2006) who concluded the loss of BamHI restriction site in the cultivated accessions which probably occurred prior to domestication. All these conclusions suggested a bottleneck effect induced by a single domestication event and explain the low genetic diversity of *V. unguiculata* var. *unguiculata* reported in several studies (Li et al. 2001; Fall et al. 2003; Badiane et al. 2004, 2012; Diouf and Hilu 2005; Asare et al. 2010). Assessing the USDA collection maintained at the USDA-ARS Plant Genetic Resources Conservation Unit (PGRCU) located at Griffin, GA, the United States, Wang et al. (2008) used 30 gene-derived markers developed from legumes for examining the genetic relationship in 48 accessions representing 12 *Vigna* species. Their studies revealed a putative misclassification accession among the collection. This assertion was in agreement with the seed morphology. Based on their results, the authors reported that some subspecies were underrepresented among the collection which needs to be corrected to expand the genetic variation. Studies based on univariate and multivariate analyses of 376 cowpea accessions showed that the cowpea from the United States were grouped with the one collected from the Mediterranean Basin (Cyprus, Egypt, Greece, Israel, Italy, and Libya), the collection from Egypt and Libya forming a subcluster in that group. In contrast, the cowpea from Nigeria and Southern Africa formed another cluster in their dendrogram. According to their studies, the Mediterranean cowpea showed the highest variability, but compared to the others, Southern African provenances shared the same characters such as shorter and broader pod, same number of locules, and low seed weight (Perrino et al. 1993). These conclusions are in agreement with the findings of Fang et al. (2007) leading to the assertion of the European origin of the cowpea cultivated in the United States. The investigators noticed that the cowpea cultivated in Asia are more closer to the European breeding lines, thus including West African provenances could increase variability and vice versa. Similar suggestions are valid for Algerian farmers as their landraces showed a low genetic variability (Ghalmi et al. 2010). However, other investigators suggested that in situ maintenance of several populations in the region from where they came can induce genetic variability due to rare outcrossing occurring between varieties (Nkongolo 2003; Tosti and Negri 2005).

Based on morphological characters, Sène (1962) was the first to survey the Senegalese landraces and identified 53 cowpea varieties. This collection was estimated to 247 in 2002 by the "Institut Sénégalais de Recherches Agricoles" (ISRA) (Cissé et al. 2005) due to the addition of new varieties from the breeding programs or coming from seed exchange activities with other germplasm. Later, physiological studies using acetylene reduction assay (ARA) described Diongoma as a high nitrogen-fixing variety while Mouride was the lowest (Fall et al. 2003). In contrast, the variety from IITA, IT81D-1137, was more sensitive to drought by reducing its lateral root numbers but 985 was the most tolerant variety and they showed genetic variability as revealed by RAPD analyses (Badiane et al. 2004). Furthermore, Diouf and Hilu (2005) reported that microsatellite markers were more efficient to detect relationship among cowpea accessions and varieties from Senegalese germplasm. Based on these suggestions, Badiane et al. (2012) analyzed the genetic variation of Senegalese cowpea germplasm using microsatellites and concluded that most of the local varieties were clustered in the same group suggesting their close relationship compared to the one from IITA. Asare et al. (2010) applied this technology to 141 cowpea accessions collected from 9 regions in Ghana. According to their conclusions, the accessions were clustered in five main groups, each of which was loosely associated with geographical regions from which they were collected. These markers were also powerful to determine the genetic relationship among 16 cowpea varieties used in breeding for *Striga* resistance in Burkina Faso by grouping the resistant accessions into two separate clades (Sawadogo et al. 2010). These findings can be used in breeding program as *Striga* is one of the major weeds affecting cowpea production in the sub-Saharan region.

Recently, Fatokun et al. (2012) made a notable work based on physiological studies of 1,288 lines selected randomly from the collection maintained at IITA cowpea germplasm. By stopping irrigation 5 weeks after sowing, they selected 142 drought-tolerant lines of which 6 are potentially usable in a breeding program to increase drought tolerance in cowpea. In addition, significant investigations were carried out during the last three decades to study the impact of salinity on cowpea which led to the conclusions that NaCl affects seed germination and plant development (Dantas et al. 2005; Murillo-Amador et al. 2006; Hussein et al. 2007). Similar results were reported by Thiam et al. (2013) while screening 15 cowpea varieties from Senegalese germplasm for their sensitivity to NaCl at germination and growth stages. They identified high salt-tolerant (Diongoma, 58–78, 58–191) and salt-sensitive (Mougne and Yacine) varieties. On the basis of dendrogram analyses, the salt-tolerant and the salt-sensitive varieties were grouped in separate clades, and chlorophyll and proline accumulation were more in the first group.

The genetic diversity of 206 accessions of 14 wild *Vigna* including 12 species of Asian *Vigna* (subgenus *Ceratotropis*), *V. vexillata* (subgenus *Plectotropis*), and *V. pilosa* (*Dolichovigna*) from India was studied using morphological characters subjected to multivariate analysis. A great homology of morphological characters and growth habit was observed in the *mungo-radiata* group of the subgenus *Ceratotropis* except *V. khandalensis*, viz., *V. radiata* var. *sublobata*, *V. radiata* var. *setulosa*, *V. mungo* var. *silvestris*, and *V. hainiana*, while the species differed by flower, pod, and seed characteristics (Bisht et al. 2005). *V. umbellata* was more similar to *V. dalzelliana* than *V. bourneae* and *V. minima* in the *angularis-umbellata* (adzuki bean) group, and the *V. trilobata* accessions were more diverse than *V. aconitifolia* in the *aconitifolia-trilobata* (moth bean) group (Bisht et al. 2005). In addition, assessing the genetic diversity of 46 accessions of *V. vexillata* from Africa, America, and Australia based on globulin protein variability, Piergiovanni (1998) revealed a high variation in the Southeastern provenances and observed variability between the African and American provenances.

5.6 Production-Related Problems

The production of cowpea is limited by a wide range of factors because the crop is sensitive to biotic factors such as virus, bacteria, fungi, insects, nematodes, and weeds, which are responsible for significant yield loss in several parts of the world. In addition, despite its high tolerance to drought, some investigations performed at different stages of plant growth showed that the behavior of the crop vis-à-vis drought depends on the genotype. Drought affects root architecture; flowering and maturity time impact cowpea yield negatively (Ehlers and Hall 1997; Matsui and Singh 2003). In the Sahelian region, high temperature late in the night is a constraint for cowpea production as it disturbs flower production and pollination (Hall 2004). Breeding cowpea tolerant to high temperature will impact positively the production of the crop in that region which is the biggest producer in the world.

Because low phosphorus availability in the soil is another major constraint in the Sudano-Sahelian region of West Africa, Saidou et al. (2007) screened 200 cowpea genotypes leading to the identification of high variation in grain and folder yield among varieties. They identified tolerant genotypes which produce high yield under low phosphorus availability conditions. In contrast, in the wet zone, soil acidity is a major constraint for cowpea yield that reduces nutrient availability due to low microorganism activities. Germplasm is the basic material for breeders to create new varieties that will be distributed to the farmers to increase their production. However, germplasm accessibility is sometimes limited due to diverse reasons. The first reason can be a lack of strong link between researchers, curators, and breeders. This assertion is supported by the report of Dumet and Fatokun (2010) showing that 40 % of cowpea germplasm in the IITA collection has never been distributed. In addition, a lot of information collected by the breeders is not very often given to the germplasm provider, suggesting a lack of sharing information within the genetic resource community. The second reason is related the low number of breeders particularly in Africa and the small amount of fund to support their activities which limits the potential use of genetic resources for the producers' benefit. In addition, the capacity of the breeding program to absorb the new varieties has to be improved. Therefore, a lot of effort should be carried out by the policymakers to meet the needs of the genetic resource community. The third reason is a lack of evaluation of genetic resources. It is relevant to know the traits of economic importance for a large number of accessions maintained worldwide and to assess their genotype-environment interaction using multilocation and replicated evaluation. Introducing molecular technique will also be useful to identify the putative duplicated samples for better germplasm management. The fourth reason is related to the nature of the varieties. The early flowering varieties producing grain in 60 days are widely planted in the tropical region, where agriculture depends on the rainy season which is usually short. In contrast, the long season genotypes producing grain in 150 days are well adapted in the wet regions (Timko et al. 2007). In addition, some cowpea genotypes are sensitive to photoperiodism leading to the selection in West Africa of cultivars adapted to specific regions (Wein and Summerfield 1980; Ehlers and Hall 1996). This constraint limits the wider usefulness of these cultivars in other areas of adaptation and the exchange between farmers from these regions.

5.7 Traits of Importance for Base Broadening

In the last decades, several breeding programs have been implemented to increase cowpea productivity and to expand the genetic base of the crop (Singh 2005; Timko et al. 2007). Despite these progresses, significant efforts need to be done due to the fact that the crop is susceptible to a wide range of parasites like virus, bacteria, fungi, nematodes, and parasitic weeds (*S. gesnerioides* and *A. vogelii*). In addition to this, insect pests are also one of the major constraints to increase cowpea production in different parts of the world; therefore, developing resistant cultivars is relevant for the low-income farmers because of the high cost of agrochemicals. Achieving these objectives is a challenge because the crop is susceptible to a wide range of insects at different stages of its growth including at seed storage level. But in the future, pyramiding resistance genes will be a good approach to develop cowpea cultivar resistant to several parasites depending on the priority of each geographical region because strong resistance genes against certain insects are not yet available. Breeding abiotic stresses including drought, heat tolerance, and deep rooting have been implemented in the past years. For this purpose, Singh and Matsui (2002) reported that 78 % of lines bred from IITA were drought tolerant at vegetative stage and susceptibility is due to a single recessive gene. Adaptation to drought may be improved when the crop is able to develop a deep rooting system to explore the far horizons within the soil. Studies performed in the Sahelian regions indicated that current cultivars can develop root system reaching a depth of 200 cm in the wet season. Thus, breeding for this character in well-adapted farmers' varieties will improve production. High temperature is also another constraint in the Sahelian region. It reduces productivity because it affects flowers, pollination, and pod production (Hall 2004). The attempt to breed a heat-tolerant variety failed in the tropical zones of Africa because probably the gene needs to be in combination with other compatible genes (Ehlers and Hall 1998; Ismail and Hall 1998). The results showed that the heat-tolerant varieties had a semidwarfing character which may be useful in intensive irrigation conditions. Attempting to separate heat-tolerant gene(s) might be another challenge in the coming years for breeders. Furthermore, photoperiod sensitivity is also a relevant character to include in a breeding program due to the fact that most of the traditional cowpea cultivars in West Africa are photosensitive and sown at the end of the rainy season in the Sahelian region.

Cowpea seed protein and other mineral content have been intensively assessed in the recent years, leading to the conclusion that some genotypes (IT97K-1042-3, IT99K-216-48-1, and IT97K-556-4, LST-IIC-12 and HC-5) contain significant quantity of these products (Timko and Singh 2008; Gupta et al. 2010). Taking into account the high consumption of cowpea dry seeds in the poorer areas in tropical regions, breeding for improved cowpea protein content will be relevant to fight malnutrition for the low-income farmers. Another challenge for cowpea breeders is related to the seed quality or salinity tolerance. In Eastern and Southern Africa and parts of Central and South America, consumers prefer red- or brown-seeded varieties with smooth seed coats while the white- and brown-seeded varieties with rough seed coats are not welcome in their dishes. In addition, cooking time is another important character, for example, in Senegal, where people prefer low cooking time varieties. The future breeding programs can include these consumer preferences for their success.

5.8 Interspecific Hybridization in Crop Species

In the genus *Vigna*, interspecific hybridization has been reported occurring with different levels of the hybrid progeny fertility in the subgenus *Ceratotropis*. These findings were supported by the clustering of *V. reflexo-pilosa* ($2n=4x=44$), a tetraploid species, with *V. trinervia* in the four plastid DNA region phylogenetic tree (Ye Tun Tun and Yamaguchi 2007; Javadi et al. 2011). This suggested that *V. trinervia* is the maternal progenitor of *V. reflexo-pilosa.* This view was also in agreement with *atpB-rbcL* spacer region analysis reported by Doi et al. (2002). Using nuclear DNA (ITS), the same authors suggested that the male progenitor of *V. reflexo-pilosa* could be *V. exilis*, *V. hirtella*, or *V. umbellata* (adzuki bean group). In addition, referring to the Ye Tun Tun and Yamaguchi (2007) conclusions, similarities of ITS sequences between *V. reflexo-pilosa* and adzuki bean group could be due to concerted evolution. On the basis of these discoveries, artificial interspecific hybridizations have been attempted in the subgenus *Ceratotropis* but limited successes were achieved (Dana and Karmakar 1990). Successful interspecific hybridization between *V. radiata* × *V. mungo*, *V. radiata* × *V. trilobata*, and *V. radiata* × *V. umbellata* was reported by Dana (1966a, b, c). Another hybrid involving *V. radiata* like *V. radiata* × *V. mungo*, *V. angularis* × *V. radiata*, and *V. radiata* × *V. umbellata* has been well documented in the literature (Boling et al. 1961; Sen and Ghosh 1960; Sawa 1974). Pandiyan et al. (2010) achieved significant success by crossing *V. radiata* with 13 wild *Vigna*. They recorded a low pod set in the cross *V. radiata* × *V. dalzelliana* while the cross *V. umbellata* × *V. radiata* exhibited a significant number of branches and number of clusters in reciprocal direction, and the hybrids from *V. radiata* × *V. umbellata* showed virus resistance for nine seasons. The results paved the way towards the development of interesting agronomic lines.

Interspecific hybridization involving the cowpea and other *Vigna* species are not well documented. Fatokun and Singh (1987) reported interspecific hybridization between *V. unguiculata* and its hairy wild relatives *V. pubescens*. The hybrid embryos generated were rescued by culturing in Murashige and Skoog medium and produced F_1 plants sharing characteristics from both the parents. In addition, interspecific hybridization was also reported between *V. unguiculata* and *V. vexillata* by Barone et al. (1992) and Gomathinayagam et al. (1998).

5.9 Barriers to Interspecific Hybridization

The need to have cowpea varieties with desirable agronomic characteristics and resistance to a wide range of diseases has led breeders to rely on interspecific hybridization. For this purpose, several attempts to hybridize cowpea with its wild relatives, which are the reservoir of resistant or tolerant genes for biotic and abiotic factors, have been performed using conventional cross techniques or in vitro embryo culture but the excepted results have not been achieved yet. The conventional hybridization techniques were constrained by pre-zygotic or post-zygotic barriers. The pre-zygotic barriers can be characterized by low germination of pollen on the stigma, slowing the growth of pollen tube in the style tissue and/or its distortion as it was described in a cross between *V. unguiculata* and *V. vexillata* that caused the reduction of fertilization between 18 and 30 % (Barone et al. 1992). The post-zygotic barriers can occur with no hybrid embryo development, partial or complete sterility of the hybrid seeds or F1 plants (Barone et al. 1992). To overcome the post-zygotic barriers, embryogenic calli were generated from immature embryos resulting from the cross between *V. vexillata* × *V. unguiculata* cultured in Murashige and Skoog medium supplemented with 2,4-D (2 mg/l) (Gomathinayagam et al. 1998). The hybrid plants regenerated from these calli inherited stem, leaf, and pod hairiness shape from *V. vexillata* which could be used as a barrier against viral vector. In addition, through biochemical analysis, they concluded that the hybrid plants exhibited high level of peroxidase, polyphenol oxidase, and phenylalanine ammonia lyase activities. They also concluded through cytological analysis that because of the univalent formation in the hybrid chromosomes, the genomes of the two parents were structurally differentiated (Gomathinayagam et al. 1998). Pre-zygotic barriers were also observed in the cross between *V. radiata* and *V. aconitifolia* (Bharathi et al. 2006). In contrast, in the interspecific crosses *Vigna umbellata* × *Vigna radiata*, *Vigna sublobata* × *Vigna radiata*, and *Vigna trilobata* × *Vigna radiata*, the high rate of seed abscission and low seed set suggested a post-zygotic barrier (Bharathi et al. 2006). At intraspecific level, pre-zygotic barriers were also described by different investigators in the cross between *V. unguiculata* and the subspecies *baoulensis*, *letouzeyi*, and *tenuis* (Echikh 2000; Kouadio et al. 2006). The findings were in agreement with the results reported by Kouadio et al. (2006) who showed that more than 4 % of the ovules fertilized in the cross *V. unguiculata* × subspecies *baoulensis*.

5.10 Conventional and Contemporary Approaches of Interspecific Gene Transfer

5.10.1 Conventional Approaches

Several studies concluded that the cultivated species (*V. unguiculata* ssp. *unguiculata* var. *unguiculata*) showed a narrow genetic variability (around 16 %) due to a bottleneck resulting from a single domestication event (Panella and Gepts 1992; Badiane et al. 2004; Diouf and Hilu 2005). On the other hand, the crop is sensitive to a wide range of parasites (virus, bacteria, fungi, insects, and weeds). However, the closest wide relative *V. unguiculata* ssp. *unguiculata* var. *spontanea* is moderately resistant to insect pests while the *V. oblongifolia* and *V. luteola* from the genus *Vigna* and *V. vexillata* have a high resistance. For this reason, the attempt was made to transfer the hairiness character of the wild *V. unguiculata* ssp. *unguiculata* var. *pubescens* (TVNu-110-3A), used as a male parent, to a cowpea improved variety (IT84S-2246-4). After 12 days pods and seeds withered and the rescued hybrid embryos produced a partially fertile F_1 population (Fatokun and Singh 1987). The objective of this work was to introduce some degree of insect resistance into cowpea. In contrast, the investigations of Mohammed et al. (2010) showed that the cross of *V. unguiculata* ssp. *unguiculata* var. *pubescens* (TVNu-110-3A), used as a pollen parent with five cowpea varieties (Dandunga, IT89KD-288, IT93K-452-1, IT97K-499-38, and IT81D-994), resulted in the production of F_1 plants which produced viable seeds and subsequent F_2 generation. These findings suggested that cowpea genotypes can play a crucial role in the success of interspecific hybridization. Kouadio et al. (2007) made a successful cross between *V. unguiculata* ssp. *unguiculata* var. *unguiculata* and *V. unguiculata* ssp. *stenophylla* with a rate of 41.7 %, but the reduction of pollen fertility suggested a genetic distance among these subspecies. These findings provide hope to fight against aphid which causes a lot of damage in cowpea production because it is well documented that *V. unguiculata* ssp. *stenophylla* contains one or two flavonol p-coumaroyl glycosides, an antifeedant product against this virus. During their studies, the authors also experimented a cross between *V. unguiculata* ssp. *unguiculata* var. *unguiculata* and the subspecies *alba*, *spontanea*, and *baoulensis*. Conventional approaches for transferring the gene of interest from others species into *Vigna unguiculata* ssp. *unguiculata* were time consuming because it requires several years for development of pure lines. In addition, there were compatibility issues between the two genomes involved in the cross. To overcome this constraint, contemporary approaches have been attempted for gene transfer into cowpea.

5.10.2 Contemporary Approaches

The advantages of genetic transformation strategies used were related to the resolution of incompatibility encountered during the classical cross between genotypes belonging to different species and to break the crossing barriers. This approach has been attempted and several investigations concluded that contemporary approaches were reliable methods to introduce alien genes in *V. unguiculata* ssp. *unguiculata* var. *unguiculata*.

For this purpose, the susceptibility of this crop to *Agrobacterium tumefaciens* has been tested (Garcia et al. 1986, 1987; Penza et al. 1991, 1992) in addition to the direct DNA delivery methods used (Akella and Lurquin 1993; Muthukumar et al. 1996; Ikea et al. 2003). These investigations led to the implementation of reliable protocols of genetic transformation of cowpea which were used by Popelka et al. (2006) to introduce successfully the *bar* gene encoding herbicide resistance. At the end they concluded on the basis of strong scientific evidence that the *bar* gene was transmitted to the progenies in the Mendelian fashion. A similar approach was used successfully by Ivo et al. (2008) to transfer the *ahas* gene coding for a mutated acetohydroxyacid synthase, under the control of the *ahas* 5′ regulatory sequence. The AHAS is the first enzyme involved in the biosynthesis of valine, leucine, and isoleucine and is the target of several herbicides. In their studies, the authors used the *herbicide imazapyr* gene under the control of *ahas5* promoter in order to develop as a selectable marker. The same gene was also introduced by Citadin et al. (2013) into the cultivar Nova Era using a biolistic method. They isolated a line (named 59) showing high herbicide tolerance of 400 g per ha of imazapyr, a concentration four times higher than the commercial recommended dose. Surprisingly, this line showed a developmental growth similar to the control plants. In addition, researchers have paid some attention to the damage caused by bruchids, a major threat to stored cowpea grains particularly for the low-income farmers, which is mainly caused by *Callosobruchus maculatus* and *C. chinensis* with a loss estimated from 20 to 60 % (Abrol 1999; Tarver et al. 2007). Presently, there is no identified cowpea variety showing strong resistance to bruchids. In contrast, high resistance was described in the wild relative *Vigna vexillata* but nonviable seeds resulting from their cross make this approach inappropriate to transfer these genes to the cultivated species (Fatokun 2002). However, artificial diet bioassay performed on cowpea weevils suggested that α-amylase inhibitor 1 (αAI-1) isolated from common bean (*Phaseolus vulgaris*) can be used to fight against these pest attacks (Ishimoto et al. 1999). These findings were confirmed by several authors as the introduction of this gene in pea (Shade et al. 1994; Schroeder et al. 1995; Morton et al. 2000; Sousa-majer et al. 2007), adzuki bean (Ishimoto et al. 1996), and chickpea (Sarmah et al. 2004; Ignacimuthu and Prakash 2006) conferred resistance against bruchid beetles. All these studies allowed Solleti et al. (2008) to introduce the *αAI-1* gene under bean phytohemagglutinin promoter, in Pusa Komal a commercially important Indian cultivar, and to develop fertile transgenic plants which strongly inhibited the development of *C. maculatus* and *C. chinensis* insect bioassay. Similar results were reported by Lüthi et al. (2013) who introduced *αAI-1* gene under the control of cotyledon-specific promoter into the breeding line IT86D-1010, developed at the International Institute of Tropical Agriculture (IITA), Nigeria, and the Japanese cultivar 'Sasaque' with 100 % larval (*C. chinensis* and *C. maculatus*) mortality in the seeds of transgenic lines (Table 5.1).

The Lepidoptera *Maruca* pod borer (*Maruca vitrata*) is one of the most devastating insects in

Table 5.1 Useful agronomic traits introduced in cowpea

Genes	Traits	Gene delivery systems	References
α-amylase inhibitor 1 (*αAI-1*)	Bruchid resistance	*Agrobacterium*	Solleti et al. (2008)
	(*Callosobruchus maculates*)		
	(*Callosobruchus chinensis*)		
Cry 1Ab	Pod borer resistance	Electroporation	Adesoye et al. (2008)
	(*Maruca vitrata*)		
Cry 1Ac	Pod borer resistance	*Agrobacterium*	Bakshi et al. (2011)
	(*Maruca vitrata*)		
Imazapyr	Herbicide tolerance	Particle bombardment	Citadin et al. (2013)

the tropical region which attacks the flower buds, flowers, and young pods and can be responsible up to 80 % yield loss in cowpea (Jackai and Adalla 1997; Sharma 1998). However, the investigations performed by Jackai et al. (1997) showed that the insect pests of cowpea can be controlled by several different forms of Bt crystal toxins. This basic information was used by Adesoye et al. (2008) and Bakshi et al. (2011) to introduce *Cry1Ab* in cultivars (TVu 201, Ife Brown, IT90K-277-2, IT90K-288, and IT90K-391) and *Cry1Ac* genes in cultivar (Pusa Komal) in different cowpea genotypes. Their experiments showed that the transgenes were transmitted in Mendelian fashion to the progenies which showed significant reduction of larvae survival and weight.

5.11 Molecular Markers, Genome Mapping, and Genomics as an Adjunct to Breeding

In the past decades, significant progress has been made on genomic research of legumes, since a large amount of databases are available for the ones considered as the models such as *Medicago truncatula*, *Glycine max*, and *Phaseolus vulgaris* (http://www.phytozome.org/). In contrast to cowpea, until recently, very little genomic data were available posing a major constraint to gene discovery. At present, the combination of technology improvement, the low cost of sequencing, and the significant progress made by several teams working on this crop allowed the identification of more than 189,593 EST (http://www.ncbi.nlm.nih.gov/nucest?term=vigna%20unguiculata). Recently, using methylation filtering technology, Chen and coworkers (2007) sequenced the hypomethylated regions of cowpea genome leading to the identification of 268,950 gene-space sequence reads (GSRs) with an average length of 610 bp covering approximately 160 MB. These GSRs were blasted against the four plant proteomes from *Arabidopsis thaliana*, *Oryza sativa*, *Medicago truncatula*, and *Populus trichocarpa* (http://cowpeagenomics.med.virginia.edu/) providing a significant tool for marker development and comparative genomics. In addition, other genomic data are available such as microRNA likely to be involved in transcription, development, growth, metabolism, and other physiological process regulation or associated to drought (Lu and Yang 2010; Barrera-Figueroa et al. 2011). Since the first attempt of mapping by Fatokun et al. (1993) consisting of 10 linkage groups based on 87 random genomic DNA fragments, 5 cDNAs, and RAPDs, others with the objective to improve the initial map were published by several authors using different techniques or markers (RAPD, RFLP, AFLP) (Menéndez et al. 1997; Ouédraogo et al. 2002). These improved maps were based on 11 linkage groups spanning a total of 2,670 cM with 18 diseases or pest resistance-related markers. The most important markers are *Striga gesnerioides* races 1 and 3 localized on the LG1 and LG6, cowpea mosaic virus (CPMV), and cowpea severe mosaic virus (CPSMV) in two distinct loci on LG2, B1CMV (black-eyed cowpea mosaic virus) in LG8, southern bean mosaic virus (SBMV) in LG6, *Fusarium* wilt in LG3, and root-knot nematode (gene *Rk*; NemR) on LG1. Candidate genes for Resistance Gene Analogs were mapped on LG2, LG3, LG5, and LG9 using RFLP markers.

A consensus genetic linkage map was developed using EST-derived SNP (single nucleotide polymorphism) assay of cowpea leading to the integration of 928 markers into a genetic map spanning 680 cM with 11 linkage groups and an average marker distance of 0.73 cM. Using comparative genomic approaches, a significant macrosynteny with *G. max*, *M. truncatula* genome, and some microsynteny with *Arabidopsis thaliana* genome (Muchero et al. 2009b) with cowpea genome were reported. Nine QTL (*Mac*-1 to 9) for resistance to *Macrophomina phaseolina* were mapped on LG2, LG3, LG5, LG6 and LG11. Analyzing these QTLs, Muchero et al. (2011) found strong synteny with proteins involved in parasite resistance in *G. max* and *M. truncatula* or with other products like Ca^{2+}-binding proteins, light-harvesting complex PSII, UDP glycosyltransferase, lipase class 3 family protein, and flowering genes. Pottorff et al. (2012b) mapped a candidate gene for resistance to *Fusarium oxysporum f.sp. tracheiphilum* (*Fot*) Race 3 in a LG6 of cowpea of a recombinant inbred line population, California Blackeye 27 (resistant)×24-125B-1 (susceptible). The locus *Fot 3–1* was found in the chromosome 6 of cowpea showing macro- and microsyntenies with *G. max* chromosome numbers 9 and 15 where four resistance genes are located. In contrast, Muchero et al. (2010a), working on the cross from the foliar trips susceptible IT93K503-1 and the resistant black-eyed cowpea cultivar 'California Blackeye No. 46' (CB46), identified three QTLs on the linkage groups 5 and 7. These QTLs' (*Thr-1*, *Thr-2*, and *Thr-3*) peaks were colocated with the AFLP markers ACC-CAT7, ACG-CTC5, and AGG-CAT1 and were associated with foliar damage caused by *Thrips tabaci* and *Frankliniella schultzei* (*Thysanoptera*: *Thripidae*). These encouraging results paved the way for genetic characterization of thrips resistance in cowpea as this disease causes 15 % yield loss in West Africa and impacts production in Asia and South America negatively.

Using synteny approaches Pottorff et al. (2012a) identified QTL for leaf shape called *Hls* (hastate leaf shape) on the linkage group 15 of a genetic map of Sanzi×Vita 7. The same QTL was localized on the LG4 of the consensus genetic map of cowpea spanning from 25.57 to 35.96 cM. This finding might be important as fresh and dry leaves are used, respectively, in human consumption and in animal feeding. In addition, other markers such as QTL related to drought were mapped on linkage groups 1, 2, 3, 5, 6, 7, 9, and 10, while the QTL for maturity were on LG7 and LG8 (Muchero et al. 2009a, b). In the same vein, Andargie et al. (2013) studying populations resulting from a cross between a short duration domesticated cowpea variety (524B) and a relatively long duration, the wild accession *V. unguiculata* var. *spontanea* (219-01), mapped 8 QTLs distributed in four linkage groups. The QTLs linked to time of flower opening (FOT) (*qFotl.1*, *qFotl.2*, *qFotl.3*) were located on the LG1, the *qfot2* on LG2, and *qfot10* on LG10. The QTLs linked to days to flowering (DTF) time *qdtf1*, *qdtf2*, and *qdtf7* were located on LG1, LG2, and LG7, respectively. The authors concluded that these results will provide a tool to restrain efficiently gene flow between cultivated and wild *Vigna*. Furthermore, fine mapping and positional cloning will be needed for the identification of the genes governing FOT and DTF as time of flower opening and the day of flowering time are two important agronomic traits playing a role in geographical adaptability, the expression of which is controlled by environmental factors. Conventional breeding was based on the selection of elite varieties showing the best morphological traits, resistance to pest and disease, yield, grain quality, or abiotic stresses. The transfer of one of these traits or many of them by pyramiding strategies by population development was time consuming, taking around 5–10 years to come up with a new variety. The selection of the best variety had to be postponed until the process reaches the F_5 to F_8 generation allowing the line to become homogeneous. Currently biotechnology based on DNA markers derived on genetic and genomic studies bring a great hope in plant breeding. The integration of DNA markers into plant breeding program known as marker-assisted selection (MAS) forms a new discipline, "molecular breeding." But these DNA markers have to meet five requirements like reliability, quantity and quality of DNA required, technical procedure for marker assay, level of polymorphism, and cost. The main advantages of MAS are linked to the fact that in some cases, a small population can only be required, the number of generation to achieve the goal is reduced, the accuracy of the evaluation is increased, and a single technology can handle the selection of multiple traits. In contrast, MAS is not necessarily useful or effective for every trait and require substantial investments for their development explaining their weak integration in the breeding programs of developing countries or in the orphan crops. On the other side the low impact of marker-assisted selection is related to the fact that their validation is not considered enough for scientific publication, their accurate localization can be problematic, their unknown interaction with environmental factors, the laboratory equipments and field trial required or intellectual property rights, the lack of strong linkage between research centers working on MAS, and the centers developing new varieties and misunderstanding between molecular biologists and plant breeders (Collard and Mackill 2008). In cowpea, an AFLP marker linked to *Striga* races 1 and 3 resistance was converted into a SCAR (sequence-characterized amplified region) called 61R detecting a single band in some resistant varieties (B301, IT82D-849, and Tvu 14676). In addition, other putative candidate MAS for insect or disease resistance in cowpea were reported (Timko et al. 2008). Presently, the recent influx of SSR and SNP markers identified in cowpea or closely related species like *Glycine max* which is sharing significant synteny constitute interesting tools for the development of MAS.

5.12 Conclusions

Improving cowpea varieties was the main objective of several programs implemented by research institutes around the world. For this purpose, different approaches have been used. The first approach was related to classical breeding technique based on the cross of two parents available in the germplasm in order to introduce new genes in a selected cultivar. This technique has led to the selection and release of several cowpea genotypes with agronomically important characters such as disease resistance. Other important traits such as drought tolerance, grain quality and size, and maturity were also included in the breeding programs but the biggest challenge still is the incorporation of several characters in individual varieties adapted to local conditions with high yield and high quality of seeds. Genetic transformation approach can be useful to achieve this goal as transgenic lines of cowpea carrying the gene of interest have been obtained. The second approach was based on the use of molecular technique leading to the identification of several markers associated with important agronomic traits and the implementation of a good linkage genetic map. Refining the available genetic map by integrating the new molecular techniques and the bioinformatic tools will be useful for gene discovery. Finally, the integration of all these approaches in combination with the use of omic datasets available in the model plant (*Arabidopsis*, *Oryza sativa*, *Glycine max*, *Medicago truncatulata*, etc.) will speed up the release of new elite varieties to the farmers.

References

Abrol DP (1999) Pulse susceptibility to *Callosobruchus chinensis* (L) (*Bruchidae*: *Coleoptera*) under field conditions. Trop Agric 76:150

Adesoye A, Machuka J, Togun A (2008) *CRY 1AB* transgenic cowpea obtained by nodal electroporation. Afr J Biotechnol 7(18):3200–3210

Akella V, Lurquin PF (1993) Expression in cowpea seedlings of chimeric transgenes after electroporation into seed-derived embryos. Plant Cell Rep 12:110–117

Andargie M, Pasquet RS, Muluvi GM, Timko MP (2013) Quantitative trait loci analysis of flowering time related traits identified in recombinant inbred lines of cowpea (*Vigna unguiculata*). Genome 56:289–294

Anderson E (1949) Introgressive hybridization. Wiley, New York

Asare AT, Gowda BS, Galyuon IKA, Aboagye LL et al (2010) Assessment of the genetic diversity in cowpea [*Vigna unguiculata* (L.) Walp.] germplasm from Ghana using simple sequence repeat markers. Plant Genet Resour Charact Util 8:142–150

Asiwe J (2009) Insect mediated out crossing and gene flow in cowpea (*Vigna unguiculata* (L.) Walp.): Implication for seed production and provision of containment structures for genetically transformed cowpea. Afr J Biotechnol 8:226–230

Badiane FA, Diouf D, Sané D, Diouf O, Goudiaby V, Diallo N (2004) Screening cowpea [*Vigna unguiculata* (L.) Walp.] varieties by inducing water deficit and RAPD analyses. Afr J Biotechnol 3:174–178

Badiane FA, Bhavani SG, Cissé N, Diouf D, Sadio O, Timko MP (2012) Genetic relationship of cowpea *Vigna unguiculata* varieties from Senegal based on SSR markers. Genet Mol Res 11(1):292–304

Bakshi S, Sadhukhan A, Mishra S, Sahoo L (2011) Improved *Agrobacterium*-mediated transformation of cowpea via sonication and vacuum infiltration. Plant Cell Rep 30:2281–2292

Barone A, Del Giudice A, Ng NQ (1992) Barriers to interspecific hybridization between *Vigna unguiculata* and *Vigna vexillata*. Sex Plant Reprod 5(3):195–200

Barrera-Figueroa BE, Gao L, Diop NN, Wu Z, Ehlers JD, Roberts PA, Close TJ, Zhu JK, Liu R (2011) Identification and comparative analysis of drought-associated microRNAs in two cowpea genotypes. BMC Plant Biol 11:127

Ba-Sylla F, Pasquet RS, Gepts P (2004) Genetic diversity in cowpea [*Vigna unguiculata* (L.) Walp.] as revealed by RAPD markers. Genet Resour Crop Evol 51:539–550

Baudoin JP, Maréchal R (1985) Cowpea taxonomy, origin and germplasm. In: Singh SR, Rachie KO (eds)

Cowpea research, production and utilization. Wiley, Chichester, pp 3–9

Bharathi A, Vijay Selvaraj KS, Veerabadhiran P, Subba Lakshmi B (2006) Crossability barriers in mungbean (*Vigna radiata* L. Wilczek): with its wild relatives. Indian J Crop Sci 1(1–2):120–124

Bisht IS, Bhat KV, Lakhanpaul S, Latha M, Jayan PK, Biswas BK, Singh AK (2005) Diversity and genetic resources of wild *Vigna* species in India. Genet Resour Crop Evol 52:53–68

Boling M, Sander DA, Matlock RS (1961) Mungbean hybridization technique. Agron J 53:54–55

Chen X, Laudeman TW, Rushton PJ, Spraggins TA, Timko MP (2007) CGKB: an annotation knowledge base for cowpea (*Vigna unguiculata* L.) methylation filtered genomic genespace sequences. BMC Bioinform 8:129

Cissé N, Wey J, Seck D, Gueye MT, Gueye M (2005) Les légumineuses à grains. In: ISRA, ITA, CIRAD (eds) Bilan de la recherche agricole et agroalimentaire au Sénégal. ISRA/CIRAD, Dakar/Paris, pp 257–266

Citadin CT, Cruz ARR, Aragão FJL (2013) Development of transgenic imazapyr-tolerant cowpea (*Vigna unguiculata*). Plant Cell Rep 32:537–543

Collard BCY, Mackill DJ (2008) Marker-assisted selection: an approach for precision plant breeding in the twenty-first century. Phil Trans R Soc Lond B 363: 557–572

Collins RO (2006) The African slave trade to Asia and the Indian Ocean Islands. Afr Asian Stud 5(3–4): 325–346(22)

Coulibaly S, Pasquet RS, Papa R, Gepts P (2002) AFLP analysis of the phenetic organization and genetic diversity of *Vigna unguiculata* L. Walp. Reveals extensive gene flow between wild and domesticated types. Theor Appl Genet 104:358–366

Craufurd PQ, Summerfield RJ, Ell RH, Roberts EH (1997) Photoperiod, temperature and the growth and development of cowpea (*Vigna unguiculata*). In: Singh BB, Mohan Raj DR, Dashiell KE, Jackai LEN (eds) Advances in cowpea research. Co-publication Int Inst Tropic Agric (IITA) and Japan Int Res Center Agric Sci (JIRCAS), Sayce, pp 75–86

Dana S (1966a) Cross between *Phaseolus attreus* Roxb. and *P. mtmgo* L. Genetics 37:259–274

Dana S (1966b) Cross between *Phase~dus aureus* Roxb. and *P. ricwardianus*. Ten Genrt Ibrr 18:141–156

Dana S (1966c) Species cross between *Phaseolus aureus* Roxb. and *P. trilobus* Ait. Cytologia 31:176–187

Dana S, Karmakar PG (1990) Species relation in *Vigna* subgenus *Ceratotropis* and its implications in breeding. In: Janick J (ed) Plant breeding review. Timber Press, Portland, pp 19–42

Dantas BF, De Sáribeiro L, Aragão CA (2005) Physiological response of cowpea seeds to salinity stress. Rev Bras Sementes 27(1):144–148

Diouf D (2011) Recent advances in cowpea [*Vigna unguiculata* (L.) Walp.] "omics" research for genetic improvement. Afr J Biotechnol 10(15):2803–2810

Diouf D, Hilu KW (2005) Microsatellites and RAPD markers to study genetic relationship among cowpea breeding lines and local varieties in Senegal. Genet Resour Crop Evol 52:1057–1067

Doebley J, Stec A, Wendel J et al (1990) Genetic and morphological analysis of maize-teosinte F2 population: implications for the origin of maize. PNAS 87(24): 9888–9892

Doi K, Kaga A, Tomooka N, Vaughan DA (2002) Molecular phylogeny of genus *Vigna* subgenus *Ceratotropis* based on rDNA ITS and *atpB-rbcL* intergenic spacer of cpDNA sequences. Genetica 114:129–145

Dumet D, Fatokun C (2010) Global strategy for the conservation of cowpea (*Vigna unguiculata* ssp. *unguiculata*). World cowpea conference, September, Saly, Senegal

Echikh N (2000) Organisation du pool génique de formes sauvages et cultivées d'une légumineuse alimentaire, *Vigna unguiculata* (L.) Walp. Thèse de doctorat, Faculté universitaire des Sciences agronomiques, Gembloux, Belgique, p 307

Ehlers JD, Hall AE (1996) Genotypic classification of cowpea based on responses to heat and photoperiod. Crop Sci 36:673–679

Ehlers JD, Hall AE (1997) Cowpea (*Vigna unguiculata* L. Walp.). Field Crops Res 53:187–204

Ehlers JD, Hall AE (1998) Heat tolerance of contrasting cowpea lines in short and long days. Field Crops Res 55:11–21

Ellstrand NC (2003) Current knowledge of gene flow in plants: implications for transgene flow. Philos Trans R Soc Lond B 358:1163–1170

Fall L, Diouf D, Fall-Ndiaye MA, Badiane FA, Gueye M (2003) Genetic diversity in cowpea [*Vigna unguiculata* (L.) Walp.] varieties determined by ARA and RAPD techniques. Afr J Biotechnol 2:48–50

Fang J, Chao CCT, Roberts PA, Ehlers JD (2007) Genetic diversity of cowpea [*Vigna unguiculata* (L.) Walp.] in four West African and USA breeding programs as determined by AFLP analysis. Genet Resour Crop Evol 54:1197–1209

Fatokun CA (1991) Wide hybridization in cowpea: problems and prospects. Euphytica 54:137–140

Fatokun CA (2002) Breeding cowpea for resistance to insect pests: attempted crosses between cowpea and *V. vexillata*. In: Fatokun CA, Tarawali SA, Singh BB, Kormawa PM, Tamo M (ed) Challenges and opportunities for enhancing sustainable cowpea production. Proceedings of the world cowpea conference III, held at the International Institute of Tropical Agriculture, IITA, Ibadan Nigeria, 4–8 2000. IITA, Ibadan

Fatokun CA, Singh BB (1987) Interspecific hybridization between *Vigna pubescens* and *V. unquiculata* (L.) Walp through embryo rescue. Plant Cell Tissue Organ Cult 9(3):229–233

Fatokun CA, Danesh D, Menancio-Hautea D, Young ND (1993) A linkage map for cowpea [*Vigna unguiculata* (L.) Walp.] based on DNA markers. In: O'Brien JS (ed) A compilation of linkage and restriction maps of genetically studied organisms, Genetic maps 1992. Cold Spring Harbor Laboratory Press, Cold Spring Harbor, pp 6.256–6.258

Fatokun CA, Ousmane B, Satoru M (2012) Evaluation of cowpea (*Vigna unguiculata* (L.) Walp.) germplasm lines for tolerance to Drought. Plant Genet Resour Charact Util 10:171–176

Feleke Y, Pasquet RS, Gepts P (2006) Development of PCR-based chloroplast DNA markers that characterize domesticated cowpea (*Vigna unguiculata* ssp. *unguiculata* var. *unguiculata*) and highlight its crop-weed complex. Plant Syst Evol 262:75–87

Fery RL (1985) The genetics of cowpea: a review of the world literature. In: Singh SR, Rachie KO (eds) Cowpea research, production and utilization. Wiley, Chichester, pp 25–62

Fotso M, Azanza JL, Pasquet RS, Raymond J (1994) Molecular heterogeneity of cowpea (*Vigna unguiculata Fabaceae*) seed storage proteins. Plant Syst Evol 191(1/2):39–56

Garcia JA, Hille J, Goldbach R (1986) Transformation of cowpea (*Vigna unguiculata*) cells with an antibiotic-resistance gene using a Ti-plasmid-derived vector. Plant Sci 44:37–46

Garcia JA, Hille J, Vos P, Goldbach R (1987) Transformation of cowpea (*Vigna unguiculata*) with a full-length DNA copy of cowpea mosaic virus M-RNA. Plant Sci 48:89–98

Ghalmi N, Malice M, Jacquemin JM, Ounane SM, Mekliche L, Baudoin JP (2010) Morphological and molecular diversity within Algerian cowpea (*Vigna unguiculata* (L.) Walp.) landraces. Genet Resour Crop Evol 57(3):371–386

Gomathinayagam P, Ganesh Ram S, Rathnaswamy R, Ramaswamy NM (1998) Interspecific hybridization between *Vigna unguiculata* (L.) Walp. and *V. vexillata* (L.) A. Rich. through in vitro embryo culture. Euphytica 120(2):203–209

Gupta P, Singh R, Malhotra S, Boora KS, Singal HR (2010) Characterization of seed storage proteins in high protein genotypes of cowpea [*Vigna unguiculata* (L.) Walp.]. Phys Mol Biol Plants 16(1):53–58

Hall AE (2004) Comparative ecophysiology of cowpea, common bean and peanut. In: Nguyenand H, Blum A (eds) Physiology and biotechnology integration for plant breeding. Marcel Dekker, New York, pp 271–325

Hall AE (2012) Phenotyping cowpeas for adaptation to drought. Front Physiol 3:1–8

Hall AE, Ismail AM, Ehlers JD, Marfo KO, Cisse N et al (2002) Breeding cowpeas for tolerance to temperature extremes and adaptation to drought. In: Fatokun CA, Tarawali SA, Singh BB, Kormawa PM, Tamo M (eds) Challenges and opportunities for enhancing sustainable cowpea production. International Institute of Tropical Agriculture, Ibadan, pp 14–21

Hancock JF (2005) Contributions of domesticated plant studies to our understanding of plant evolution. Ann Bot 96:953–963

Huesing J, Romeis J, Ellstrand N, Raybould A, Hellmich R, Wolf J, Ehlers J, Dabire C, Fatokun C, Hokanson K et al (2011) Regulatory considerations surrounding the deployment of Bt-expressing cowpea in Africa: report of the deliberations of an expert panel. GM Crops 2(3):211–214

Hussein MM, Balbaa LK, Gaballah MS (2007) Developing a salt tolerant cowpea using alpha tocopherol. J Appl Sci Res 3(10):1234–1239

Huynh BL, Close TJ, Roberts PA, Hu Z, Wannamaker S, Lucas MR, Chiulele R, Cissé N, David A, Hearne S, Fatokum C, Diop NN, Ehlers JD (2013) Genepools and the genetic architecture of domesticated cowpea [*Vigna unguiculata* (L.) Walp.]. Plant Genome. doi:10.3835/plantgenome2013.03.0005

Ignacimuthu S, Prakash S (2006) *Agrobacterium-mediated* transformation of chickpea with α-amylase inhibitor gene for insect resistance. J Biosci 31:339–345

Ikea J, Ingelbrecht I, Uwaifo A, Thotttappilly G (2003) Stable gene transformation in cowpea (*Vigna unguiculata* L. Walp.) using particle gun method. Afr J Biotechnol 2:211–218

Ishimoto M, Sato T, Chrispeels MJ, Kitamura K (1996) Bruchid resistance of transgenic azuki bean expressing seeds α-amylase inhibitor of the common bean. Entomol Exp Appl 79:309–315

Ishimoto M, Yamada T, Kaga A (1999) Insecticidal activity of an α-amylase inhibitor-like protein resembling a putative precursor of α-amylase inhibitor in the common bean, *Phaseolus vulgaris* L. Biochim Biophys Acta 1432:104–112

Ismail AM, Hall AE (1998) Positive and potential negative effects of heat-tolerance genes in cowpea lines. Crop Sci 38:381–390

Ivo NL, Nascimento CP, Vieira LS, Campos FAP, Aragão FJL (2008) Biolistic-mediated genetic transformation of cowpea (*Vigna unguiculata*) and stable Mendelian inheritance of transgenes. Plant Cell Rep 27:1475–1483

Jackai LEN, Adalla CB (1997) Pest management practices in cowpea: a review. In: Singh BB, Mohan Raj DR, Dashiell KE, Jackai LEN (eds) Advances in cowpea research. IITA and Japan International Research Centre for Agricultural Sciences (JIRCAS) IITA, Ibadan, pp 240–258

Jackai LEN, Moar W, Thottappilly G (1997) Evaluation of insecticidal crystal proteins form common soil bacterium, *Bacillus thuringiensis* (Bt) for efficacy against the maruca pod borer, *Maruca testulalis*. Crop Improvement Division and Biotechnology Research Unit, Archival report (1992–1994), pp 55–57

Javadi F, Tun YT, Kawase M, Guan K, Yamaguchi H (2011) Molecular phylogeny of the subgenus *Ceratotropis* (genus *Vigna*, Leguminosae) reveals three eco-geographical groups and Late Pliocene–Pleistocene diversification: evidence from four plastid DNA region sequences. Ann Bot 108(2):367–380

Kouadio D, Toussaint A, Pasquet RS, Baudoin JP (2006) Barrières pré-zygotiques chez les hybrides entre formes sauvages du niébé, *Vigna unguilata* (L.) Walp. Biotechnol Agron Soc Environ 10(1):33–41

Kouadio D, Echikh N, Toussaint A, Pasquet RS, Baudoin JP (2007) Organisation du pool génique de *Vigna unguiculata* (L.) Walp.: croisements entre les formes sauvages et cultivées du niébé. Biotechnol Agron Soc Environ 11:47–57

Kouam EB, Pasquet RS, Campagne P, Tignegre JB, Thoen K, Gaudin R, Ouedraogo JT, Salifu AB, Muluvi GM, Gepts P (2012) Genetic structure and mating system of wild cowpea populations in West Africa. BMC Plant Biol 12:113

Langyintuo AS, Lowenberg-DeBoer J, Faye M, Lambert D, Ibro G, Moussa B, Kergna A, Kushwaha S, Musa S, Ntoukam G (2003) Cowpea supply and demand in West Africa. Field Crops Res 82:215–231

Lelou B, Diatewa M, Van Damme P (2011) A study of intraspecific hybrid lines derived from the reciprocal crosses between wild accessions and cultivated cowpeas (*Vigna unguiculata* (L.) Walp.). Afr J Plant Sci 5(6):337–348

Levin DA, Kerster HW (1974) Gene flow in seed plants. Evol Biol 7:139–220

Li CD, Fatokun CA, Ubi B, Singh BB, Scoles G (2001) Determining genetic similarities among cowpea breeding lines and cultivars by microsatellite markers. Crop Sci 41:189–197

Linné C (1760) Disquisito de sexu plantarum. Amoenitates Acad 10:100–131

Lu Y, Yang X (2010) Computational identification of novel MicroRNAs and their targets in *Vigna unguiculata*. Comp Funct Genom 2010:1–17. doi:10.1155/2010/128297

Lüthi C, Álvarez-Alfageme F, Ehlers JD, Higgins TJV, Romeis J (2013) Resistance of αAI-1 transgenic chickpea (*Cicer arietinum*) and cowpea (*Vigna unguiculata*) dry grains to bruchid beetles (*Coleoptera: Chrysomelidae*). Bull Entomol Res 103:373–381

Maréchal R, Mascherpa JM, Stainer F (1978) Etude taxonomique d'un group complexe d'espèces des genres *Phaseolus* et *Vigna* (*Papillionaceae*) sur la base de données morphologiques et polliniques traitées par l'analyse informatique. Boissiera 28:1–273

Matsui T, Singh BB (2003) Root characteristics in cowpea related to drought tolerance at the seedling stage. Exp Agric 39:29–38

Maxted N, Mabuza-Diamini P, Moss H, Padulosi S, Jarvis A, Guarino L (2004) An ecogeographic study: African *Vigna*. International Plant Genetic Resources Institute, Rome

Menéndez CM, Hall AE, Gepts P (1997) A genetic linkage map of cowpea (*Vigna unguiculata*) developed from a cross between two inbred, domesticated lines. Theor Appl Genet 95:1210–1217

Mohammed MS, Russom Z, Abdul SD (2010) Studies on crossability between cultivated cowpea (*Vigna unguiculata* [L.] Walp.) varieties and their wild relative (var. *pubescens* TVNu110-3A). Int Res J Plant Sci 1(5): 133–135

Morton RL, Schroeder HE, Bateman KS, Chrispeels MJ, Armstrong E, Higgins TJV (2000) Bean α-amylase inhibitor 1 in transgenic peas (*Pisum sativum*) provides complete protection from pea weevil (*Bruchus pisorum*) under field conditions. PNAS 97:3820–3825

Muchero W, Diop NN, Bhat PR, Fenton RD, Wanamaker S, Pottorff M, Hearne S, Cisse N, Fatokun C, Ehlers JD, Roberts PA, Close TJ (2009a) A consensus genetic map of cowpea [*Vigna unguiculata* (L) Walp.] and synteny based on EST-derived SNPs. PNAS 106(43):18159–18164

Muchero W, Ehlers JD, Close TJ, Roberts PA (2009b) Mapping QTL for drought stress-induced premature senescence and maturity in cowpea [*Vigna unguiculata* (L.) Walp.]. Theor Appl Genet 118(5):849–863

Muchero W, Ehlers JD, Roberts PA (2010a) QTL analysis for resistance to foliar damage caused by *Thrips tabaci* and *Frankliniella schultzei* (*Thysanoptera*: *Thripidae*) feeding in cowpea [*Vigna unguiculata* (L.) Walp.]. Mol Breed 25(1):47–56

Muchero W, Ehlers JD, Roberts PA (2010b) Restriction site polymorphism-based candidate gene mapping for seedling drought tolerance in cowpea [*Vigna unguiculata* (L.) Walp.]. Theor Appl Genet 120(3):509–518

Muchero W, Ehlers JD, Close TJ, Roberts PA (2011) Genic SNP markers and legume synteny reveal candidate genes underlying QTL for *Macrophomina phaseolina* resistance and maturity in cowpea [*Vigna unguiculata* (L) Walp.]. BMC Genomics 12:8

Murillo-Amador B, Troyo-Diéguez E, García-Hernández JL et al (2006) Effect of NaCl salinity in the genotypic variation of cowpea (*Vigna unguiculata*) during early vegetative growth. Sci Hortic 108(4):423–431

Muthukumar B, Mariamma M, Veluthambi K, Gnanam A (1996) Genetic transformation of cotyledon explants of cowpea (*Vigna unguiculata* L. Walp) using *Agrobacterium tumefaciens*. Plant Cell Rep 15:980–985

Ng NQ (1995) Cowpea *Vigna* unguiculata (*Leguminosea-Papilionoidaea*). In: Smartt J, Simmonds N (eds) Evolution of crop plants. Longman, London, pp 326–332

Ng NQ, Marechal R (1985) Cowpea taxonomy, origin and germplasm. In: Singh SR, Rachie KO (eds) Cowpea research, production and utilization. Wiley, New York, pp 11–21

Nkongolo KK (2003) Genetic characterization of Malawian cowpea (*Vigna unguiculata* (L.) Walp) landraces: diversity and gene flow among accessions. Euphytica 129:219–228

Ouédraogo JT, Gowda BS, Jean M, Close TJ, Ehlers JD, Hall AE, Gillespie AG, Roberts PA, Ismail AM, Bruening G, Gepts P, Timko MP, Belzile FJ (2002) An improved genetic linkage map for cowpea (*Vigna unguiculata* L.) combining AFLP, RFLP, RAPD, biochemical markers and biological resistance traits. Genome 45:175–188

Padulosi S (1993) Genetic diversity, taxonomy and ecogeographic survey of the wild relatives of cowpea (*Vigna unguiculata* (L.) Walp.). Thèse de doctorat, Université Catholique de Louvain, Louvain-La-Neuve, Belgique, p 228

Padulosi S, Ng N (1997) Origin, taxonomy, and morphology of *Vigna unguiculata* (L.) Walp. In: Singh B, Mohan Raj D, Dashiel K, Jackai L (eds) Advances in cowpea research. International Institute of Tropical Agriculture (IITA) and Japan International Research Center for Agricultural Sciences (JIRCAS), Ibadan, pp 1–12

Pandiyan M, Senthil N, Ramamoorthi N, Muthiah AR, Tomooka N, Duncan V, Jayaraj T (2010) Interspecific hybridization of *Vigna radiata x* 13 wild *Vigna* species for developing MYMV donar. Electron J Plant Breed 1(4):600–610

Panella L, Gepts P (1992) Genetic relationship within *Vigna unguiculata* (L.) Walp. based on isozyme analyses. Genet Resour Crop Evol 39:71–88

Pasquet RS (1996a) Wild cowpea (*Vigna unguiculata*) evolution. In: Pickersgill B, Lock J (eds) Legumes of economic importance, vol 8, Advances in legume systematics. Royal Botanic Gardens, Kew, pp 95–100

Pasquet RS (1996b) Cultivated cowpea (*Vigna unguiculata*): genetic organization and domestication. In: Pickersgill B, Lock J (eds) Legumes of economic importance, vol 8, Advances in legumes systematics. Royal Botanic Gardens, Kew, pp 101–108

Pasquet RS (1999) Genetic relationships among subspecies of *Vigna unguiculata* (L.) Walp based on allozyme variation. Theor Appl Genet 98:1104–1119

Pasquet RS (2000) Allozyme diversity of cultivated cowpea Vigna unguiculata (L.) Walp. Theor Appl Genet 101:211–219

Pasquet RS, Baudoin JP (1997) Le niébé, *Vigna unguiculata*. In: Charrier A, Jacquot M, Hammon S, Nicolas D (eds) L'amélioration des plantes tropicales. CIRAD-ORSTOM, Montpellier, pp 483–505

Pasquet RS, Fotso M (1994) Répartition des cultivars de niébé, *Vigna unguiculata* (L.) Walp., du Cameroun: influence du milieu et des facteurs humains. J Agric Tradit Bot Appl 36:93–143

Pasquet RS, Peltier A, Hufford MB, Oudin E, Saulnier J, Paul L, Knudsen JT, Herren HR, Gepts P (2008) Long-distance pollen flow assessment through evaluation of pollinator foraging range suggests transgene escape distances. PNAS 105(36):13456–13461

Penza R, Lurquin PF, Filippone E (1991) Gene transfer by co cultivation of mature embryos with *Agrobacterium tumefaciens*-application to cowpea (*Vigna unguiculata* Walp). J Plant Physiol 138:39–43

Penza R, Akella V, Lurquin PF (1992) Transient expression and histological localization of a *gus* chimeric gene after direct transfer to mature cowpea embryos. Biotechniques 13:576–580

Perrino P, Laghetti G, Spagnoletti Zeuli PL, Monti LM (1993) Diversification of cowpea in the Mediterranean and other centres of cultivation. Genet Resour Crop Evol 40:121–132

Piergiovanni AR (1998) *Vigna vexillata* (L.) A. Rich seed proteins: heterogeneity in subunits of globulin fraction. Genet Resour Crop Evol 45:97–103

Popelka JC, Gollasch S, Moore A, Molvig L, Higgins TJV (2006) Genetic transformation of cowpea (*Vigna unguiculata* L.) and stable transmission of the transgenes to progeny. Plant Cell Rep 25:304–312

Pottorff M, Ehlers JD, Fatokun C, Roberts PA, Close TJ (2012a) Leaf morphology in Cowpea [*Vigna unguiculata* (L.) Walp]: QTL analysis, physical mapping and identifying a candidate gene using synteny with model legume species. BMC Genomics 13:234

Pottorff M, Wanamaker S, Ma YQ, Ehlers JD, Roberts PA et al (2012b) Genetic and physical mapping of candidate genes for resistance to *Fusarium oxysporum* f.sp. *tracheiphilum* race 3 in cowpea [*Vigna unguiculata* (L.) Walp]. PLoS One 7(7):e41600

Rawal KM, Rachie KO, Franckowiak JD (1976) Reduction in seed size in crosses between wild and cultivated cowpeas. J Hered 67:253–254

Saidou AK, Abaidoo RC, Singh BB, Iwuafor ENO, Sanginga N (2007) Variability of cowpea breeding lines to low phosphorus tolerance and response to external application of Phosphorus. In: Advances in integrated soil fertility management in sub-Saharan Africa: challenges and opportunities. Springer, Dordrecht, pp 413–422

Sakupwanya S, Mithen R, Mutangaundura-Mhlanga (1990) Studies on the African *Vigna* genepool II. Hybridization studies with *Vigna unguiculata* var. *tenuis* and var. *stenophylla. IBPGR*. Plant Genet Res Newsl 78(79):5–10

Sarmah BK, Moore A, Tate W, Molvig L, Morton RL, Rees DP, Chiaiese P, Chrispeels MJ, Tabe LM, Higgins TJV (2004) Transgenic chickpea seeds expressing high levels of a bean α-amylase inhibitor. Mol Breed 14:73–82

Sawa M (1974) On the interspecific hybridization between the adzuki bean *Phaseolus angularis* (Willd.) and the green gram *Phaseolus radiatus* L. II. On the characteristics of amphiploid in C ~ generation from the cross green gram rice bean, *Phaseolus calcaratus* Roxb. Jpn J Breed 24:282–286

Sawadogo M, Ouedraogo JT, Gowda BS, Timko MP (2010) Genetic diversity of cowpea [*Vigna unguiculata* (L.) Walp.] cultivars in Burkina Faso resistance to *Striga gesnerioides*. Afr J Biotechnol 9:8146–8153

Schroeder HE, Gollash S, Moore A (1995) Bean α-amylase inhibitor confers resistance to the pea weevil (*Bruchus pisorum*) in transgenic peas (*Pisum sativum* L.). Plant Physiol 107:1233–1239

Sen NK, Ghosh AK (1960) Interspecific hybridization between *Phaseolus aurens* Roxb. (green gram) and *Ph. mungo* (black gram). Bull Bot Soc Bengal 14:1–4

Sène D (1962) Inventaires des principales variétés de niébé (*Vigna unguiculata* W.) cultivées au Sénégal. Agron Trop 8:927–933

Shade RE, Schroeder RE, Poueyo JJ, Tabe LM, Murdock LI, Higgins TJV, Chrispeels MJ (1994) Transgenic pea seeds expressing the α-amylase inhibitor of the common bean are resistant to bruchid beetles. Biotechnology 12:793–796

Sharma HC (1998) Bionomics, host plant resistance, and management of the legume pod borer, *Maruca vitrata*-a review. Crop Prot 17(5):373–386

Simon MV, Benko-Iseppon AM, Resende LV, Winter P, Kahl G (2007) Genetic diversity and phylogenetic relationships in *Vigna Savi* germplasm revealed by DNA amplification fingerprinting. Genome 50(6):538–547

Singh BB (2005) Cowpea [*Vigna unguiculata* (L.) Walp. In: Singh RJ, Jauhar PP (eds) Genetic resources, chromosome engineering and crop improvement, vol 1. CRC Press, Boca Raton, pp 117–162

Singh BB, Matsui T (2002) Cowpea varieties for drought tolerance. In: Fatokun CA, Tarawali SA, Singh BB, Kormawa PM, Tamò M (eds) Challenges and opportunities for enhancing sustainable cowpea production. IITA, Ibadan, pp 287–300

Singh BB, N'tar (1985) Development of improved cowpea varieties in Africa. In: Singh SR, Rachie KO (eds) Cowpea research, production and utilization. Wiley, Chichester, pp 105–115

Solleti SK, Bakshi S, Purkayastha J, Panda SK, Sahoo L (2008) Transgenic cowpea (*Vigna unguiculata*) seeds expressing a bean α-amylase inhibitor 1 confer resistance to storage pests, bruchid beetles. Plant Cell Rep 27:1841–1850

Sousa-majer MJD, Hardie DC, Turner NC, Higgins TJV (2007) Bean α-Amylase inhibitors in transgenic peas inhibit development of pea weevil larvae. J Econ Entomol 100:1416–1422

Tarver MR, Shade RE, Shukle RH, Moar WJ, Muir WM, Murdock LM, Pittendrigh BR (2007) Pyramiding of insecticidal compounds for control of the cowpea bruchid (*Callosobruchus maculatus* F.). Pest Manag Sci 63:440–446

Thiam M, Champion A, Diouf D, SY MO (2013) NaCl effects on *In Vitro* germination and growth of some Senegalese cowpea (*Vigna unguiculata* (L.) Walp.) cultivars. ISRN Biotechnol 2013:1–11 http://dx.doi.org/10.5402/2013/382417

Thulin M, Lavin M, Pasquet R, Delgado-Salinas A (2004) Phylogeny and biography of *Wajira* (*Leguminosae*): a monophyletic segregate of *Vigna* centered in the Horn of Africa region. Syst Bot 29(4):903–920

Timko MP, Singh BB (2008) Cowpea, a multifunctional legume. In: Moore PH, Ming R (eds) Genomics of tropical crop plants. Springer, New York, pp 227–258

Timko MP, Ehlers JD, Roberts PA (2007) Cowpea. In: Kole C (ed) Genome mapping and molecular breeding in plants, pulses, sugar and tuber crops, vol 3. Springer, Berlin/Heidelberg, pp 49–67

Timko MP, Rushton PJ, Laudeman TW, Bokowiec MT, Chipumuro E, Cheung F, Town CD, Xf C (2008) Sequencing and analysis of the gene-rich space of cowpea. BMC Genomics 9:103

Tosti N, Negri V (2005) On-going on-farm micro evolutionary processes in neighboring cowpea landraces revealed by molecular markers. Theor Appl Genet 110:1275–1283

Tun YT, Yamaguchi H (2007) Phylogenetic relationship of wild and cultivated *Vigna* (Subgenus *Ceratotropis*, *Fabaceae*) from Myanmar based on sequence variations in non-coding regions of *trnT-F*. Breed Sci 57:271–280

Vaillancourt RE, Weeden NF (1992) Chloroplast DNA polymorphism suggests Nigerian center of domestication for the cowpea, *Vigna unguiculata*, Leguminosae. Am J Bot 79:1194–1199

Vaillancourt RE, Weeden NF, Barnard J (1993) Isozyme diversity in the cowpea species complex. Crop Sci 33:606–613

Vanderborght T, Baudoin JP (2001) Cowpea [*Vigna unguiculata* (L.) Walpers]. In: Raemaekers RH (ed) Crop production in tropical Africa. Directorate General for International Co-operation, Brussels, pp 334–348

Wang ML, Barkley NA, Gillaspie GA, Pederson GA (2008) Phylogenetic relationships and genetic diversity of the USDA *Vigna* germplasm collection revealed by gene-derived markers and sequencing. Genet Res 90:467–480

Watt EE, Kueneman EA, de Araujo JPP (1985) Achievement in breeding cowpea in Latin America. In: Singh SR, Rachie KO (eds) Cowpea research, production and utilization. Wiley, Chichester, pp 125–135

Wein HC, Summerfield RJ (1980) Adaptation of cowpeas in West Africa: effects of photoperiod and temperature responses in cultivars of diverse origin. In: Summerfield RJ, Bunting AH (eds) Advances in legume science. Royal Botanic Gardens, Kew, pp 405–417

Williams CB, Chambliss OL (1980) Out crossing in southern pea. Hortic Sci 15:179

Lentil

6

Richard Mario Fratini, Marcelino Pérez de la Vega, and Mª Luisa Ruiz Sánchez

Abstract

Among the cool season legume crops, lentil production has displayed the greatest increase, while cropping of many other grain legumes is actually in decline. This increase is most likely due to a convenient fast cooking coupled to a saving of fuel and time – dehulled lentils cook even faster than milled rice. Worldwide lentil consumption has augmented more than twice the rate of human population growth, with lentil consumption over the past 40 years having increased more than any other food crop. Therefore, continued genetic improvement of lentil is essential in all production regions; nevertheless, without access to and use of diverse germplasm, prospects for yield genetic gain in lentil may be difficult to accomplish. Wild *Lens* accessions represent less than 1 % of the world germplasm collection of this genus, this small fraction likely representing the greatest untapped pool of genetic variation in lentil. Germplasm evaluation together with a successful and targeted hybridization allows for an efficient breeding and selection of varieties adapted to specific environments. The aim of the present chapter is to describe the evolutionary aspects of lentil and assess the gene pool and flow for crop improvement, considering levels of diversity and agronomic traits of importance, with emphasis placed on wild species and interspecific *Lens* hybridization, to finally conclude with molecular aspects and future prospects of lentil breeding.

R.M. Fratini (✉) • M. Pérez de la Vega •
M.L. Ruiz Sánchez
Área de Genética, Departamento de Biología
Molecular, Universidad de León,
Campus de Vegazana SN, 24071 León, Spain
e-mail: richard.fratini@unileon.es;
m.perez.delavega@unileon.es; ml.ruiz@unileon.es

M. Singh et al. (eds.), *Broadening the Genetic Base of Grain Legumes*,
DOI 10.1007/978-81-322-2023-7_6, © Springer India 2014

6.1 Introduction

The global economic position of lentil (*Lens culinaris* Medik.) among grain legumes has increased in international trade and currently ranks sixth in terms of production after dry bean, pea, chickpea, faba bean, and cowpea. During the period 2003–2006, lentil constituted 6 % of the total world dry pulse production. Lentil production increased more than fourfold (413 %) in a period of 40 years, from 917,000 t in 1961–1963, with an average yield of 560 kg/ha, to a world harvest in 2004–2006 of 3,787,000 t with a mean yield of 950 kg/ha (Erskine 2009). According to FAOSTAT (http://faostat.fao.org), in (2007), lentils were being produced in 52 countries with approximately 3.8 million hectares under cultivation. The spread of lentil from its center of origin in the "Fertile Crescent" after its early domestication about 10,000 years ago (Cubero et al. 2009) has been accompanied by selection for traits that enhance adaptation to a wide range of agroecological environments. The crop is now being cultivated from cold-winter continental climates to subtropical areas and high altitudes in tropical regions. With an initial selection against pod dehiscence, the main objective of human selection has remained seed size together with flowering response to photoperiod and temperature, as well as resistance or tolerance to abiotic and biotic stresses. If lentils have been maintained by farmers through ages, it is most likely because they grow in poor soils, harsh climates, and conditions unfavorable for humans, animals, and crops (Cubero et al. 2009). In many cases, lentils may have been the only source of protein available to farmers. Lentil cultivation and adaptation to different environments and human preferences have resulted in a wealth of phenotypic seed variation together with the development of different culinary customs. Lentils are used worldwide as a starter, as a main dish, as a side dish, or in salads. For instance, according to Sarker and Kumar (2011), in West Asia and North Africa, *Mujaddarah* is a popular dish made out of whole lentil and immature wheat seed. *Koshary* is a commonly served dish in Egypt, made out of a mixture of rice and red lentil. In North Africa, lentil is prepared with vegetables and the recipe is known as *Lentil Tagine*. Of course, red lentil soup is popular all over West Asia, but most particularly in Turkey, Lebanon, Jordan, Palestine, and Syria. *Wot* is a traditional dish of Ethiopia. In South Asia, as indicated by Saha and Muehlbauer (2011), lentil is an important source of protein and is consumed almost daily as *Dal*, a typical stew seasoned with turmeric, ginger, onion, and other spices, by all strata of people irrespective of social and economic status. While in Western Europe, particularly in Spain, the brown lentil typified by variety pardina (greenish-brown seed coat with some dark speckling and mottling with yellow cotyledons) is preferred to be cooked as lentil stew. In addition, lentil may be deep fried and eaten as a snack or combined with cereal flour in the preparation of bread and cake. Large-seeded green lentils are commonly used in salads. The newly developed product *Genki Energy Bar* is an example of a potential wider use of the wholesome lentil (Vandenberg 2011). On a global scale, lentil consumption has augmented more than twice the rate of human population growth, with lentil consumption over the past 40 years having increased more than any other food crop. Among the cool season pulses, lentil is by far the fastest-growing crop, while many of the other grain legumes are actually in decline. This is most likely due to the convenient fast cooking coupled with a saving of fuel and time; dehulled lentils cook even faster than milled rice. Moreover, lentil requires less processing when compared to soya beans and cereals. As part of a diet, lentils are rich in protein (20–30 %), complex carbohydrates, and dietary fiber and are an excellent source of a large range of micronutrients (Thavarajah and Thavarajah 2011).

In order to introduce genetic variation, there remains no doubt as to the need to undertake manual crosses in a highly autogamous species such as lentil, allowing for segregation and selection, while introgression and genetic-based broadening go hand in hand. Germplasm evaluation together with a successful and targeted hybridization allows for an efficient breeding and selection of varieties adapted to specific environments.

Therefore, knowledge on germplasm and specific practical skills regarding ways to optimize hybridization constitute vital factors to advance breeding programs. Our aim here is to provide insight into these issues; simultaneously we describe the evolutionary aspects of lentil and assess the gene pool and flow for crop improvement; consider levels of diversity and agronomic traits of importance, with emphasis placed on wild species and interspecific *Lens* hybridization; and finally conclude with molecular aspects and future prospects of lentil breeding.

6.2 Crop Gene Pool, Evolutionary Relationships, and Systematics

6.2.1 The Genus *Lens* Miller

The genus *Lens* Miller is taxonomically classified within the tribe Vicieae DC. (syn. Fabeae Rchb.) (syn. Fabaceae, Papilionoideae) together with three other genera: *Lathyrus* L., *Pisum* L., and *Vicia* L. The tribe Vicieae is included in the inverted repeat-lacking clade (IRLC) (Wojciechowski et al. 2000), accordingly denominated because the chloroplast genome is uniquely marked by the loss of one copy of the large inverted repeat (approx. 25 kb). The clade includes all the members of the tribes Cicereae, Hedysareae, Trifolieae, and Vicieae, as well as at least three other genera (Wojciechowski 2006; Jansen et al. 2008). Thus, the interesting crop legume genera such as *Trifolium*, *Medicago*, and *Cicer* all hold a close phylogenetic relation to lentil and other Vicieae species. Although all botanical texts since the sixteenth century have used the name *Lens* for the species to which lentil belongs, the first botanist to truly assign the status of the genus was Tournefort in 1700. Miller in 1740 verified the designation and became the authority on the genus, while he is also responsible for producing the oldest available botanical description. In 1787, the German botanist and physician Medikus assigned lentil the scientific name *Lens culinaris* and the nomenclature is currently accepted and used. Numerous taxonomic treatments during the nineteenth century were published based on similarities of *Lens* with other taxa such as *Ervum*, *Ervilia*, *Vicia*, *Lathyrus*, *Orobus*, and even *Cicer* (all of which belong to a relatively young group of plants that are still in active evolution). Nonetheless, by the end of the nineteenth century, the genus *Lens* was finally well established (for historical references and synonyms, see Cubero et al. 2009).

The taxonomy of the genus *Lens* is far from easy given the close relationships among species, and the phylogenetic relationships between *Lens* species continue to be a subject of scientific discussion (Cubero et al. 2009). The classical structure of the genus based on morphological analyses and/or biochemical markers included four species: *Lens culinaris*, *L. orientalis*, *L. nigricans*, and *L. ervoides* (Ladizinsky 1979). In 1984, accessions of *nigricans* were reclassified as *odemensis*, and a new taxonomy of the genus was proposed, *Lens culinaris* with cultigen subspecies *culinaris* and wild ssp. *orientalis* and *odemensis*, and *Lens nigricans* with two subspecies, *nigricans* and *ervoides* (Ladizinsky et al. 1984). However, Ladizinsky later recognized the following taxa: *L. culinaris*, with subspecies *culinaris* and *orientalis*, *Lens odemensis*, *L. ervoides*, and *L. nigricans* (Ladizinsky 1993). In 1997, two new species were added to the genus, *Lens tomentosus* (ex *L. orientalis*; Ladizinsky 1997) and *Lens lamottei* (ex *L. nigricans*; Van Oss et al. 1997). Ferguson et al. (2000), using morphological as well as molecular markers, considered both *odemensis* and *tomentosus* as subspecies of *L. culinaris*. The following classification of the genus *Lens* was proposed by these authors: *L. culinaris*, with four subspecies, namely, *culinaris*, *orientalis*, *tomentosus*, and *odemensis*; *L. ervoides*; *L. nigricans*; and *L. lamottei*. This nomenclature is currently being used by the National Center for Biotechnology Information (NCBI). A phylogenetic nomenclature based on SDS-PAGE analysis of seed proteins undertaken by Zimniak-Przybylska et al. (2001) clustered *culinaris*, *orientalis*, *tomentosus*, and *odemensis* in one group; *L. ervoides* and *L. lamottei* in a second group; and *L. nigricans* in a third group; this result conforms in grosso modo with the previous study of Ferguson et al. (2000).

To summarize, taxonomic analyses based on morphological and/or biochemical markers ranged from four species in 1979, namely, *L. culinaris*, *L. orientalis*, *L. nigricans*, and *L. ervoides*, to two species in 1984: *L. culinaris* (with subspecies *culinaris* (the cultigen), *orientalis*, and *odemensis*) and *L. nigricans* (with ssp. *nigricans* and *ervoides*); again to four species in 1993, i.e., *L. culinaris* (ssp. *culinaris* and *orientalis*), *L. odemensis*, *L. nigricans*, and *L. ervoides*; and to the previous four species plus the addition of *L. tomentosus* and *L. lamottei* to comprise a total of six species in 1997. The number of six species is now widely accepted: *L. culinaris* [ssp. *culinaris*, the cultigen and ssp. *orientalis* (Boiss.) Ponert, the wild ancestor of the cultivated forms], *L. odemensis* (Godr.) Ladiz., *L. tomentosus* Ladiz., *L. nigricans* (Bieb.) Godr., *L. ervoides* (Bring.) Grande., and *L. lamottei* Czfr. The geographical distribution of the *Lens* species which are all self-pollinated is described by Cubero et al. (2009).

6.2.2 Genome Size, Ploidy, Chromosomes, and Karyotypes

The C value (unreplicated haploid) of the lentil genome was determined by flow cytometry to amount to 4.41 pg, which is equivalent to 4,063 Mbp (Arumuganathan and Earle 1991). This result agrees with a previous evaluation undertaken by Bennet and Smith (1976) using Feulgen microdensitometry of 4.6 pg, comparable also to the genome size of closely related Vicieae species such as pea (3,947–4,397 Mbp). Neither the mitochondrial nor the chloroplast genomes of lentil have been sequenced to date according to the data available in the NCBI Eukaryote Organelles and the Organelle Genome Megasequencing Program websites. However, Van Oss et al. (1997) carried out a comparative restriction fragment length polymorphism (RFLP) analysis of the *Lens* species chloroplasts.

All *Lens* species possess the same chromosome number ($2n=14$) and share similar karyotypes, which consists of three pairs of metacentric and submetacentric chromosomes, three pairs of acrocentric chromosomes, and a pair of metacentric chromosomes with the nucleolar organizing region (NOR) secondary constriction site proximal to the centromere location (Ladizinsky 1993), although some karyotype variants have been described (Balyan et al. 2002). In addition to the standard ssp. *culinaris* karyogram, Ladizinsky (1993) also described other karyotypes in ssp. *orientalis*: a single accession (No 113) showed a karyotype in which about three-quarters of the satellite had been translocated to another chromosome and the metacentric-satellited chromosome had become acrocentric. Accessions from Iran differed by a chromosomal rearrangement involving two metacentric chromosomes, and accessions from Turkey differed by a paracentric inversion and inclusive of two or even three chromosome rearrangements. Accordingly, ssp. *orientalis* appeared to be the most chromosomally variable *Lens* taxon (Ladizinsky 1993). Balyan et al. (2002) using fluorescence in situ hybridization (FISH) identified the NOR region to be located in chromosome 3; chromosome 2 consisted of a submetacentric chromosome with a 5S locus distally located on the long arm (the major site of hybridization), whereas chromosome 6 comprised an acrocentric chromosome with the other 5S locus located proximal on the short arm (minor site). Further data based on *Lens* homologous probes, which differentially hybridized to the 5S non-transcribed spacer (NTS), allowed a better differentiation between the two 5S loci (Fernández et al. 2005). Locus 5S on chromosome 2 corresponded to the longer NTS family sequence, while the second 5S locus located on chromosome 6 corresponded to the shorter NTS family. Furthermore, FISH data indicated that in some wild *orientalis* accessions (e.g., ILWL-7), three-quarters of the satellite had been transferred to another chromosome, the metacentric-satellited chromosome had become acrocentric, and one of the submetacentric chromosomes had been lengthened. In this *orientalis* accession, only a single 5S locus was detected using the longer NTS probe on chromosome 6; thus, it is possible that some ssp. *orientalis* karyotypes lack one of the 5S loci (Fernández et al. 2005).

6.2.3 Gene Pools

The previous discussion regarding the genus *Lens* and their different karyotypes can be confusing; in addition, as will be described further on, no clear hybridization barriers have been defined between the accepted species. The levels of fertility and sterility so far found have depended on the taxa involved in a specific cross but also, to a greater or lesser extent, on the particular populations used within these taxa. This phenomenon is common as it happens in *Lens*, since the taxa are close relatives. *Lens* is a genus that belongs to a very active group from an evolutionary point of view. Even with the exceptions described above, of which more details are provided in the next sections, hybridization experiments have supported the idea of six differentiated species. In line with the above and particularly with hybridization barriers in mind, support for the idea of six differentiated *Lens* species has gained. Besides of the forms of the cultigen (*L. culinaris* ssp. *culinaris*), *L. culinaris* ssp. *orientalis* obviously belongs to the primary gene pool and *L. odemensis* to the second, although success of *L. odemensis* crosses with the cultigen may or may not require embryo rescue depending on the specific accessions used. The tertiary gene pool is composed of *L. tomentosus*, *L. lamottei*, *L. nigricans*, and *L. ervoides*, but these can become part of the secondary gene pool by means of embryo rescue. In any case, further hybridization studies are needed to truly establish whether *L. tomentosus* and *L. lamottei* belong definitely to the secondary or tertiary gene pool (Cubero et al. 2009).

6.2.4 Origin and Evolution

Seed size seems to be the only character which has allowed domestication events to be traced in archaeological remains. The oldest remains of wild lentils have been found in Mureybit (Syria) dated around 10,000 BP, while those of the cultigens have been dated around 9,000 BP and were discovered in aceramic Neolithic layers located in the Near East (see Cubero et al. 2009 for archaeological data). Given the coexistence of the wild and domesticated forms which is not found elsewhere, coupled with archaeological evidence, the Fertile Crescent is the most likely candidate to be the center of origin of the cultivated lentil. In addition, lentil diversity in the center of origin is still very high for both the cultivated primitive forms and the wild relatives. Using molecular markers, Ferguson et al. (1998a, b) located areas of high diversity for ssp. *orientalis* alongside the border between Turkey and Syria as well as in areas of southern Syria and Jordan; this data supports the archaeological evidence and indicates the Near East region as the most likely center of lentil domestication. Ladizinsky (1999) also suggested the Near East as the center of origin based on the polymorphism found in wild accessions of ssp. *orientalis* and the monomorphism of *culinaris*. Indicators of human activity suggest that some populations of *orientalis* were unconsciously subjected to selection (Harlan 1992) which resulted in the crop that we know today as lentil. According to Zohary (1999) and based on chromosome and DNA polymorphisms, the domestication event of lentil happened once or only a few times. Toklu et al. (2009) used ISSRs and AFLPs to undertake a variability analysis of lentil landraces from Southeast Turkey concluding that they exhibited a high genetic variation index (He=0.180); nonetheless one particular landrace from the Karacadag/Diyarbakir region was found to be significantly different to the rest of the germplasm evaluated. The Karacadag mountain range located in Southeast Turkey is placed within the core area of plant domestication of the Fertile Crescent. Lev-Yadun et al. (2000) suggested a lentil domestication site close to or overlapping the area where einkorn and emmer wheats had been domesticated in the Fertile Crescent, concretely, the Karacadag/Diyarbakir region as the domestication center of lentil. Furthermore, two centers of diversity for *L. culinaris* ssp. *orientalis* have been described: (1) Southeast Turkey and Northwest Syria and (2) West and North Jordan and southern Syria (Ferguson et al. 1998a).

Compared to *L. orientalis*, cultivated lentils have a greater stem and rachis length, more leaflets per leaf, a greater leaf area, an increased number of flowers per peduncle, as well as an increased number of pods and seeds. In addition, peduncles of the cultivated forms are generally shorter or equal in length to the rachis when compared to the wild forms. Aside from all of the above, pod indehiscence of the cultivated forms in contrast to the dehiscent mode of the wild species, together with the erect growth habit of the cultivated lentil compared to the wild procumbent and prostrate plant structure, typifies the most noticeable differences between the domesticated and wild relatives, being traits of great economic value and with many implications on yield and ease of harvesting. All of the abovementioned characters are strongly associated with increased yield levels, in a similar way as observed for the other domesticated food legumes. Lentil cultigens show a higher frequency of white flowers compared to the dominant purple-colored inflorescences of the wild relatives, most probably a character associated to a better culinary quality and fixed by indirect selection of lighter-colored seed coats (Fratini et al. 2011). As is also the case of chickpea and faba bean, there exists a clear regional distribution pattern of the cultivated lentil forms (displayed in Cubero et al. 2009). The trend from eastern to western lentils comprises an increase in seed size, a higher number and size of leaflets in addition to the length of the calyx teeth relative to the corolla length. To explain this cline, it has been postulated that the introgression into western forms of lentil came from *odemensis* (more likely than from *nigricans* as *odemensis* was given its specific status because of its crossability with *culinaris*), while an introgression from *orientalis* played the leading role with regard to the eastern forms. For the short calyx of the prevalent lentil forms found in East Africa and mainly Ethiopia, the genetic influence of *ervoides* has also been postulated. A comparison between the geographical patterns of the wild species and the cultivated forms (see Cubero et al. 2009) seems to verify the introgression hypothesis as *orientalis* is the only wild form spreading eastward, *ervoides* to the Ethiopian region, and both *nigricans* and *ervoides* to the west. Nonetheless, the latter two species do not readily cross with *culinaris* as hybrids result in embryo abortion (Ladizinsky et al. 1984; Fratini and Ruiz 2006), but sporadic crosses through a long period of time cannot be readily dismissed. Besides, in the same way as *odemensis* was separated from *nigricans* because of the differential fertility level with cultigens, other strains of *nigricans* and *ervoides* could have behaved in a similar way.

Thus, although crosses between *culinaris* and *odemensis* are feasible and produced longer (sometimes branched) tendrils than those observed for *culinaris* (Ladizinsky 1979; Ladizinsky et al. 1984), more experimental work, including molecular biology analysis, is necessary to prove the historical introgression of traits derived from *odemensis* into the cultigens. The geographical pattern could simply be an indirect (correlated) response to different human selection approaches in different parts of the world accompanied with the usual sources of genetic variation (mutation, migration, and genetic drift) and crosses with companion weeds. In fact, molecular marker analyses have indicated that the genetic variability within cultivated lentil forms is relatively low (Durán and Pérez de la Vega 2004; Sonnante et al. 2003), suggesting that the two great groups of cultivated lentils, *microsperma* and *macrosperma*, could only be variants for quantitative traits resulting from disruptive selection.

Summing up the available information on crop evolution, lentils were domesticated in the foothills of the mountains of southern Turkey and northern Syria (Ferguson et al. 1998a, b, c; Ladizinsky 1999; Lev-Yadun et al. 2000) most likely by unconscious selection (Harlan 1992) within populations of ssp. *orientalis*. Nonetheless, the influence of other wild relatives cannot be excluded as has been shown to have occurred in the origin of most cultivated crops. In the concrete case of lentil, the similarity among wild species could have been a factor in favor of producing companion weeds and maintaining them in cultivated stocks. The ssp. *orientalis* and *L. odemensis* are the most likely candidates to have constituted the main origin of extra-specific variability for the cultigens; however, more experimental proof is further needed. The genetic variability found with molecular markers seems to be low (Durán and Pérez de la Vega 2004; Sonnante et al. 2003), suggesting a common origin for all the cultivated forms of lentil and a narrow range of achieved artificial selection. The observed differences detected among different geographical groups could be the plain result of a correlated response exerted by selection for higher yield, compared to a direct consequence of a more basic genetic difference. All in all, the role of introgression from wild forms, however, requires further study.

6.3 Assessment of Gene Flow for Crop Improvement

Ladizinsky (1979) made crosses between *culinaris* (three accessions representing extreme values for seed characteristics and other traits), *orientalis* (four), and *nigricans* (one). The F_1 generation of the *culinaris* × *orientalis* crosses grew normally, and the F_2 generation segregated for growth habit, flower color, pod dehiscence, and seed coat color. Meiosis was almost normal with the presence of a quadrivalent in most of the cells examined and, in very few cases, a trivalent and one univalent. This result supported previous proposals considering *orientalis* as a subspecies of *culinaris*. Other hybridization experiments have indicated that *culinaris* × *orientalis* crossing successes are equal or even higher than cross combinations within *L. c. culinaris* (Fratini et al. 2004), so far not having observed embryo abortion using different *orientalis* accessions, although clear signs of varying degrees of plant deterioration, such as chlorosis and other weakness signs, are clearly detectable in recombinant inbred populations (RILs) most likely due to the fixing of unfavorable allele associations (Fratini, unpublished). Altogether, the *orientalis* taxon remains the best source among the wild species for the introgression of traits to the cultigen. According to Ladizinsky (1979), *culinaris* and *orientalis* share the same karyotype, although karyotype variants in *orientalis* have been described (Ladizinsky 1993; Balyan et al. 2002; Fernández et al. 2005). Likewise, some extent of intraspecific karyotype variations in several *Lens* species has also been described (Galasso 2003; Fernández et al. 2005). Since fertility of the inter-subspecific hybrids between *culinaris* and *orientalis* depends on the chromosome arrangement of parents (Ladizinsky 1979; Ladizinsky et al. 1984), three crossability groups have been identified in the wild ssp. *orientalis*: a common, a unique, and an intermediate. Crosses between members of the common and unique groups yield aborted seeds which can be rescued by embryo culture, while members of the intermediate group are cross compatible with the other two groups (Ladizinsky 1993; van Oss et al. 1997). Subspecies *orientalis* is readily crossed with *L. odemensis*, the hybrids are vegetatively normal but partially sterile due to meiotic irregularities resulting from three chromosome rearrangements between the parental strains (Ladizinsky 1993). The first hybrids between *culinaris* and *nigricans* developed normally (Ladizinsky 1979). However, later on crosses with other accessions of *nigricans* were observed to be unviable (Ladizinsky et al. 1984). Therefore, in 1984, the former *nigricans* accession was reclassified as *L. odemensis*. *Lens tomentosus* is morphologically closer to *L. c.* ssp. *orientalis* than to any other *Lens* taxon; nevertheless, they are isolated by hybrid embryo breakdown, complete sterility, and five chromosomal rearrangements (van Oss et al. 1997), which supports the idea of a specific status for *tomentosus*. Likewise, *L. tomentosus* is reproductively isolated from *L. lamottei* and *L. odemensis* by hybrid embryo abortion (van Oss et al. 1997). Linkage studies have revealed chromosomal rearrangements between *L. culinaris* and *L. odemensis* (Ladizinsky 1993), which could possibly explain why certain accessions of *odemensis* are freely crossed with *culinaris* (Ladizinsky 1979) while others need embryo rescue (Fratini and Ruiz 2006). *Lens odemensis* is cross incompatible with *nigricans* and *ervoides* due to hybrid embryo abortion (Ladizinsky et al. 1984; Ladizinsky 1993).

The morphological differences between *L. nigricans* and *L. lamottei* are limited to stipule shape; however, the two species differ by four reciprocal translocations and one paracentric inversion, resulting in the complete sterility of their hybrids (Ladizinsky et al. 1984). *Lens nigricans* × *L. ervoides* interspecific hybrids are vegetatively normal but completely sterile (Ladizinsky 1993). Fratini and Ruiz (2006) carried out extensive crosses between *L. c.* ssp. *culinaris* and

L. nigricans and *L. ervoides* and *L. odemensis*. Hybrids between the cultigens and the other species were viable only through embryo rescue; the rates of adult plants obtained were low: 9 % with *odemensis* and 3 % with *nigricans* and *ervoides*. Previously, it had been already shown that crosses between *culinaris* and *nigricans* or *ervoides* needed embryo rescue to recover interspecific hybrids (Ladizinsky et al. 1984; Ladizinsky 1993). In summary, hybridization barriers support the idea of six differentiated species. Together with *Lens c. culinaris*, the cultivated form, *Lens c. orientalis* obviously belongs to the primary gene pool, whereas *L. odemensis* to the second. Although the success in such crosses (needing embryo rescue or not) depends on the particular *odemensis* accessions used in the study, *L. tomentosus, L. nigricans, L. ervoides*, and *L. lamottei* belong to the tertiary gene pool, but can become part of the secondary gene pool by means of embryo rescue; however, further hybridization studies are needed to establish whether *L. tomentosus* and *L. lamottei* truly belong to the secondary or tertiary gene pool (Cubero et al. 2009).

6.3.1 Gene Flow Constraints

Mating system is a major determinant of the genetic structure of plant populations and of the breeding methods in crops. Lentil is a highly self-pollinating species in which outcrossing rates are usually low. Wilson and Law (1972) reported less than 0.8 % of natural cross-pollination as discerned by morphological markers. More recently, the outcrossing rate of lentils was established to range from 0.06 to 5.12 % among three varieties as measured in Central Europe using a dominant orange cotyledon allele as a genetic marker (Horneburg 2006). The outcrossing rate of individual plants ranged from 0 to 22.2 %. Results were strongly influenced by cultivar, year, and location; thus, for instance, the outcrossing rates of the cultivar Pisarecka Perla in the same locality were 0.89, 5.12, and 3.01 % in the years 1999, 2000, and 2001, respectively (Horneburg 2006).

The large variation detected in natural outcrossing rates of lentil, which for a given cultivar depend upon year and location (Horneburg 2006), indicates that in addition to the genotype, the environmental conditions can play a determinant role with regard to hybridization success. The effect of environmental conditions on intra- and inter-subspecific hybridization success was analyzed by Fratini et al. (2004). Fall crossing success under controlled greenhouse conditions among lentil cultivars or with subspecies *orientalis* was substantially higher with regard to the same field crosses carried out in the spring. In the first published literature on intraspecific lentil hybridization (Wilson 1972), it was observed that lentil crosses in the greenhouse were most successful during the winter months, when the temperatures were maintained in a range of 18–24 °C, and the relative humidity was above 50 %, while manual crossing in the field was unsuccessful. The study of Fratini et al. (2004) established that the fall crossing success in the greenhouse was favored by temperatures of 20–25 °C, high humidity, and sunny mornings; crosses performed in the morning and early afternoon (10–14 h) usually gave better results than crosses late in the afternoon (16–17 h); light conditions also influenced crossing success as crosses carried out on cloudy days (with lower temperatures around 15 °C) yielded a significantly lower pod and seed set; also a temperature range from 20 to 25 °C under high humidity conditions with an optimum temperature at 24 °C was found to be the most suitable for crossing lentil. Previously, Malhorta et al. (1978) had also determined that morning pollinations (9–11 a.m.) in India produced the highest percentage of pod set in intraspecific crosses. The field results of the spring crosses undertaken by Fratini et al. (2004) agreed with previous field data obtained for such crosses (Wilson 1972; Malhorta et al. 1978; Solh et al. 1980; Mera and Erskine 1982; Malaviya and Shukla 1990; Kumar and Singh 1998); all studies pointing out that conditions of high humidity and mild temperatures favored crossing success. In addition, for spring crosses in the field, it has been determined that the bagging of pollinated flowers or plants, together with a lower illumination, such as applying artificial shading after pollination, increased the intraspecific field hybridization success (Solh et al. 1980; Mera and Erskine 1982). For instance, Mera and Erskine (1982) obtained an average percentage of pod set of 27.1 %, ranging from 60.9 % with plastic bags left on shaded plants until harvest to 7.7 % for the unshaded control. During winter in India, Kumar and Singh (1998) found an overall pod set of 23 %, observing that evening crosses, with temperatures of 23–26 °C and a relative humidity of 85 %, were the most successful.

Besides environmental factors determining hybridization success, genotype differences related to crossing efficiency have also been detected. For instance, Fratini et al. (2004) found that the cultivar Lupa was on average the worst male parent (5.9 % of pod and 7.1 % of seed set per pollinated flower); however, this cultivar was on average to be the best female parent (29.1 and 39.4 %, respectively). Under controlled greenhouse conditions, it was concluded that the higher significance of the variable *Male* relative to the *Female* pointed out the importance of pollen quality (Fratini et al. 2004). As to specific genotype influences, it has been noted that certain intraspecific cross combinations are easier to obtain than others, inclusive in the case of reciprocal crosses, especially with regard to crosses between *microsperma* and *macrosperma* varieties of lentil (Fratini, unpublished). Previously three crossability groups had been identified in the wild ssp. *orientalis*: a common, a unique, and an intermediate; only members of the intermediate group were cross compatible with the other two groups (Ladizinsky 1993; van Oss et al. 1997). Therefore, it seems that lentil genotypes influence crossing success, although further studies are needed to clarify the relative influence of the two main factors, genotype and environment. Besides from all of the above considerations regarding the mating system, environmental conditions, and parental genotype cross combinations, all of which constrain gene flow in lentil breeding; the operator efficiency undertaking manual hybridizations is also a determinant factor. A skilled operator needs to possess the experience to discern highly fertile female lentil plants and appropriate male pollen donors. Identifying flowers at the correct physiological stage of ¾ sepals to petals to emasculate and pollinate or to gather pollen is of vital importance. For crosses undertaken in the greenhouse under controlled conditions, out of 247 emasculations and intra- and inter-subspecific pollinations, a total of 60 hybrids were obtained as discerned by morphological and molecular markers, equal to an overall hybridization efficiency of 24 % (Fratini et al. 2004).

As a summary, gene flow in lentil breeding is constrained by (1) the mating system with the inherent complexity to undertake crosses by hand emasculation and pollination (which means, for instance, that gene introgression using the backcross method of breeding is difficult and tedious); (2) the need to dispose of controlled environmental greenhouse conditions to assure hybridization success; (3) certain parental cross combinations result in higher crossing efficiencies than other genotype combinations; (4) need of skilled manual hybridization operators; and (5) in the case of certain inter-subspecific crosses and in most interspecific crosses, embryo abortion must be overcome by in vitro embryo rescue, although this may not preclude a future hybrid breakdown during the juvenile stage or sterility during the adult phase. More data on this aspect will be provided in the following sections. On the other hand, traditional lentil breeding is still limited by: (1) the apparent low genetic variability and the relatively low number of described Mendelian genes; (2) the lack of information on genetic, physiological, and other aspects of the crop; and (3) the lack of screening methods under controlled conditions. Traditional breeding has often been directed to address major production constraints, which in turn has limited to take into consideration other possible breeding goals.

6.4 Level of Diversity in Crop Germplasm

6.4.1 Morphological Diversity

The first detailed and complete study of the cultivated lentil was made by Barulina (1930), a disciple and subsequently wife of N. I. Vavilov. She considered two subspecies, *microsperma* and *macrosperma*, according to seed size, the main objective of human selection. Barulina also considered the geographical distribution of a cluster of characters, defining regional groups or *greges*. The main characters used to define the subspecies consisted of pod and seed traits together with distinct differences in flower length. Important characters used to define the groups within the subspecies included dehiscence, length of calyx teeth, and number of flowers per peduncle (Cubero et al. 2009). More than half a century later, a study of the quantitative agro-morphological variation found in the germplasm of 13 major lentil-producing countries identified three major regional groups (Erskine et al. 1989). The close resemblance detected between germplasm from adjacent climatically similar countries indicated that environmental conditions were the major evolutionary driving force in the case of the cultivated lentil. Furthermore, it appeared that the phenological adaptation to the environment was central to crop evolution. Landrace accessions from India and Ethiopia are grouped together with striking similar phenotypes for the nine quantitative traits surveyed. The group was characterized by early flowering and maturity, low biological yield, short stature, low pod height, and small seeds. This similarity for quantitative traits had already been noted by Barulina (1930).

South Asia is the world's largest lentil-growing region; nevertheless, indigenous lentils (with specific ecotype *pilosae*) show a noticeable lack of variability. According to Erskine et al. (1998), this lack of variability is the result of a genetic bottleneck (most likely a founder effect) derived from lentil introduction into the Indian subcontinent from Afghanistan around 2,000 BC. Further evidence of the variation paucity within Indian germplasm resides in a sensitivity study of a world lentil collection to temperature and photoperiod (Erskine et al. 1990). In the agro-morphological study (Erskine et al. 1989), germplasm from Afghanistan was found markedly different to the Indo-Ethiopian group and had a greater similarity to lentil germplasm from Iran and Turkey. Of all the accessions tested, the Afghan accessions were among the latest to mature. Germplasm from Pakistan was also of the *pilosae* group for qualitative characters, yet intermediate between Afghan and Indian material for quantitative traits. Remarkably, countries with a much smaller area dedicated to lentil production (e.g., Lebanon and Chile) compared to South Asian countries (e.g., India and Pakistan) showed a higher mean coefficient of variation for the nine morpho-agronomic characters scored (Erskine et al. 1998).

6.4.2 Biochemical and Molecular Diversity

The introduction of biochemical and molecular markers in *Lens* diversity studies represented a leap forward, overcoming the previous limitations imposed by the relative low number of morphological characteristics recorded. Isozymes and storage proteins were first used in the 1980s to study genetic control and differences between *Lens* species and among lentil germplasm collections and to lay out geographical distribution (Skibinski and Warren 1984; Zamir and Ladizinsky 1984; Pinkas et al. 1985; Hoffman et al. 1986; Muehlbauer et al. 1989; Rosa and Jouve 1992). The use of isoenzymatic and protein SDS-electrophoresis procedures rapidly declined when DNA markers were introduced. Restriction fragment length polymorphism (RFLP) markers were the first molecular markers used to undertake lentil genetic and linkage analyses (Havey and Muehlbauer 1989). Subsequently, random amplified polymorphic DNA (RAPD) was used, among other purposes, to study the diversity, phylogeny, and taxonomy of *Lens* (Ford et al. 1997; Ferguson et al. 1998a, b, c). The most extensive analysis of the genetic variation in cultivated lentil was carried out by Ferguson et al. (1998c) who analyzed isozyme and RAPD variation in accessions derived from 16 different countries. The germplasm of Afghanistan clustered with that of South Asian countries. Accessions from these two areas were strikingly different to all other countries studied and showed low levels of variability. Another remarkable result was that the classification into *macrosperma* or *microsperma* types did not reflect overall country relationships, thus suggesting that seed size variation has little relation to between-country genetic similarities. However, different results in relation to seed size were observed by Duran and Pérez de la Vega (2004). Using RAPD and ISSR markers, they found that continental Spanish *macrosperma* and *microsperma* were more similar between them than with *microsperma* from different origins. Most of the studies on genetic variability after 1998 have been limited to local or regional germplasm samples (Sonnante and Pignone 2001, 2007; Durán and Pérez de la Vega 2004; Babayeva et al. 2009; Fiocchetti et al. 2009; Toklu et al. 2009). The general overview is that genetic variability in lentil is higher in or near the area where the domestication of lentil occurred in the Fertile Crescent.

In relation to the wild *Lens* species, the analysis of molecular variation with isozyme and RAPD data has revealed that between 78 and 99 % of the variation is attributable to between-population differences (Ferguson et al. 1998b). Areas of high diversity and unique diversity were located for four wild species, thus indicating regions where further germplasm collection is likely to yield novel genetic variation (Ferguson et al. 1998a). A systematic comparative study on the extent of genetic variation in *Lens* is still lacking, but in general it is considered that wild species hold more genetic variation compared to cultivated materials. In addition, when compared to other diploid sexually reproduced and cultivated crop species, lentils have less variation but show similar levels to that of the other cool season grain legumes.

Arbitrarily produced amplified fragment length polymorphism (AFLP) markers have also been used to study genetic diversity (Sharma et al. 1996) and to differentiate cultivars (Závodná et al. 2000; Fiocchetti et al. 2009). More recently, simple sequence repeat (SSR or microsatellite) and inter-simple sequence repeat (ISSR) markers have been included in the list of markers managed in lentils. Sets of microsatellite markers have been developed by Závodná et al. (2000) and Hamwieh et al. (2005, 2009) at the ICARDA, while other sets (also functional markers) are currently being produced at the Universidad de León (de la Puente 2012). In addition, cross-genera SSRs have been used in a diversity analysis of *Lens* species (Reddy et al. 2010). Muehlbauer et al. (2006) reviewed the molecular marker use in lentil. During the first decade of the twenty-first century, numerous studies regarding lentil diversity using biochemical and/or molecular markers have analyzed local or regional germplasm collections (Sonnante and Pignone 2001, 2007; Durán and Pérez de la Vega 2004; Scippa et al. 2008; Yuzbasioglu et al. 2008; Inder et al. 2008; Fiocchetti et al. 2009; Babayeva et al. 2009; Hamwieh et al. 2009; Toklu et al. 2009). The extensive use of single nucleotide polymorphisms (SNPs) will further improve the usefulness of molecular markers applied to lentil for basic genetic studies and breeding.

6.5 Germplasm Collection

Managed germplasm collections are available for many legume species, including lentil; characterization of the genetic diversity within these collections is a necessary prelude to their efficient use (Varshney et al. 2009). At least 43 lentil germplasm collections are maintained across the globe for breeding and research purposes. Muehlbauer et al. (1995) reviewed the most important institutions holding lentil collections. With a conservation capacity of more than one thousand accessions, the International Center for Agricultural Research in the Dry Areas (ICARDA; http://www.icarda.org), the Indian Agricultural Research Institute (http://www.iari.res.in), the N I Vavilov Research Institute of Plant Industry (http://www.vir.nw.ru), the National Bureau of Plant Genetic Resources, India, and finally the USDA collection at the Regional Plant Introduction Station, Pullman, Washington (http://www.ars.usda.gov) stand out. Another national germplasm bank holding an important collection of lentils is the Spanish Plant Genetic Resource Center (http://www.inia.es) with more than 600 accessions. A complete picture of the genetic resources available can be obtained at the germplasm collection directory web page of Bioversity International (http://www.bioversityinternational.org/), which integrates information regarding most of the world germplasm collections. For instance, the query on *Lens culinaris* landraces and traditional cultivars generated in the year 2012 a list of 43 gene banks and institutions holding a total of 12,097 accessions, whereas the query on wild or weedy forms of *Lens* yielded a list of six institutions maintaining only a total of 296 accessions (Pérez de la Vega et al. 2012). The GapAnalysis tool (http://gisweb.ciat.cgiar.org/GapAnalysis/) under a wild lentil search describes gaps in the collected *Lens* gene pool and provides recommendations for new recollection priorities, particularly *L. tomentosus*, which is given a high priority and of which only 4 gene bank accessions exist worldwide. Aside from the relatively low number of less than 300 wild accessions indexed by the germplasm collection directory web page of Bioversity International and maintained at six institutions, it is also remarkable that no genetic stocks, introgressed forms, or mutants were listed in 2012 among the available materials (Pérez de la Vega et al. 2012). Nevertheless, obtaining new useful mutants by artificial mutagenesis has been attended to in lentils. Chemical mutagens and ionizing radiation have been used to produce mutant lentil lines released for commercial production (review by Toker et al. 2007), and gamma ray-induced mutations have been used to obtain and release new varieties of outstanding performance ("NIA- MASOOR-05") in Pakistan (Ali and Shaikh 2007). Dixit and Dubey (1986) described the effect of alkylating agents to generate chromosome mutations, mainly translocations. Moreover, the world lentil core collections have been mainly developed from the ICARDA and USDA-ARS collections (Varshney et al. 2009), implying a great extent of overlapping; in addition, the possibility might exist misclassification for certain wild accessions given their great morphological similarity.

Ladizinsky et al. (1990) described tertiary trisomic lines of lentils with *L. orientalis* and *L. ervoides*. Trisomic plants were weaker than euploid plants, and many did not reach maturity. There are no further records of their use in genetic analyses or breeding. Colchicine-induced lentil polyploids have been reported (Singh et al. 1992). Polyploidy results generally in gigantism for most determinate plant parts yet causes reduced fertility and yield; nonetheless, fertility and yield increased dramatically after selection from C1 to C4 generations. However, no commercial tetraploid lentil varieties have so far been released.

Wild *Lens* species can constitute an important source of donor genes. For instance, resistance to *Ascochyta* blight was found in an ICARDA wild germplasm collection: 24 out of 86 accessions of *L. culinaris* ssp. *orientalis* were resistant, as were 12 out of 35 accessions of *L. odemensis*, 3 out of 35 accessions of *L. nigricans*, and 36 out of 89 accessions of *L. ervoides* (Bayaa et al. 1994). Ye et al. (2002) listed a series of cultivated and wild lentil materials as sources of partial resistance to *Ascochyta* blight, and Muehlbauer and Chen (2007) reported that the most prominent among the partially resistant germplasm accessions consisted of lentil accessions PI 339283, PI 374118, ILL5588, ILL5684, PR86-360, and ILL7537. Wild materials have also been used as a source of genes for winter hardiness. Assessment of 245 wild lentil accessions and 10 cultivated lentil lines in Turkey showed that, on average, *L. orientalis* exhibited the highest level of winter hardiness, whereas accessions of *L. ervoides* were the most susceptible (Hamdi et al. 1996). Cultivated materials with large seeds were found in West Asia to be less susceptible to winter cold than those with small seeds, with average temperatures below 10 °C from January to April; the yield of the large-seeded group was higher than the group of the small-seeded lentils, while the converse was true at higher temperatures (Erskine 1996).

L. nigricans provided the maximum number of resistant accessions for biotic (wilt, powdery mildew, and rust) and tolerance to abiotic stresses (Gupta and Sharma 2006). Tullu et al. (2006) tested all six wild *Lens* species for resistance to *Colletotrichum truncatum*. Accessions of *L. ervoides* revealed the highest level of resistance (3–5) to races Ct1 and Ct0 followed by *L. lamottei* in both field and greenhouse assays. *L. orientalis*, *L. odemensis*, *L. nigricans*, and *L. tomentosus* were highly susceptible (8–9) to race Ct0 in the greenhouse. The highest levels of resistance were found among *L. ervoides* accessions originating from Syria and Turkey.

Resistance against the parasitic plant *Orobanche crenata* was found in wild *Lens* materials (Fernández-Aparicio et al. 2007, 2009). The screening of 23 wild accessions for resistance to *O. crenata* under field conditions showed a wide range of responses from complete resistance to high susceptibility. The highest levels of resistance were observed in accessions of *L. ervoides*, *L. odemensis*, and *L. orientalis*. Likewise, resistance against *Sitona crinitus* Herbst, a major insect pest limiting lentil productivity mainly in the countries of West Asia and the North African region, was found in wild species (*L. ervoides*, *L. odemensis*, *L. nigricans*, and *L. c.* ssp. *orientalis*) (El-Bouhssini et al. 2008). A lentil germplasm collection of 571 accessions from 27 countries including wild species was screened for susceptibility to seed bruchids under natural field conditions in Central Spain (Laserna-Ruiz et al. 2012); a total of 32 accessions including cultivated landraces, *L. c.* ssp. orientalis, *L. nigricans*, and *L. lamottei* showed lower infestation rates than the susceptible check and were selected as potential sources of resistance to seed weevil (*Bruchus* spp.).

Resistance to fusarium wilt has been described in germplasm from India (Kamboj et al. 1990); and Bayaa et al. (1997) screened a core collection of 577 germplasm accessions from 33 countries and rescreened a subset of the 88 mostly resistant accessions. A total of 1,771 lentil accessions from the US lentil collection (Pullman, WA) and the Institut für Pflanzengenetik und Kulturpflanzenforschung (Gatersleben, Germany) were screened for resistance to *C. truncatum*; about 95 % of the accessions were susceptible when inoculated with a single isolate in the field, and 16 accessions displayed resistance (Buchwaldt et al. 2004). Tullu et al. (2006) searched for sources of resistance to anthracnose in cultivated and wild materials.

6.6 Production-Related Problems

Muehlbauer et al. (2006) pointed out that several abiotic stresses (drought, heat, salinity, iron deficiency, among others) are important constrains to lentil production worldwide. The severity of these stresses is unpredictable in field experiments; hence, field trials are increasingly being supplemented with controlled environment testing and physiological screening. Lentils are mainly grown as a rainfed crop and often are seeded in marginal areas where limited rainfall and deficient soil moisture conditions are encountered. Thus, lentil breeding for improved drought resistance is a major goal for most regions where lentils are grown and particularly in areas of limited and erratic rainfall such as the Middle East and areas of the Mediterranean climate. Drought-tolerant cultivars are required to stabilize lentil production and also to extend cultivation to further dry areas. Muehlbauer et al. (1995) indicated that screening for drought tolerance was difficult due to the lack of knowledge of what to screen for and the absence of efficient screening techniques.

Screening for drought tolerance has improved with controlled environment testing and physiological screening. In a recent review on screening techniques and sources of resistance to abiotic stresses in cool season food legumes (Stoddard et al. 2006), it was stated that over the last decade, there has been considerable progress to increase the adaptation of lentil to autumn sowing, and a large body of literature exists regarding drought responses, particularly in chickpea and faba bean. Yet the available literature on salinity tolerance and waterlogging is minimal. The review of Stoddard et al. (2006) also summarizes information on identified lentil sources of resistance or tolerance to drought and heat stress [e.g., MI-30, a mutant line showing high yields under limited water supply identified by Salam and Islam (1994)], to chilling or freezing stress (Hamdi et al. 1996; Ali et al. 1999; Kahraman et al. 2004a), to salinity stress (Mamo et al. 1996; Ashraf and Zafar 1997; Rai and Singh 1999), to waterlogging (Ashraf and Chishti 1993), and to high boron concentrations (Hobson et al. 2006). Another very important component of drought resistance is the plant root system. Sarker et al. (2005) analyzed the variation in shoot and root characteristics and their association with drought tolerance in several lentil landraces. Stem length, taproot length, and lateral root number were highly correlated, both among themselves and with yield. High heritability estimates provided reliability for screening based on these traits.

Among a total of 40 genotypes, only one line (ILL 6002) was strikingly different to all other tested genotypes. Another approach to drought resistance screening is to select for drought avoidance through early maturity, particularly when the major moisture stress occurs at the end of the growing season. Early planting can contribute to this avoidance since lentil plants usually develop and initiate flowering before serious drought conditions prevail. This is the case for fall versus spring lentil sowing in cold areas. On the other hand, among cool season legumes, lentils seem to be the least tolerant to transient waterlogging in greenhouse experiments (Solaiman et al. 2007; Singh et al. 2013a).

A factor for increasing production in the Mediterranean region and other areas with cool winters consists of earlier sowing in the fall rather than at the end of the winter. Available winter-hardy lentil germplasm has prompted interest in the development and use of cultivars that can be sown in the fall in cold highland areas. This change in production of lentils from normal spring sowing to fall sowing increases yield potential (Kahraman et al. 2004a; Barrios 2012). Early sowing has proved to increase lentil production in several countries, for instance, by shifting sowing from spring to early spring or to autumn sowing in areas such as the highlands of Turkey, Iran, Afghanistan, the Baluchistan area in Pakistan, and also Spain. Early fall-seeded lentil increased yield from 480 to 590 kg ha^{-1} when compared to early spring sowing of non-winter-hardy varieties in the US Pacific Northwest and northern Great Plains of Canada. Fall sowing in Australia allowed for a longer period of vegetative and reproductive growth, rapid canopy development, greater absorption of photosynthetically active radiation, as well as a better water use efficiency, increased dry matter, and improved seed production when compared to a later sowing. Early sown lentils began flowering and forming seeds ahead in the growing season, at a time when vapor pressure deficits and air temperatures were lower, using more water in the post-flowering period compared to the delayed sowing treatments (Siddique et al. 1998; Chen et al. 2006). Delaying spring seeding in the northern Great Plains of Canada by 4 weeks negatively affected crop development parameters due to the exacerbation of summer drought stress. Seed yields decreased at the majority of site years and the production reduction averaged 38 %. Seeding delays of 2 and 4 weeks decreased average seed size by 10 % (Miller et al. 2006). However, early sowing increased the incidence of some fungal and bacterial diseases (Davidson and Kimber 2007). According to Muehlbauer et al. (1995), diseases of lentil are in general less damaging than those of most other food legume crops. However, there are some important and potentially devastating diseases (Taylor et al. 2007). Several reviews on screening techniques and sources of resistance to root and foliar diseases in cool season grain legumes have been published (Infantino et al. 2006; Sillero et al. 2006; Tivoli et al. 2006).

The root rot/wilt complex is probably the most important lentil disease problem worldwide (Muehlbauer et al. 1995). This set of diseases is caused by several fungi of the genera *Fusarium*, *Pythium*, *Rhizoctonia*, and *Sclerotinia*, in particular *F. oxysporum* f. sp. *lentis*, *R. solani*, and *S. sclerotiorum*. Another fungus which can be included in this complex is *Aphanomyces euteiches*, a soil-borne pathogen causing *Aphanomyces* root rot of several legumes, including alfalfa, bean, lentil, and pea. Relatively little is known about the population biology of this legume pathogen (Grunwald and Hoheisel 2006). The inheritance of resistance to *Fusarium* vascular wilt was investigated by Eujayl et al. (1998a, b) who found several random amplified polymorphic DNA (RAPD) markers linked to the *Fw* locus. Wang et al. (2006) found small but consistent differences among 14 lentil cultivars with regard to their reaction to *Rhizoctonia* seedling blight and root rot caused by *Rhizoctonia solani* in greenhouse trials and in field experiments conducted in Canada. *Fusarium redolens* has also been described as a causal agent of *Fusarium* wilt in lentil (Riccioni et al. 2008).

Rust caused by *Uromyces viciae-fabae* Pers. is a serious problem in areas where mild temperatures and humid conditions favor the development of this fungus (Muehlbauer et al. 1995). Incomplete (partial) resistance to *Uromyces* has been described in lentil cultivars (Negussie et al. 2005). These authors analyzed the response to four components of resistance, namely, latent period, infection efficiency, pustule size, and spore production and concluded that, since there was an interdependence of the components, selection based on more than one component should help to obtain lines with higher levels of partial resistance. *Uromyces*-resistant cultivars, such as Laird or Calpun-INIA derived from Laird, have been described (Penazola et al. 2007). *Ascochyta* blight, a seed-borne disease described in at least 16 countries across five continents, is probably one of the most widely distributed and devastating diseases. The fungus, *Ascochyta lentis* (syn. *A. fabae* f.sp. *lentis;* teleomorph: *Didymella lentis*), attacks leaves, stems, and pods causing necrotic spots. In some cases it kills plants before harvest or infects seeds to such a degree that lentils are unmarketable. Ascospores of *Didymella lentis* can be found on lentil stubble under natural conditions, and the presence of the teleomorph has implications in the long-distance dispersal of *A. lentis* (Galloway et al. 2004). Species of the genus *Ascochyta* cause most of the major diseases of many cool season grain legumes, and a considerable amount of information about the pathogens, plant hosts, and disease resistance responses is available. Recently, *Phytopathologia Mediterranea* (Millan 2013) has brought out a special issue on *Ascochyta.*

The mating type (MAT) locus of *Ascochyta lentis* was cloned and characterized, and a multiplex PCR assay for mating type was developed based on the MAT idiomorph and flanking sequences (Cherif et al. 2006). Molecular diagnostic techniques have been developed to differentiate the *Ascochyta* pathogens that infect cool season food and feed legumes, as well as to improve the sensitivity of detecting latent infection in plant tissues (Taylor and Ford 2007). Recent multilocus phylogenetic analyses of a worldwide sample of *Ascochyta* fungi causing *Ascochyta* blights of chickpea (*Cicer arietinum*), faba bean (*Vicia faba*), lentil (*Lens culinaris*), and pea (*Pisum sativum*) have revealed that the fungi causing disease of each host formed a monophyletic group. Host inoculations of these fungi demonstrated that they were host-specific, causing disease only on the host species from which they were isolated (Hernández-Bello et al. 2006; Peever 2007). Tivoli et al. (2006) provided a thorough review on the sources of resistance and screening techniques for *Ascochyta* blight in lentil, and Ye et al. (2002) gave an account of suitable breeding techniques for the selection of lentils with resistance to *Ascochyta* blight; partial resistance to the disease is available in germplasm (Muehlbauer and Chen 2007).

Anthracnose is another important disease in lentil caused by the fungus *Colletotrichum truncatum* (Schwein) Andrus and Moore [teleomorph *Glomerella truncate*, described by Armstrong-Cho and Banniza (2006)]. The presence of two distinct pathogen races (Ct0 and Ct1) was demonstrated (Buchwaldt et al. 2004), and Tullu et al. (2006) searched for sources of resistance to anthracnose in cultivated and wild materials, localizing two resistant cultivars (Indianhead and PI 320937) to race Ct1 but none to race Ct0. Some accessions of wild *Lens* species are potential sources of genes for resistance to both races. Transfer of resistance to both races, Ct1 and Ct0, from *L. ervoides* accession L-01-827A to the cultivar Eston by single-seed descent from F_2 individuals to $F_{7:8}$ RILs was described by Fiala et al. (2009). Following rooting and hybrid clonal propagation procedures developed for lentil (Fratini and Ruiz 2003, 2008), Vail and Vandenberg (2010) produced a clonal propagation protocol to evaluate disease data of lentil infections. These authors tested the feasibility of cloning individual lentil plants to generate replicated ratings in order to evaluate the response to anthracnose, proposing that this method is more dependable than using single-plant evaluations. Powdery mildew is a minor disease in lentil, predominantly present under controlled greenhouse conditions, even though it can severely affect certain lentil cultivars in some parts of the world. *Erysiphe pisi* is the pathogen which has been detected in several countries, although *E. trifolii* has recently been added as a causal agent of this disease in the Pacific Northwest of the United States (Attanayake et al. 2009).

Genotypes with resistance to various biotic stresses, particularly resistance to vascular wilt, rust, and *Ascochyta* blight, have been identified and directly exploited or used in breeding programs (Sarker and Erskine 2006). Inclusive multiple resistant materials exist among both breeding lines and landraces. For instance, Laird is a macrosperma multiresistant cultivar, and Assano is a recently released cultivar highly resistant to rust caused by *Uromyces viciae-fabae* with also moderate resistance to *Ascochyta* blight and to the wilt-root rot complex (*Fusarium* sp. and *Rhizoctonia* sp.) (Fikru et al. 2007). A Pakistani accession, 66013-6, was found to have a high level of resistance to blight, rust, and pea seed-borne mosaic virus (Hussain et al. 2008).

Lentils have been reported to be susceptible to more than 25 viruses according to the list supplied by the Plant Virus Online (http://www.agls.uidaho.edu/ebi/vdie/). In a field experiment, 12 out of 16 lentil genotypes were ranked as highly susceptible and four as susceptible to the pea seed-borne mosaic virus (PSbMV). Once infected, plant sensitivities (symptom severities) were low in most lentil genotypes. Seed lots harvested from PSbMV-infected plants of lentil were usually less affected than other cool season grain legumes. Seed transmission of PSbMV (6 %) was detected in seed from infected lentil plants (Coutts et al. 2008). In a study (Bashir et al. 2005) to assess the amount of variation among lentil germplasm with regard to infection reaction against several viruses and also to identify sources of resistance, most of the lentil genotypes were susceptible to highly susceptible, and none was found to be highly resistant although some genotypes were found to be resistant and moderately resistant. Some differences in the sequence of the *eIF4E* lentil gene have been found in lentil (Saenz de Miera, pers. comm.); this gene is related to virus resistance in pea and other plants species (Robaglia and Caranta 2006).

Another major threat with regard to lentil cultivation in the Mediterranean area and Western Asia consists of crenata broomrape (*Orobanche crenata*). This weedy root parasite genus constitutes a serious threat to lentils. Broomrape control is difficult due to the large number of wind-blown seeds which are produced and dispersed each year and which may thereafter remain dormant in soils for several years. The control strategy has consisted in applying different agronomic practices and herbicides. Resistance breeding so far is hampered by a scarcity of proper resistance sources in addition to a reliable and practical screening procedure. A Spanish germplasm collection of 234 accessions of lentils was screened for resistance to crenate broomrape under field conditions (Fernández-Aparicio et al. 2007), followed by a germplasm screening of 23 wild accessions (Fernández-Aparicio et al. 2009). A wide range of responses was observed; however, complete resistance was not detected. Other *Orobanche* species can also parasitize lentil plants (Muehlbauer et al. 1995).

Lentil is damaged by many types of insects and other pests. Among insects, the major pests in the field are aphids (*Aphis craccivora*, *Acyrthosiphon pisum*), leaf weevil (*Sitona* spp.), lygus bugs (*Lygus* spp.), and the cutworm (*Agrotis ipsilon*). Another major pest problem causing great seed losses are the seed insect species: *Bruchus ervi* and *B. lentis* and also *Callosobruchus chinensis* and *C. maculatus* (Stevenson et al. 2007). Little progress in breeding for insect resistance was described by Muehlbauer et al. (1995), and compared to breeding for diseases, the situation continues to be similar nowadays. No sources of resistance to *Sitona* were detected, and ICARDA materials showed only to be effective against the adult insect while not to larvae (Stevenson et al. 2007). However, resistance to *S. crinitus* Herbst has been described in wild materials (El-Bouhssini et al. 2008). An extensive search on bruchid species infecting lentil was carried out in 10 years by the National Bureau of Plant Genetic Resources, New Delhi (Bhalla et al. 2004). A total of 2,517 imported lentil samples from 40 countries were X-ray screened for hidden insects in seeds. About 30 % of the samples were infected, and the species found consisted of *Bruchus ervi* Froelich, *B. lentis* Froelich, *B. tristiculurs* Fahraeus, *Callosobruchus analis* (Fabr.), *C. chinensis* (L.), and *C. maculatus* (Fabr.). After screening 571 accessions including wild species of the Spanish germplasm collection for susceptibility to seed bruchids under natural field conditions in Central Spain, a total of

32 accessions including cultivated landraces, *L. c.* ssp. *orientalis*, *L. nigricans*, and *L. lamottei* were selected as potential sources of resistance to Seed weevil (Laserna-Ruiz et al. 2012).

6.7 Traits of Importance for Base Broadening

6.7.1 Pod and Seed Characteristics

Reduced shattering is a trait of great economic value as pod dehiscence can cause significant losses before or during harvest; the trait is considered to have played a major role in lentil domestication. Pod dehiscence was found to be completely dominant over indehiscence and was assigned the gene symbol *Pi* (Ladizinsky 1979; Vaillancourt and Slinkard 1992); nonetheless, response to selection for pod indehiscence indicates the presence of quantitative variation (Erskine 1985), which was confirmed by quantitative trait loci (QTL) analysis that determined one major recessive QTL and two minor dominant QTLs which together accounted for 81 % of the observed variation (Fratini et al. 2007). Pod dehiscence in segregating populations shows several degrees between completely dehiscent and non-dehiscent plants, therefore classified since 1985 by the IBPGR/ICARDA lentil descriptor into four classes on a scale from 0 to 9 when scored 1 week after maturity.

Seed size is an important economic trait with special attributes in lentil consumption and trade, considering that wholesale prices of small-seeded (*microsperma*) and large-seeded (*macrosperma*) varieties differ by a large margin depending on consumer preferences and farmers' choice. Cultivated lentils are divided on the basis of seed size differences into *microsperma* (<4.5 mm) and *macrosperma* (>4.5 mm and sometimes over 7 mm). Seed size has a continuous distri bution in F_2 progenies following crossing of large- and small-seeded types (Ladizinsky 1979), and at least two major additive QTLs and one minor dominant QTL have been described to determine seed size variation (Fratini et al. 2007). Recombinant inbred populations (RILs) derived from the interspecific cross with *L. ervoides* (Fiala et al. 2009) revealed transgressive segregants with an 8 % increase in seed size (Tullu et al. 2011). With regard to seed mass, QTL analysis has shown a polygenic control with partial dominance for low seed weight alleles (Abbo et al. 1991), which has also been confirmed in a second study which describes two recessive and one additive QTL associated to low seed weight (Fratini et al. 2007).

Imbibition is one of the first steps in breaking seed dormancy to initiate seed germination; dormancy and viability can be maintained over long periods of time in hard-seeded legumes because seed coats are impermeable to water, some legume seeds being capable of remaining viable for more than 100 years (Rolston 1978). This is important ecologically in wild populations and economically in cultivated legumes, since seed coat hardness conditions cooking and palatability. Typically, wild *Lens* accessions are hard seeded, whereas cultivated varieties are not. Seed hardness is an undesired agronomic character, when lentil was domesticated from wild populations seed hardness most probably was selected against – especially considering that among pulse crops, lentil stands out for their fast imbibition and cooking time. Reports on the inheritance of seed hardness in lentil are highly conflicting, particularly considering that the quantitative nature of the trait has been thoroughly demonstrated since 1990 in the closely related grain legume *Glycine max.* (Keim et al. 1990); however, in lentil Ladizinsky (1985) intially described a single recessive gene in crosses with ssp. orientalis compared to a single dominant gene in crosses with L. ervoides, nonetheless thereafter it has persistently been stated to be a monogenic trait and seed hardness was designated the gene symbol *Hsc* which in addition is assured to be linked to the *Pi* locus for pod dehiscence (Sharma 2011).

Lentil is a rich source of dietary proteins among crops that are consumed without industrial processing. Furthermore, lentil is possibly the richest source of proteins among edible pulses that are cooked and directly consumed without prior processing for quality alterations. The usual protein content in dry lentil is around 26 %, although a germplasm range from 20.4 to 29.8 % has been reported (Sharma 2011). Studies to analyze the inheritance of seed protein concentration in lentil (Hamdi et al. 1991; Chauhan and Singh 1995; Tyagi and Sharma 1995) have revealed the quantitative nature of this trait and a nonsignificant correlation with grain yield and seed size (Hamdi et al. 1991; Tyagi and Sharma 1995). According to Sharma (2011), in most crops, especially cereals, protein content is invariably negatively correlated to seed size; however, lentil is possibly an exception and protein content has been claimed to be positively (although mildly) correlated to seed size.

6.7.2 Flowering and Plant Architecture

Flowering time is particularly important for adaptation and yield; it determines the length of the vegetative phase and conditions crop exposure during reproductive growth to climatic settings. Selection for response to specific regional balances between photoperiod and temperature for the onset of flowering has played a vital role in the adaptation of lentil to different regions and conditions around the globe (Erskine et al. 1990). Earliness is a desirable trait ensuring completion of the crop cycle in a relatively short period of time, thereby making more efficient use of resources as well as avoiding losses due to high temperatures during crop maturation. The inheritance of flowering time was described as monogenic, with earliness being recessive, and the gene symbol *Sn* was assigned (Sarker et al. 1999); in spite of the authors also observing transgressive segregation for early flowering, they considered the transgressive phenotypes to be a consequence of the interaction between the recessive *Sn* locus and a polygenic system of minor earliness genes.

Continuous polygenic variation had already been described in another analysis regarding flowering response in crosses between early maturing *microsperma* varieties from India and an early maturing *macrosperma* variety from Latin America (Tyagi and Sharma 1989). Transgressive segregation was observed for these crosses hinting strongly to the quantitative nature of flowering time. As crosses between *microsperma* varieties did not produce transgressive flowering segregants, they concluded that *microsperma* and *macrosperma* lentils possess different sets of genes controlling flowering time, while the Indian *microsperma* genotypes all share the same gene pool with regard to flowering response. Finally, three QTLs that accounted for more than 90 % of the observed variation in flowering response were detected, one was a major recessive gene while the remaining two were minor and dominant (Fratini et al. 2007). As the *Sn* locus was shown to be linked to *Scp* (seed coat pattern) and the latter major recessive QTL was located in the linkage group (LG) containing *Gs* (green stem), it seems likely that flowering response is under a complex genetic control with several qualitative and quantitative genes segregating differentially in diverse crosses. As a matter of fact, Tullu et al. (2008) reported two QTLs affecting earliness which accounted for 37–46 % of the variation observed in number of days to flower. Recently, a functional SSR marker in the *TFL1* gene related to earliness has been described (de la Puente et al. 2013).

Plant structure holds many implications regarding yield and ease of harvesting. Incomplete dominance of a bushy/erect growth habit was reported and the gene symbol *Gh* proposed (Ladizinsky 1979), later, the recessive *Ert* gene was discovered by Emami and Sharma (1999) and mapped in the LG containing *Gs* together with *Rdp* (red pod) and *Bl* (brown leaf). The erect phenotype *Ert* is recessive to the most prevalent growth habit in cultivated lentil in which plants remain procumbent for about a month, and thereafter branches grow in a semi-erect or semi-spreading fashion, while the wild *Lens* species retain prostrate stems much longer. Although according to Sharma (2011) there is a gradation between spreading and completely erect growth habit among the cultivated lentil varieties, genotypes with spreading growth habit can be grouped into several categories from highly prostrate to more upright; the spreading genotypes are generally endowed with profuse branching (basal as well as secondary). With regard to other plant structure attributes, QTL analysis has revealed that the number of branches at the first node is mainly controlled by a dominant quantitative locus together with two minor loci of opposing effects which together explain 92 % of the observed variation. Two additive QTLs explained one-third of the variance for the height of the first node. Finally, one recessive and one dominant QTL jointly explained half of the variation encountered for the total number of branches (Fratini et al. 2007).

Plant height is strongly associated to yield index. Plants of erect and a very tall habit tend to lodge as they approach maturity; thus, erect varieties of medium height should be expected to be more lodging resistant, which results in higher yields and also to a more amenable mechanical harvesting. Tallness in plants has almost universally been shown to be dominant over dwarfness since Mendel's times. In lentil, the gene *Ph* for plant height was first reported (Tahir et al. 1994) and found to be dominant over dwarfness. *Ph* is located in an LG comprising eight morphological (Kumar et al. 2005) and more than a dozen isozyme markers (Tahir et al. 1994). In addition, plant height was also found to be quantitatively inherited (Haddad et al. 1982), while another study described one additive QTL and two recessive QTLs, but the three QTLs explained only 38 % of the observed variation in plant height (Fratini et al. 2007); likewise, five QTL for plant height accounted for 31–40 % of the observed variation (Tullu et al. 2008).

Flower number per peduncle in lentil may retain significance in relation to productivity potential. This agronomic trait is known as prolificacy and is defined as the ability of a genotype to produce many flowers and ultimately pods on each peduncle (Sharma 2011). Genetic analysis is complicated because of an inconsistent expression of the trait which is highly influenced by environmental conditions and declines as plants get older. Nonetheless, monogenic inheritance with the two-flower phenotype dominant over the three flowered was initially described (Gill and Malhotra 1980), followed by contrasting results from different studies that concluded that a higher flower number was dominant over a low flower number per peduncle (Sharma 2009).

It can be concluded from this section that contrasting results have been obtained for several morphological markers; some studies considering monogenic inheritance while other studies have determined quantitative variation. These contrasting results are not unexpected since many characters are controlled by both qualitative (e.g., dwarf plants insensible to phytohormones) and quantitative genes (generating variation within dwarf and "normal" plants). Anyway, an extensive revision of the status of several morphological traits is necessary. A very comprehensive list of the gene symbols used for all of the lentil traits of simple monogenic inheritance described so far, some of which are oligogenic instead of being strictly monogenic, together with the different linkage studies in which they have been used can be found in Pérez de la Vega et al. (2012); the list also provides references of linked markers to biotic disease resistance loci and to the abiotic radiation-frost tolerance gene.

6.7.3 Abiotic Stresses

Tolerance to frost injury is an essential requirement when lentils are sown during the winter and cultivated in cooler climates. Monogenic inheritance of radiation-frost tolerance was reported and assigned gene symbol *Frt*; the gene was also tagged with a random amplified polymorphic DNA (RAPD) marker at a distance of 9.1 cm (Eujayl et al. 1999). On the other hand, it has been concluded that winter hardiness in lentil is a polygenic trait, and several QTLs together accounted for 42 % of the variation observed in recombinant inbred lines (RILs) (Kahraman et al. 2004a, b). At least four QTLs were detected under controlled frost conditions and field conditions in two RIL populations; furthermore, two QTLs related to frost response were also related to yield under winter-sown conditions (Barrios et al. 2007); QTLs with a major effect for winter hardiness and yield seem to be closely located within a single linkage group, and some molecular markers are potentially useful for their tracking (Barrios 2012). SuperSAGE expression profiling in response to some biotic and abiotic stresses has been carried out within the LEGRESIST and other projects, for instance, lentil SuperSAGE genomic analysis was used to analyze the allele-specific differential expression of transcripts potentially involved in frost tolerance by bulk segregant analysis among 90 F_9 RILs derived from the Precoz × WA8649041 lentil cross (Barrios et al. 2010).

Lentil has been traditionally grown in semiarid regions under rainfed conditions; thus it combines a high degree of drought resistance and a low water requirement; in fact, excessive water supply is damaging to the crop. The genetics of drought tolerance or waterlogging are still to be explored.

6.7.4 Biotic Stresses

Most genetic studies regarding rust (*Uromyces fabae*) resistance in lentil have reported that resistance is under a monogenic control, resistance being dominant over susceptibility. However, reports of incomplete resistance (Negussie et al. 2005), as well as duplicate dominant genes controlling resistance, have frequently emerged; only one unconfirmed report has suggested rust resistance to be a recessive trait (see Sharma 2009 for a review). Furthermore, it has been observed that the dominant gene for resistance of a *macrosperma* variety from Latin America differs from that of an Indian *microsperma* variety (Sharma 2009, 2011). Therefore, it seems that at least two separate genes controlling rust resistance have evolved in spatially and temporally isolated lentil groups. Gene symbols *urf1*, *urf2*, and the unconfirmed *urf3* have been proposed (Sharma 2009). *Uromyces*-resistant cultivars, such as Laird or Calpun-INIA derived from Laird, have been described (Penazola et al. 2007); in addition, Assano is a cultivar highly resistant to *Uromyces* with moderate resistance to *Ascochyta* blight and to the wilt-root rot complex (Fikru et al. 2007).

Resistance to *Fusarium* wilt was described in a germplasm collection from India (Kamboj et al. 1990). Allelism tests concluded that *Fusarium* resistance in lentil is conferred by five dominant genes (Kamboj et al. 1990). However, subsequently only one dominant gene (*Fw*) for wilt resistance was reported. It was tagged with a RAPD marker at 10.8 cm (Eujayl et al. 1998b); a microsatellite marker and an AFLP marker were further linked to the *Fw* locus at distances of 8.0 and 3.5 cm, respectively (Hamwieh et al. 2005). Moreover, a 3-year screening of lentil germplasm at the ICARDA yielded a total of 34 strains confirmed for resistance to *Fusarium* wilt. Evaluations from the F_5 to F_8 successfully identified 753 resistant lines. Among the small-seeded lines, 72 % were wilt resistant compared with 41 % of the large-seeded lines, suggesting that genes for small seed size might be loosely associated with genes for wilt resistance (Sarker et al. 2004).

Resistance to *Ascochyta* blight, screening techniques, and breeding methods have been reported (Bayaa et al. 1994; Ye et al. 2002; Muehlbauer and Chen 2007; Tivoli et al. 2006). The genetic control of *Ascochyta* blight resistance was described to be monogenic recessive in spite of having observed transgressive segregations (Tay and Slinkard 1989). However, two complementary dominant genes were further described in a cross between *L. ervoides* and *L. odemensis* (Ahmad et al. 1997), while only one dominant gene was found in crosses between *L. culinaris* accessions (Ahmad et al. 1997). The existence of two complementary dominant genes within cultivated lentils was thereafter established (Nguen et al. 2001). Two flanking RAPD markers at distances of 8.0 and 3.5 cm from the designated resistance locus *Ral1* (*Abr1*) have been mapped (Ford et al. 1999); likewise, two additional RAPD markers have been located in flanking positions of the recessive gene for resistance *ral1* at distances of 6.4 and 10.5 cm (Chowdhury et al. 2001). The first report on the use of the microarray technology to study gene expression in lentil was recently published (Mustafa et al. 2009). Highly resistant (ILL7537) and highly susceptible (ILL6002) lentil germplasm accessions were inoculated with *Ascochyta lentis*. Ninety genes were differentially expressed in ILL7537, and 95 genes were differentially expressed in ILL6002. The expression profiles of the two accessions showed substantial difference in the type of genes and the time of expression in response to the pathogen. The resistant variety showed early upregulation of *PR4* and 10 proteins and other defense-related genes. The susceptible genotype showed early downregulation of defense-related genes. Real-time PCR (RT-PCR) was used to verify microarray expression ratios. The resistant and susceptible lentil accessions differed not only in the type of genes expressed but also in the time and level of expression in response to *A. lentis* inoculation (Mustafa et al. 2009).

Tullu et al. (2006) searched for sources of resistance to anthracnose caused by *Colletotrichum truncatum* in cultivated and wild materials and found resistant cultivars (Indianhead and PI 320937) to race Ct1 but none to race Ct0. Some accessions of wild *Lens* species are potential sources of genes for resistance to both the races. Transfer of resistance to both races, Ct1 and Ct0, from *L. ervoides* accession L-01-827A to the cultivar Eston through embryo rescue followed by single-seed descent from F_2 individuals to $F_{7:8}$ RILs was described (Fiala et al. 2009); authors indicated that resistance to race Ct1 and race Ct0 may be conferred by two recessive genes. Resistance to Ct1 race was reported to be under the control of one recessive gene (*lct-1*) in one cultivar, while two dominant genes designated *LCt-2* and *LCt-3* were respectively responsible for resistance in two additional cultivars (Buchwaldt et al. 2004). The *LCt-2* resistance locus was tagged with two flanking RAPD markers at 6.4 and 10.5 cm (Tullu et al. 2003); in addition, three AFLP markers were also found to be linked to this locus (Tar'an et al. 2003). However, most recently Buchwaldt et al. (2013) have established that resistance to Ct1 race is controlled by two recessive genes (*lct-1* and *lct-2*) and three closely linked dominant genes (*LCt-3*, *LCt-4* and *LCt-5*).

Pea seed-borne mosaic virus (PSbMV) is a major disease in lentil transmitted through seed as well as aphids. Monogenic recessive inheritance of viral immunity has been confirmed in four crosses, and the gene symbol *sbv* was proposed to denote PsbMV resistance in lentil (Haddad et al. 1978). Pakistan accession 66013-6 was found to have a high level of resistance to blight, rust, and pea seed-borne mosaic virus (Hussain et al. 2008).

6.8 Interspecific Hybridization in Crop Species

Artificial cross-pollination in a highly self-pollinated crop species, such as lentil, is important to increase genetic variability. Wide crosses to yield interspecific hybrids allow for the introgression of important alleles of agricultural interest from wild species to cultivars, as, for instance, the resistance or tolerance to abiotic and biotic stresses (Erskine et al. 1994; Ocampo et al. 2000; Davis et al. 2007; Cubero et al. 2009; Pérez de la Vega et al. 2012). Lentil is generally described as a strictly self-pollinated species holding cleistogamous flowers (Wilson 1972; Kumar and Singh 1998); nonetheless, recent studies point out that natural outcrossing rates of lentil can be relatively high (Horneburg 2006). Lentil flowers are complete with a typical structure of the subfamily Papilionaceae of the Leguminosae family (Muehlbauer et al. 1980), and as it was previously mentioned, emasculation and artificial crossing in lentil is a demanding task. In the case of interspecific hybridization, crossing efficiencies are much lower than intraspecific ones, for instance, out of a total of 1,707 pollinations, six interspecific hybrids with *L. odemensis*, two with *L. nigricans*, and one with *L. ervoides* were recovered using embryo rescue (Fratini and Ruiz 2006). Therefore, a large number of manual pollinations are needed to recover each single interspecific hybrid.

As has already been discussed previously, intraspecific crosses between cultivated lentils produce viable descendants (Malhorta et al. 1978; Solh et al. 1980; Mera and Erskine 1982; Kumar and Singh 1998; Fratini et al. 2004; Singh et al. 2013b). With regard to inter-subspecific hybrids of lentil, it has been reported that the domesticated lentil is readily crossable with subspecies *orientalis* (Ladizinsky 1979; Muehlbauer et al. 1989; Vandenberg and Slinkard 1989; Vaillancourt and Slinkard 1992; Fratini et al. 2004), although the fertility of the hybrids depends on the chromosome arrangement of the wild parent (Ladizinsky 1979; Ladizinsky et al. 1984).

Lens odemensis produces partially fertile hybrids with both subspecies of *L. culinaris* depending on the accessions used (Cubero et al. 2009). Interspecific embryos between cultivated lentils and either *L. ervoides* or *L. nigricans* abort (Abbo and Ladizinsky 1991, 1994; van Oss et al. 1997; Fratini and Ruiz 2006) and embryo rescue techniques are necessary to recover hybrids (Cohen et al. 1984; Ladizinsky et al. 1985; Fratini and Ruiz 2006; Fiala et al. 2009). Nonetheless, gibberellic acid (GA_3) application after pollination aided to develop viable pods and interspecific *Lens* hybrids were obtained without the need of in vitro embryo culture (Ahmad et al. 1995). To enhance the gene pool and the hybridization range of lentil, hand-pollination followed by embryo rescue is in most cases the only practical method to recover interspecific hybrids of *Lens*; thus, it is necessary to dispose of an efficient embryo rescue protocol.

6.9 Barriers to Interspecific Hybridization

Pollination followed by fertilization normally leads to the production of an embryo which in the intact plant is linked with normal seed development. Crossability was defined by Ladizinsky (1992) as the potential for intercrossing individuals belonging to the same or different taxa to produce embryos or seeds that can give rise to an F_1 plant. Crossability is either limited by incompatibility or by incongruity barriers; the sexual barriers that belong to the second aspect have been further divided into pre- and postfertilization barriers (Stebbins 1958). Part of the postfertilization barriers may be overcome by using in vitro embryo rescue methods (Yeung et al. 1981; Fratini and Ruiz 2011), although depending on plant species, the process can entail the culture of ovaries immediately after pollination and/or in ovulo embryo culture and/or embryo culture. Ladizinsky (1992) explained that the success in lentil crosses depends on the interaction between the parental genomes in the hybrid zygote, embryo, and endosperm and between the hybrid tissue and the surrounding maternal tissue. The crossability between the cultivated lentils and the wild *Lens* relatives has been described to be hampered by pre- as well as by postfertilization barriers (Abbo and Ladizinsky 1991, 1994; Ladizinsky 1993; Ladizinsky and Abbo 1993; van Oss et al. 1997). It is now clear that epigenetic processes during embryo-endosperm development can also influence crossability at a post-zygotic level (Law and Jacobsen 2010).

Cohen et al. (1984) and Ladizinsky et al. (1985) were the first to describe hybrid embryo abortion after interspecific hybridization in lentil. Interspecific hybridization between the cultivated lentil and *L. ervoides* and *L. nigricans* resulted initially in pod development followed by an arrest 10–16 days after pollination with a consequent production of shriveled and nonviable seeds. This postfertilization barrier has prevented gene flow from the wild *L. ervoides* and *L. nigricans* species into the cultivated lentil, hindering the use of these wild forms for breeding purposes. The anatomical aspects of hybrid embryo abortion in the genus *Lens* were described by Abbo and Ladizinsky (1991); more recently Fratini and Ruiz (2006) also have provided a description of the appearance and frequency of aborted pods and ovules 14 days after pollination. So far the only study which describes the genetic aspects leading to embryo abortion of wide *Lens* hybridizations was carried out by Abbo and Ladizinsky (1994).

6.10 Conventional and Contemporary Approaches of Interspecific Gene Transfer

The first lentil embryo rescue protocol published (Cohen et al. 1984) allowed to recover interspecific hybrids between the cultivated lentil and *L. ervoides* and *L. nigricans*. This method consisted of a two-step culture procedure to recover interspecific hybrids: first, ovule-embryos were collected 2 weeks after pollination and cultured on MS medium supplemented with 10 % sucrose and 0.2 mg/l IAA + 0.5 mg/l ZEA + 0.5 mg/l GA_3 and 0.9 % agar. After 1 week in culture, embryos were excised from the ovular integuments and placed on a second MS medium, supplemented with 3 % sucrose and 0.2 mg/l IAA+0.2 mg/l ZEA and 0.9 % agar. Later, using the same embryo culture technique, Ladizinsky et al. (1985) again obtained hybrids of the cultivated lentil with *L. ervoides*.

On the other hand, viable interspecific hybrids were obtained without the use of embryo rescue between lentil cultivars of New Zealand crossed with *L. odemensis*, *L. ervoides*, and *L. nigricans*, applying GA_3 after pollination (Ahmad et al. 1995). Fratini and Ruiz (2006) in a first hybridization experiment carried out 630 pollinations with *L. odemensis* and *L. nigricans*, performing no embryo rescue, and no hybrid seed was recovered; in a second hybridization experiment, 218 pollinations were carried out with *L. odemensis*, *L. nigricans*, and *L. ervoides*, applying GA_3 after fertilization, increasing the percentage of ovules obtained; however, ovules started to turn brown and pods initiated to dry out 18–21 days after pollination (DAP). When ovules of both hybridization experiments were dissected, either no embryos were located or the embryos had dried out. This outcome is in disagreement with the results of Ahmad et al. (1995); nonetheless, it has been described in the literature that certain cultivars and accessions of a species are better combiners to recover interspecific hybrids than other cross combinations (Pitarelli and Stavely 1975; Harlan and de Wet 1977; Williams 1987; Pickersgill 1993). Based on the results of interspecific crosses of Spanish lentil cultivars, Fratini and Ruiz (2006) concluded that embryo rescue is necessary to obtain lentil interspecific hybrids, as had been previously indicated by other authors (Cohen et al. 1984; Ladizinsky et al. 1985; Abbo and Ladizinsky 1991, 1994).

Gupta and Sharma (2005) used a two-step in vitro method of embryo rescue analogous to Cohen et al. (1984) to overcome the postfertilization interspecific barrier. The hybrid embryos developed multiple shoots which subsequently did not allow induction of root differentiation to recover whole plants. Fratini and Ruiz (2006) applying the rescue protocol of Cohen et al. (1984) observed similar multiple shoot results recovering a single hybrid with *L. odemensis* out of 1,350 hybridizations. In view of the above, Fratini and Ruiz (2006) developed an interspecific embryo rescue protocol suited for Spanish lentil cultivars consisting of four different stages: (1) in ovule-embryo culture, (2) embryo culture, (3) plantlet development, and finally, (4) the gradual habituation to ex vitro conditions of the recovered interspecific hybrid plantlets (see also Fratini and Ruiz 2011). Ovule-embryos of 18 DAP were cultured on MS salts with 1 % sucrose and 1 μM IAA+0.8 μM KN; after 2 weeks, embryos were released from the ovular integuments and cultured on the same medium for another 2 weeks in upright position. Afterward, the embryos were transferred to test tubes containing the same medium, and 1 month later interspecific plantlets were obtained (Fratini and Ruiz 2006). Using this rescue protocol, the efficiency of interspecific hybrids with *L. odemensis*, *L. nigricans*, and *L. ervoides* recovered per number of ovules cultured ranged between 50 and 100 % (Fratini and Ruiz 2006). Authors noted that interspecific hybrids were always obtained in the spring season with increasing daylight hours and an average temperature of 20°C, albeit an approximate equal number of crosses were carried out in the fall under nearly the same environmental conditions.

The main differences between the embryo rescue protocol of Cohen et al. (1984) and that developed by Fratini and Ruiz (2006) comprise the number of culture media used and the amount of carbohydrate added to the media. Whereas in the case of Cohen et al. (1984) two different media were used for ovule and embryo culture supplemented respectively with 10 and 3 % of sucrose, the single medium used by Fratini and Ruiz (2006) for ovule, embryo, and plantlet culture contained only 1 % sucrose. Embryo rescue protocols in other legumes have also used sucrose concentrations lower than 3 %, such as 2.5 % for *Medicago* (McCoy and Smith 1986), 2 % for *Vicia faba* and *Vicia narbonensis* (Lazaridou et al. 1993), and 2 % for *Phaseolus* (Mejia-Jimenez et al. 1994). With respect to phytohormones, the embryo rescue procedure of Cohen et al. (1984) compared to that of Fratini and Ruiz (2006) contains approximately equivalent concentrations of auxins and cytokinins, even though ZEA (Cohen et al. 1984) and KN (Fratini and Ruiz 2006) were respectively used.

The most recent embryo rescue technique based on Cohen et al. (1984) for interspecies transfer of anthracnose resistance from *L. ervoides* to cultivated lentils allowed to recover only one F_1 hybrid (Fiala et al. 2009). Some of the interspecific hybrids recovered by Fratini and Ruiz (2006) were cloned in vitro in order to increase the number of segregating descendant seeds obtained to be used for breeding and genetic analysis purposes (Fratini and Ruiz 2008). Contemporary approaches to interspecific gene transfer have been reviewed by Davis et al. (2007) and Fratini and Ruiz (2011).

6.11 Molecular Markers, Genome Mapping, and Genomics as an Adjunct to Breeding

Lentil is a relatively minor crop compared to common bean, pea, and chickpea; as a result, genomic information regarding lentil is still limited by a relatively large genome size together with scarce information available on gene sequences, constituting a major obstacle to undertake genomic studies in lentil. So far for lentil, there exist no descriptions of a bacterial artificial chromosome (BAC) library, BAC-end sequence, or a physical map (Ford et al. 2007, 2009; Varshney et al. 2009; Pérez de la Vega et al. 2012). In comparison to other crop species, the number of *Lens* data indexed continues to be scarce, although this situation is rapidly changing, in particular for nucleotide-expressed sequence tags (ESTs) (Kaur et al. 2011; Bhadauria et al. 2013; Sharpe et al. 2013; Verma et al. 2013). Reports regarding the characterization of the nucleotide binding site-leucine-rich repeat (NBS-LRR) disease resistance genes of several legume species, including lentils, and on transcriptome sequencing of cool season food legumes using 454 GS-FLX Titanium technology (Penmetsa et al. 2010; Kaur et al. 2010) also reveal a change of tendency.

The first report on the use of the microarray technology to study gene expression in lentil was published by Mustafa et al. (2009). The highly resistant (ILL7537) and highly susceptible (ILL6002) lentil germplasm accessions were inoculated with *Ascochyta lentis*; a total of 90 genes were differentially expressed in ILL7537 compared to 95 genes differentially expressed in ILL6002. The expression profiles of the two accessions showed substantial difference in the type of genes and the times of expression in response to the pathogen, whereas the resistant variety showed an early upregulation of PR4 and ten proteins plus other defense-related genes; the susceptible genotype showed an early downregulation of defense-related genes. Real-time PCR (RT-PCR) was used to verify the microarray expression ratios. The resistant and susceptible lentil accessions differed not only in the type of genes expressed but also in the time and level of expression in response to *A. lentis* inoculation.

Lentil proteomics is also still underdeveloped. Two-dimensional electrophoretic maps of proteins from mature lentil seed have been described (Scippa et al. 2008, 2010). Among the hundreds of protein species, 122 were further identified by MALDI-TOF PMF and/or LC-ESI-LIT-MS/MS. Map comparisons revealed that 103 protein spots were differentially expressed within and between populations (Scippa et al. 2010).

Comparative mapping was able to show extensive macrosynteny between *Medicago truncatula* and *L. culinaris* (Phan et al. 2007). These authors generated the first gene-based genetic linkage map of lentil (*L. culinaris* ssp. *culinaris*) using an F_5 population derived from a cross between cultivar Digger (ILL5722) and Northfield (ILL5588) composed of 79 ITAP and 18 genomic SSR markers. Evidence for the macrosynteny between lentil and *M. truncatula* was found as 66 out of 71 gene-based markers, which had previously been assigned to the *M. truncatula* genetic and physical maps, were found in regions syntenic between the *L. c.* ssp. *culinaris* and *M. truncatula* genomes. However, moderate evidence of chromosomal rearrangements was found, which may account for the difference in chromosome numbers between these two legume species. Comparative mapping showed a direct and simple relationship between the *M. truncatula* and

Table 6.1 Molecular markers closely linked to desirable lentil breeding characteristics for use in marker-assisted selection

Trait mapped	Linked molecular markers	Reference
Fusarium wilt resistance (*Fw*)	OPK15	Eujayl et al. (1998b)
Ascochyta blight resistance (*AbR*1)	RV01, RB18, SCARW19	Ford et al. (1999)
Frost tolerance (*Frt*)	OPS-16	Eujayl et al. (1999)
Ascochyta blight resistance (*ral*2)	UBC227, OPD-10	Chowdhury et al. (2001)
Anthracnose resistance (*Lct2*)	OPE06, UBC704	Tullu et al. (2003)
Winter hardiness	UBC808-12	Kahraman et al. (2004b)
Fusarium wilt resistance (*Fw*)	SSR59-2B, p17m30710	Hamwieh et al. (2005)
Ascochyta blight resistance (mapped as a QTL)	C-TTA/M-AC (QTL1 and QTL2), M20 (QTL3)	Rubeena et al. (2006)
Stemphylium resistance	SRAPME5XR10 and ME4XR16c	Saha et al. (2010)
Vegetative to flowering (*TFL1a*)	SSR-TFL1	de la Puente et al. (2013)

L. culinaris ssp. *culinaris* chromosomes, with complete homology evident between LGs. Thus, lentil LG 1, 2, 3, 4, 5, 6, and 7 were proposed to be syntenic to *M. truncatula* LG 4+7, 8, 1+6, 2, 5, 3, and 3, respectively; next, considering lentil or pea LGs syntenic to the same *M. truncatula* or *M. sativa* LGs, lentil LG 1, 2, 3, 4, 5, 6, and 7 were proposed to be syntenic to pea LG VII+V, IV, II+III+VI, III+IV, I, III, and III, respectively (see Pérez de la Vega et al. 2012).

Eighteen common SSR markers were used by Phan et al. (2007) to connect their map to another comprehensive map of lentil (Hamwieh et al. 2005), providing a syntenic context for four important domestication traits. Ellwood et al. (2008) used 151 markers out of 796 ITAP markers screened to construct a comparative genetic map between *L. culinaris* and *Vicia faba*. A simple and direct macrosyntenic relationship between faba bean and *M. truncatula* was evident, while faba bean and lentil shared a common rearrangement relative to *M. truncatula*. This study also provided a preliminary indication for a finer scale macrosynteny between *M. truncatula*, lentil, and faba bean. Markers originally designed from genes on the same *M. truncatula* bacterial artificial chromosomes (BACs) were found to be grouped together in corresponding syntenic areas in lentil and faba bean. Previous research had already demonstrated marker and/or gene order conservation between species closely related to lentil such as *Pisum sativum* and *Medicago sativa* (Kalo et al. 2004), *P. sativum* and *M. truncatula* (Aubert et al. 2006), and *P. sativum*, *M. truncatula*, *M. sativa*, *Vigna radiata*, *Glycine max*, and *Phaseolus vulgaris* (Choi et al. 2004).

Tar'an et al. (2003) using 156 RILs were able to pyramid into lentil breeding lines two genes for resistance to *Ascochyta* blight caused by *A. lentis,* together with the gene for resistance to anthracnose caused by *Colletotrichum truncatum.* Genetic analysis of the disease reactions demonstrated that the observed segregation ratios of resistant versus susceptible fitted a two gene model for resistance to *Ascochyta* blight and a single gene model for resistance to anthracnose. Using molecular markers linked to the two genes conferring resistance to *A. lentis* (UBC 2271290) to *ral*1 and (RB18680) to *AbR1*, in conjunction with the major gene for resistance to anthracnose (OPO61250), they found 11 RILs that retained all of the three resistance genes providing a practical evidence of progress toward pyramiding resistance in a lentil population. Table 6.1 displays other molecular markers linked to agronomic traits of interest useful for marker-assisted selection; most recently a polymorphic microsatellite in the first intron of the *TFL1a* gene involved in the transition from vegetative to flowering stages was detected (de la Puente et al. 2013).

Detailed information regarding molecular markers, genome mapping, and genomics as an adjunct to lentil breeding can be found in Ford et al. (2007, 2009), and more recently Pérez de la Vega et al. (2012). Lastly, the development of a deep and diverse transcriptome resource for lentil using next-generation sequencing technology allowed to generate data in multiple-cultivated (*L. culinaris*) and wild (*L. ervoides*) genotypes, which together with the use of a bioinformatics workflow enabled for the identification of a large collection of SNPs and SSR markers for the subsequent development of a genotyping platform that was used to establish the first comprehensive genetic map of the *L. culinaris* genome, comprising seven linkage groups corresponding to the number of chromosome pairs of lentil (Sharpe et al. 2013). Extensive collinearity with *M. truncatula* was evident on the basis of sequence homology between mapped markers and the model genome, and large translocations and inversions relative to *M. truncatula* were identified. Synonymous changes that were only observed between *L. culinaris* and *L. ervoides* enabled to estimate the time of divergence to 677,000 years ago, whereas synonymous changes observed between *L. culinaris* and *M. truncatula* led to estimate divergence to 38 million years ago (Sharpe et al. 2013).

6.12 Conclusions

The exact taxonomy of the genus *Lens* still needs to be established, especially with regard to *L. tomentosus* and *L. lamottei*; further hybridization experiments need to be undertaken to establish if these wild species belong to the secondary or tertiary gene pool of *Lens*. With regard to wild germplasm, further recollections are needed as, for instance, only four *L. tomentosus* and 39 *L. odemensis* accessions are maintained in gene banks worldwide. More variable and valuable alleles of cultivated landraces must be conserved in ex situ collections for genotyping and phenotyping to discover new and useful variants. One example is the recent finding of a high genetic variability among accessions from Azerbaijan which suggests that this gene pool needs to be increased through additional accessions. Another example, Chinese land races are not represented in the ICARDA nor USDA core collections; however, evidence from other Chinese pulse landraces strongly indicates that collected Chinese landraces of lentil, from west and central China, will constitute a very interesting germplasm for trait and allelic variation assessment.

Wild and cultivated accessions need to be further evaluated with regard to biotic and abiotic stresses, placing emphasis on developing more accurate screening methods. In addition, further inheritance studies of resistance/tolerance to biotic and abiotic stresses are required for a better understanding of the respective genetic systems controlling these responses. This knowledge will be useful to design adequate breeding programs based on regional requirements. There is a need to include more morphological and molecular markers and to develop a comprehensive consensus genetic linkage map in lentil, allowing for molecular tagging of resistance genes against biotic and abiotic stresses in order to exploit them in breeding with increased selection efficiency. However, many morphological markers so far considered of monogenic inheritance need to be reassessed as they most likely better fit an oligogenic inheritance.

The development of fine genetic maps that include direct gene markers (functional markers) is expected to revolutionize the use of lentil genetic resources. Microsatellite markers have been developed and deployed to characterize composite and core collections at ICARDA; 109 accessions of lentil including *culinaris* (57), *orientalis* (30), *tomentosus* (4), and *odemensis* (18) were genotyped using 14 microsatellite (SSR) markers; the study of Hamwieh et al. (2009) revealed that the wild accessions were rich in alleles (151 alleles) compared to cultigens (114 alleles). New molecular tools will increase the speed and precision of introgression of these newly identified alleles from both the adapted and wild lentil species and subspecies into the advanced breeding populations. For example, a lentil pyrosequencing and SNP discovery project has recently been concluded at the University of Saskatchewan (Sharpe et al. 2013). The successful completion of this project has led to a dense linkage map of SSR and SNP in seven linkage groups constituting the seven chromosome pairs of lentil (Sharpe et al. 2013), thus allowing for a future great reduction of gene/QTL discovery timelines. The high throughput genotyping conducted by the CGIAR Challenge Program and other national programs will characterize the world's ex situ germplasm resources, leading to an understanding of the population structure from a statistical genetics perspective (Furman et al. 2009). This information combined with genome sequencing, SNP variation studies (haplotype mapping), and detailed phenotyping of the lentil germplasm will lead to successful genome-wide association studies. The understanding of allele values from the adapted and wild lentil gene pool will dramatically increase both the efficiency and worth of germplasm utilization in lentil breeding programs (Coyne et al. 2011).

References

Abbo S, Ladizinsky G (1991) Anatomical aspects of hybrid embryo abortion in the genus *Lens L.* Bot Gaz 152:316–320

Abbo S, Ladizinsky G (1994) Genetical aspects of hybrid embryo abortion in the genus *Lens L.* Heredity 72:193–200

Abbo S, Ladizinsky G, Weeden NF (1991) Genetic analysis and linkage study of seed weight in lentil. Euphytica 58:259–266

Ahmad M, Fautrier AG, McNeil DL, Burritt DJ, Hill GD (1995) Attempts to overcome postfertilization barrier in interspecific crosses of the genus *Lens*. Plant Breed 114:558–560

Ahmad M, Russell AC, McNeil DL (1997) Identification and genetic characterization of different resistance sources to *Ascochyta* blight within the genus *Lens*. Euphytica 97:311–315

Ali JAF, Shaikh NA (2007) Genetic exploitation of lentil through induced mutations. Pak J Bot 39:2379–2388

Ali A, Johnson DL, Stushnoff C (1999) Screening lentil (*Lens culinaris*) for cold hardiness under controlled conditions. J Agric Sci 133:313–319

Armstrong-Cho CL, Banniza S (2006) *Glomerella truncata* sp. nov, the teleomorph of *Colletotrichum truncatum*. Mycol Res 110:951–956

Arumuganathan K, Earle ED (1991) Nuclear DNA content of some important plant species. Plant Mol Biol Report 9:208–218

Ashraf M, Chishti SN (1993) Waterlogging tolerance of some accessions of lentil (*Lens culinaris* Medik.). Trop Agric 70:60–67

Ashraf M, Zafar ZU (1997) Effect of potassium on growth and some biochemical characteristics in two lines of lentil (*Lens culinaris* Medik.). Acta Physiol Plant 19:9–15

Attanayake RN, Glawe DA, Dugan FM, Chen W (2009) *Erysiphe trifolii* causing powdery mildew of lentil (*Lens culinaris*). Plant Dis 93:797–803

Aubert G, Morin J, Jacquin F, Loridon K, Quillet M, Petit A, Rameau C, Lejeune-Henaut I, Huguet T, Burstin J (2006) Functional mapping in pea, as an aid to the candidate gene selection and for investigating synteny with the model legume *Medicago truncatula*. Theor Appl Genet 112:1024–1041

Babayeva S, Akparov Z, Abbasov M, Mammadov A, Zaifizadeh M, Street K (2009) Diversity analysis of Central Asia and Caucasian lentil (*Lens culinaris* Medik.) germplasm using SSR fingerprinting. Genet Resour Crop Evol 56:293–298

Balyan HS, Houben A, Ahne R (2002) Karyotype analysis and physical mapping of 18S–5.8S–25S and a 5S ribosomal RNA loci in species of genus *Lens* Miller (Fabaceae). Caryologia 55:121–128

Barrios A (2012) Adaptación a la siembra invernal y tolerancia al frío en lenteja (*Lens culinaris* Medik.) mapeo de QTLs involucrados. Ph.D. Dissertation, Universidad de León. http://hdl.handle.net/10612/276

Barrios A, Kahraman A, Aparicio T, Rodríguez M, Mosquera P, García P, McPhee K, Pérez de la Vega M, Caminero C (2007) Preliminary identification of QTLS for winter hardiness, frost tolerance and other agronomic characters in lentil (*Lens culinaris* Medik.) for Castilla y León (SPAIN) region. In: Proceedings of the 6th European conference on gain legumes, Lisbon, 12–16 Nov 2007

Barrios A, Martin-Sanz A, Ramos S, Rotter B, Horres R, Plotner A, Kretzdorn N, Constantino Saldaña C, Perez de la Vega M, Winter P (2010) Allele-specific differential expression of transcripts potentially involved in cold tolerance QTL in lentil revealed by bulked extremes SuperTag digital gene expression (BE-STDGE) profiling and qRT-PCR. In: Proceedings of the 5th international food legumes research conference & 7th European conference on grain legumes, Antalya, 26–30 Apr 2010

Barulina H (1930) Lentils of the USSR and other countries. Bull Appl Bot Genet Plant Breed 40:265–304. (in Russian with English summary)

Bashir M, Azeem T, Ali A, Ijaz M (2005) Variation in lentil germplasm for reaction to virus infection. Pak J Bot 37:53–60

Bayaa B, Erskine W, Hamdi A (1994) Response of wild lentil to *Ascochyta fabae* f. sp. *lentis* from Syria. Genet Resour Crop Evol 41:61–65

Bayaa B, Erskine W, Singh M (1997) Screening lentil for resistance to fusarium wilt: methodology and sources of resistance. Euphytica 98:69–74

Bennet MD, Smith JB (1976) Nuclear DNA amounts in angiosperms. Phil Trans R Soc Lond B 274:227–274

Bhadauria V, Bett KE, Zhou T, Vandenberg A, Wei Y, Banniza S (2013) Identification of *Lens culinaris* defense genes responsive to the anthracnose pathogen *Colletotrichum truncatum*. BMC Genet 14:31

Bhalla S, Kapur ML, Singh C, Gupta K, Kumar N, Lal B (2004) Interception of bruchids in imported lentil (*Lens* spp) germplasm. Indian J Agric Sci 74:332–333

Buchwaldt L, Anderson KL, Morrall RAA, Gossen BD, Bernier CC (2004) Identification of lentil germplasm resistant to *Colletotrichum truncatum* and characterization of two pathogen races. Phytopathology 94: 236–243

Buchwaldt L, Rubeena S, Adam J, Tullu A, Slinkard AE (2013) Recessive and dominant genes confer resistance to *Colletotrichum truncatum* in cultivated lentil. Can J Plant Pathol 35:222–231

Chauhan MP, Singh IS (1995) Inheritance of protein content in lentil (*Lens culinaris* Medik.). Legum Res 18:5–8

Chen CC, Miller P, Muehlbauer F, Neill K, Wichman D, McPhee K (2006) Winter pea and lentil response to seeding date and micro- and macro-environments. Agron J 98:1655–1663

Cherif M, Chilvers MI, Akamatsu H, Peever TL, Kaiser WJ (2006) Cloning of the mating type locus from *Ascochyta lentis* (teleomorph: *Didymella lentis*) and development of a multiplex PCR mating assay for *Ascochyta* species. Curr Genet 50:203–215

Choi HK, Mun JH, Kim DJ, Zhu H, Baek JM, Mudge J, Roe B, Ellis N, Doyle J, Kiss GB, Young ND, Cook DR (2004) Estimating genome conservation between crop and model legume species. Proc Natl Acad Sci U S A 101:15289–15294

Chowdhury MA, Andrahennadi CP, Slinkard AE, Vandenberg A (2001) RAPD and SCAR markers for resistance to Ascochyta blight in lentil. Euphytica 118:331–337

Cohen D, Ladizinsky G, Ziv M, Muehlbauer FJ (1984) Rescue of interspecific *Lens* hybrids by means of embryo culture. Plant Cell Tissue Org Cult 3:343–347

Coutts BA, Prince RT, Jones RAC (2008) Further studies on Pea seed-borne mosaic virus in cool-season crop legumes: responses to infection and seed quality defects. Aust J Agric Res 59:1130–1145

Coyne CJ, McGee RJ, Redden R (2011) Lentil germplasm: a basis for improvement. Grain Legum 57:11–12

Cubero JI, Pérez de la Vega M, Fratini R (2009) Origin, phylogeny, domestication and spread. In: Erskine W, Muehlbauer F, Sarker A, Sharma B (eds) The lentil: botany, production and uses. CABI, Wallingford

Davidson JA, Kimber RBE (2007) Integrated disease management of ascochyta blight in pulse crops. Eur J Plant Pathol 119:99–110

Davis PA, Lülsdorf MM, Ahmad M (2007) Wild relatives and biotechnological approaches. In: Yadav SS, McNeil D, Stevenson PC (eds) Lentil: an ancient crop for modern times. Springer, Dordrecht

de la Puente R (2012) Mapa genético y análisis de secuencias nucleotídicas de interés en *Lens*. Ph.D. Dissertation, Universidad de León. http://hdl.handle.net/10612/2328

de la Puente R, García P, Polanco C, Pérez de la Vega M (2013) Short communication. An improved intersubspecific genetic map in *Lens* including functional markers. Span J Agric Res 11:132–136

Dixit P, Dubey DK (1986) Alkylating agent induced chromosomal aberrations in lentil (*Lens culinaris* Med. Var T 36). Cytologia 51:325–331

Durán Y, Pérez de la Vega M (2004) Assessment of genetic variation and species relationships in a collection of *Lens* using RAPD and ISSR. Span J Agric Res 4:538–544

El-Bouhssini M, Sarker A, Erskine W, Joubi A (2008) First sources of resistance to Sitona weevil (*Sitona crinitus* Herbst) in wild *Lens* species. Genet Resour Crop Evol 55:1–4

Ellwood SR, Phan HTT, Jordan M, Hane J, Torres AM, Avila CM, Cruz-Izquierdo S, Oliver RP (2008) Construction of a comparative genetic map in faba bean (*Vicia faba* L.); conservation of genome structure with *Lens culinaris*. BMC Genomics 9:380

Emami MK, Sharma B (1999) Linkage between three morphological markers in lentil. Plant Breed 118:579–581

Erskine W (1985) Selection for pod retention and pod indehiscence in lentils. Euphytica 34:105–112

Erskine W (1996) Seed-size effects on lentil (*Lens culinaris*) yield potential and adaptation to temperature and rainfall in West Asia. J Agri Sci 126:335–341

Erskine W (2009) Global production, supply and demand. In: Erskine W, Muehlbauer F, Sarker A, Sharma B (eds) The lentil: botany, production and uses. CABI, Wallingford

Erskine W, Adham Y, Holly L (1989) Geographic distribution of variation in quantitative traits in a world lentil collection. Euphytica 43:97–103

Erskine W, Ellis RH, Summerfield RJ, Roberts EH, Hussain A (1990) Characterization of responses to temperature and photoperiod for time to flowering in a world lentil collection. Theor Appl Genet 80:193–199

Erskine W, Tufail M, Russell A, Tyagi MC, Rahman MM, Saxena MC (1994) Current and future strategies in breeding lentil for resistance to biotic and abiotic stresses. Euphytica 73:127–135

Erskine W, Chandra S, Chaudary M, Malik IA, Sarker A, Sharma B, Tufail M, Tyagi MC (1998) A bottleneck in lentil: widening the genetic base in South Asia. Euphytica 101:207–211

Eujayl I, Baum M, Powell W, Erskine W, Pehu E (1998a) A genetic linkage map of lentil (*Lens* sp.) based on RAPD and AFLP markers using recombinant inbred lines. Theor Appl Genet 97:83–89

Eujayl I, Erskine W, Bayaa B, Baum M, Pehu E (1998b) Fusarium vascular wilt in lentil: inheritance and identification of DNA markers for resistance. Plant Breed 117:497–499

Eujayl I, Erskine W, Baum M, Pehu E (1999) Inheritance and linkage analysis of frost injury in a lentil population of recombinant inbred lines. Crop Sci 39:639–642

FAOSTAT (2007) FAOSTAT statistical database of the United Nations Food and Agriculture Organization (FAO), Rome. Available at: http://faostat.fao.org/site

Ferguson ME, Ford-Lloyd BV, Robertson LD, Maxted N, Newbury HJ (1998a) Mapping the geographical distribution of genetic variation in the genus *Lens* for the enhanced conservation of plant genetic diversity. Mol Ecol 7:1743–1755

Ferguson ME, Newbury HJ, Maxted N, Ford-Lloyd BV, Robertson LD (1998b) Population genetic structure in *Lens* taxa revealed by isozyme and RAPD analysis. Genet Resour Crop Evol 45:549–559

Ferguson ME, Robertson LD, Ford-Lloyd BV, Newbury HJ, Maxted N (1998c) Contrasting genetic variation amongst lentil landraces from different geographical origins. Euphytica 102:265–273

Ferguson ME, Maxted N, Van Slageren M, Robertson LD (2000) A reassessment of the taxonomy of *Lens* Miller (Leguminosae, Papilionoideae, Vicieae). Bot J Linn Soc 133:41–59

Fernández M, Ruiz ML, Linares C, Fominaya A, Pérez de la Vega M (2005) 5S rDNA genome regions of *Lens* species. Genome 48:937–942

Fernández-Aparicio M, Sillero JC, Pérez-de-Luque A, Rubiales D (2007) Identification of sources of resistance to crenate broomrape (*Orobanche crenata*) in Spanish lentil (*Lens culinaris*) germplasm. Weed Res 48:85–94

Fernández-Aparicio M, Sillero JC, Rubiales D (2009) Resistance to broomrape in wild lentils (*Lens* spp.). Plant Breed 128:266–270

Fiala JV, Tullu A, Banniza S, Séguin-Swartz G, Vandenberg A (2009) Interspecies transfer of resistance to anthracnose in lentil (*Lens culinaris* Medik.). Crop Sci 49:825–830

Fikru E, Sarker A, Fikre A, Ahmed A, Ali K, Erskine W (2007) Registration of Assano lentil. J Plant Regist 1:41

Fiocchetti F, Laddomada B, Roselli M, Crino P, Lucretti S (2009) Fingerprinting of three typical macrosperma Italian lentil (*Lens culinaris* Medik.) landraces using fluorescence-based AFLP markers. Sci Hortic 121: 383–387

Ford R, Pang ECK, Taylor PWJ (1997) Diversity analysis and species identification in *Lens* using PCR generated markers. Euphytica 96:247–255

Ford R, Pang ECK, Taylor PWJ (1999) Genetics of resistance to Ascochyta blight (*Ascochyta lentis*) of lentil and the identification of closely linked RAPD markers. Theor Appl Genet 98:93–98

Ford R, Mustafa B, Inder P, Rubeena S, Materne M, Taylor P (2007) Lensomics: advances in genomics and molecular techniques for lentil breeding and management. In: Yadav SS, McNeil D, Stevenson PC (eds) Lentil: an ancient crop for modern times. Springer, Dordrecht

Ford R, Mustafa B, Baum M, Rajesh PN (2009) Advances in molecular research. In: Erskine W, Muehlbauer F, Sarker A, Sharma B (eds) The lentil: botany, production and uses. CABI, Wallingford

Fratini R, Ruiz ML (2003) A rooting procedure for lentil (*Lens culinaris* Medik.) and other hypogeous legumes (pea, chickpea and *Lathyrus*) based on explant polarity. Plant Cell Rep 21:726–732

Fratini R, Ruiz ML (2006) Interspecific hybridization in the genus *Lens* applying *in vitro* embryo rescue. Euphytica 150:271–280

Fratini R, Ruiz ML (2008) Micropropagation of intra and interspecific *Lens* hybrids. Euphytica 159:199–206

Fratini R, Ruiz ML (2011) Wide crossing in lentil through embryo rescue. In: Thorpe TAA, Yeung ECC (eds) Plant embryo culture: methods and protocols. Methods in molecular biology series, vol 710, Walker J (ed). Humana Press, New York

Fratini R, Ruiz ML, Pérez de la Vega M (2004) Intraspecific and inter-subspecific crossing in lentil (*Lens culinaris* Medik.). Can J Plant Sci 84:981–986

Fratini R, Durán Y, García P, Pérez de la Vega M (2007) Identification of quantitative trait loci (QTL) for plant structure, growth habit and yield in lentil. Span J Agric Res 5:348–356

Fratini R, Pérez de la Vega M, Cubero JI (2011) Lentil origin and domestication. Grain Legum 57:5–9

Furman BJ, Coyne C, Redden B, Sharma SK, Vishnyakova M (2009) Genetic resources: collection, characterization, conservation and documentation. In: Erskine W, Muehlbauer FJ, Sarker A, Sharma B (eds) The lentil: botany, production and uses. CABI, Wallingford

Galasso I (2003) Distribution of highly repeated DNA sequences in species of the genus *Lens* Miller. Genome 46:1118–1124

Galloway J, MacLeod WJ, Lindbeck KD (2004) Formation of *Didymella lentis*, the teleomorph of *Ascochyta lentis*, on lentil stubble in the field in Victoria and Western Australia. Australas Plant Pathol 33:449–450

Gill AS, Malhotra RS (1980) Inheritance of flower color and flower number per inflorescence in lentils. Lens Newsl 7:15–19

Grunwald NJ, Hoheisel GA (2006) Hierarchical analysis of diversity, selfing, and genetic differentiation in populations of the oomycete *Aphanomyces euteiches*. Phytopathology 96:1134–1141

Gupta D, Sharma SK (2005) Embryo ovule rescue – *in vitro* technique for overcoming post fertilization barriers in *Lens*. J Lentil Res 2:27–30

Gupta D, Sharma SK (2006) Evaluation of wild *Lens* taxa for agro-morphological traits, fungal diseases and moisture stress in North Western Indian hills. Genet Resour Crop Evol 53:1233–1241

Haddad N, Muehlbauer FJ, Hampton RO (1978) Inheritance of resistance to pea seed-borne mosaic virus in lentils. Crop Sci 18:613–615

Haddad NI, Bogyo TP, Muehlbauer FJ (1982) Genetic variance of six quantitative characters in three lentil (*Lens culinaris* Medik.) crosses. Euphytica 31:113–120

Hamdi A, Erskine W, Gates P (1991) Relationship among economic characters in lentil. Euphytica 57:109–116

Hamdi A, Küsmenoglu I, Erskine W (1996) Sources of winter hardiness in wild lentil. Genet Resour Crop Evol 43:63–67

Hamwieh A, Udapa SM, Choumane W, Sarker A, Dreyer F, Jung C, Baum M (2005) A genetic linkage map of lentil based on microsatellite and AFLP markers and localization of fusarium vascular wilt resistance. Theor Appl Genet 110:669–677

Hamwieh A, Udupa SM, Sarker A, Jung C, Baum M (2009) Development of new microsatellite markers and their application in the analysis of genetic diversity in lentils. Breed Sci 59:77–86

Harlan JR (1992) Crops and man. American Society of Agronomy, Madison

Harlan JR, de Wet JMJ (1977) Pathways of genetic transfer from *Tripsacum* to *Zea Mays*. Proc Natl Acad Sci U S A 74:3494–3497

Havey MJ, Muehlbauer FJ (1989) Linkages between restriction fragment length, isozyme, and morphological markers in lentil. Theor Appl Genet 77:395–401

Hernández-Bello MA, Chilvers MI, Akamatsu H, Peever TL (2006) Host specificity of *Ascochyta* spp. infecting legumes of the Viciae and Cicerae tribes and pathogenicity of an interspecific hybrid. Phytopathology 96: 1148–1156

Hobson K, Armstrong R, Nicolas M, Connor D, Materne M (2006) Response of lentil (*Lens culinaris*) germplasm to high concentrations of soil boron. Euphytica 151:371–382

Hoffman DL, Soltis DE, Muehlbauer FJ, Ladizinsky G (1986) Isozyme polymorphism in *Lens* (Leguminosae). Syst Bot 11:392–402

Horneburg B (2006) Outcrossing in lentil (*Lens culinaris*) depends on cultivar, location and year, and varies within cultivars. Plant Breed 125:638–640

Hussain S, Ali A, Zahid MA, Iqbal SM (2008) Multiple disease resistance in lentil germplasm from Pakistan. Pak J Bot 40:897–903

Inder P, Materne M, Taylor PWJ, Ford R (2008) Genotyping elite genotypes within the Australian lentil breeding program with lentil-specific sequenced tagged microsatellite site (STMS) markers. Aust J Agric Res 59:222–225

Infantino A, Kharrat M, Riccioni L, Coyne CJ, McPhee KE, Grunwald NJ (2006) Screening techniques and sources of resistance to root diseases in cool season food legumes. Euphytica 147:201–221

Jansen RK, Wojciechowski MF, Sanniyasi E, Lee S-B, Daniell H (2008) Complete plastid genome sequence of the chickpea (*Cicer arietinum*) and the phylogenetic distribution of *rps12* and *clpP* intron losses among legumes (Leguminosae). Mol Phylogenet Evol 48:1204–1217

Kahraman A, Kusmenoglu I, Aydin N, Aydogun A, Erskine W, Muehlbauer FJ (2004a) Genetics of winter hardiness in 10 lentil recombinant inbred line populations. Crop Sci 44:5–12

Kahraman A, Kusmenoglu I, Aydin N, Aydogan A, Erskine W, Muehlbauer FJ (2004b) QTL mapping of winter hardiness genes in lentil. Crop Sci 44:13–22

Kalo P, Seres A, Taylor S, Jakab J, Kevei Z, Kereszt A, Endre G, Ellis T, Kiss G (2004) Comparative mapping between *Medicago sativa* and *Pisum sativum*. Mol Genet Genomics 272:235–246

Kamboj RK, Pandey MP, Chaube HS (1990) Inheritance of resistance to Fusarium wilt in India lentil germplasm. Euphytica 50:113–117

Kaur S, Pembleton L, Cogan N, Hobson K, Leonforte T, Paull J, Forster J, Materne M (2010) Transcriptome sequencing from cool season food legumes (lentil, chickpea, pea and faba bean) using 454 GS-FLX Titanium technology. In: Proceedings of the 5th international food legumes research conference & 7th European conference on grain legumes, Antalya, 26–30 Apr

Kaur S, Cogan NO, Pembleton LW, Shinozuka M, Savin KW, Materne M, Forster JW (2011) Transcriptome sequencing of lentil based on second-generation technology permits large-scale unigene assembly and SSR marker discovery. BMC Genomics 12:265

Keim P, Diers BW, Shoemaker RC (1990) Genetic analysis of soybean hard seededness with molecular markers. Theor Appl Genet 79:465–469

Kumar A, Singh DP (1998) Hybridization technique in lentil under field conditions. Lens Newsl 25:1–3

Kumar Y, Mishra SK, Tyagi MC, Singh SP, Sharma B (2005) Linkage between genes for leaf colour, plant pubescence, number of leaflets and plant height in lentil (*Lens culinaris* Medik.). Euphytica 145:41–48

Ladizinsky G (1979) The origin of lentil and its wild gene pool. Euphytica 28:179–187

Ladizinsky G (1985) The genetics of hard seed coat in the genus Lens. Euphytica 34: 539–543

Ladizinsky G (1992) Crossability relations. In: Kalloo G, Chowdhury JB (eds) Distant hybridization of crop plants, Monographs on theoretical and applied genetics, 16. Springer, Dordrecht

Ladizinsky G (1993) Wild lentils. Crit Rev Plant Sci 12:169–184

Ladizinsky G (1997) A new species of *Lens* from southeast Turkey. Bot J Linn Soc 123:257–260

Ladizinsky G (1999) Identification of the lentil wild genetic stock. Genet Resour Crop Evol 46:115–118

Ladizinsky G, Abbo S (1993) Cryptic speciation in *Lens culinaris*. Genet Resour Crop Evol 40:1–5

Ladizinsky G, Braun D, Goshen D, Muehlbauer FJ (1984) The biological species of the genus *Lens*. Bot Gaz 145:253–261

Ladizinsky G, Cohen D, Muehlbauer FJ (1985) Hybridization in the genus *Lens* by means of embryo culture. Theor Appl Genet 70:97–101

Ladizinsky G, Weeden NF, Muehlbauer FJ (1990) Tertiary trisomic in lentil. Euphytica 51:179–184

Laserna-Ruiz I, De-Los-Mozos-Pascual M, Santana-Méridas O, Sánchez-Vioque R, Rodríguez-Conde MF (2012) Screening and selection of lentil (*Lens* Miller) germplasm resistant to seed bruchids (*Bruchus* spp.). Euphytica 188:153–162

Law JA, Jacobsen SE (2010) Establishing, maintaining and modifying DNA methylation patterns in plants and animals. Nat Rev Genet 11:204–220

Lazaridou TB, Roupakias DG, Economou AS (1993) Embryo rescue in *Vicia faba* and *Vicia narbonensis*. Plant Cell Tissue Org Cult 33:297–301

Lev-Yadun S, Gopher A, Abbo S (2000) The cradle of agriculture. Science 288:1602–1603

Malaviya DR, Shukla RS (1990) Crossability among *macrosperma* and *microsperma* varieties of *Lens culinaris* Medik. Indian J Genet 50:63–69

Malhorta RS, Balyan HS, Gupta PK (1978) Crossing technique in lentils. Lens Newsl 5:7–8

Mamo T, Richter C, Heiligtag B (1996) Salinity effects on the growth and ion contents of some chickpea (*Cicer arietinum* L.) and lentil (*Lens culinaris* Medic.) varieties. J Agron Crop Sci 176:235–247

McCoy TJ, Smith LY (1986) Interspecific hybridization of perennial *Medicago* species using ovule-embryo culture. Theor Appl Genet 71:772–783

Mejia-Jimenez A, Muñoz C, Jacobsen HJ, Roca WM, Singh SP (1994) Interspecific hybridization between common and tepary beans: increased hybrid embryo growth, fertility, and efficiency through recurrent and congruity backcrossing. Theor Appl Genet 88: 324–331

Mera M, Erskine W (1982) Crossing techniques for lentil under field conditions. Lens Newsl 9:11–15

Millan T (2013) Abstracts of oral and poster presentations given at the Ascochyta 2012 workshop, Phytopathologia Mediterranea 52, Córdoba, Spain, 22–26 Apr 2012. Available at: http://www.fupress.net/index.php/pm/article/view/12383

Miller PR, Brandt SA, McDonald CL, Waddington J (2006) Chickpea, lentil, and pea response to delayed spring seeding on the Northern Great Plains. Can J Plant Sci 86:1059–1070

Muehlbauer FJ, Chen W (2007) Resistance to ascochyta blights of cool season food legumes. Eur J Plant Pathol 119:135–141

Muehlbauer FJ, Slinkard AE, Wilson VE (1980) Lentil. In: Fehr WR, Hadley HH (eds) Hybridization of crop plants. American Society of Agronomy and Crop Science Society of America, Madison

Muehlbauer FJ, Weeden NF, Hoffman DL (1989) Inheritance and linkage relationships of morphological and isozyme loci in lentil (*Lens* Miller). J Hered 80:298–303

Muehlbauer FJ, Kaiser WJ, Clement SL, Summerfield RJ (1995) Production and breeding of lentil. Adv Agron 54:283–332

Muehlbauer FJ, Cho S, Sarker A, McPhee KE, Coyne CJ, Rajesh PN, Ford R (2006) Application of biotechnology in breeding lentil for resistance to biotic and abiotic stress. Euphytica 147:149–165

Mustafa BM, Coram TE, Pang ECK, Taylor PWJ, Ford R (2009) A cDNA microarray approach to decipher lentil (*Lens culinaris*) responses to *Ascochyta lentis*. Aust Plant Pathol 38:617–631

Negussie T, Pretorius ZA, Bender CM (2005) Components of rust resistance in lentil. Euphytica 142:55–64

Nguen TT, Taylor PWJ, Brouwer JB, Peng ECK, Ford R (2001) A novel source of resistance in lentil (*Lens culinaris* ssp. *culinaris*) to Ascochyta blight caused by *Ascochyta lentis*. Australas Plant Pathol 30:211–215

Ocampo B, Conicella C, Moss JP (2000) Wide crossing: opportunities and progress. In: Knight R (ed) Linking research and marketing opportunities for pulses in the 21st century, vol 34, Current plant science and biotechnology in agriculture. Kluwer Academic Publishers, Dordrecht

Peever TL (2007) Role of host specificity in the speciation of *Ascochyta* pathogens of cool season food legumes. Eur J Plant Pathol 119:119–126

Penazola E, Tay J, France A (2007) Calpun-INIA, a large seed size and rust-resistant lentil (*Lens culinaris* Medik.) cultivar. Agri Téc 67:68–71

Penmetsa RV, Carrasquilla-Garcia N, Rosen B, Gao J, Sarma BK et al. (2010) Studies of genome evolution in Papilionoid legumes. In: Proceedings of the 5th international food legumes research conference & 7th European conference on grain legumes, Antalya, 26–30 Apr

Pérez de la Vega M, Fratini R, Muehlbauer FJ (2012) Lentil. In: Pérez de la Vega M, Torres AM, Cubero JI, Kole C (eds) Genetics, genomics and breeding of cool season grain legumes. CRC Press, Boca Raton

Phan HTT, Ellwood SR, Hane JK, Ford R, Materne M, Oliver RP (2007) Extensive macrosynteny between *Medicago truncatula* and *Lens culinaris* ssp. *culinaris*. Theor Appl Genet 114:549–558

Pickersgill B (1993) Interspecific hybridization by sexual means. In: Hayward MD, Bosemark NO, Romagosa I (eds) Plant breeding: principles and prospects. Chapman & Hill, London

Pinkas R, Zamir D, Ladizinsky G (1985) Allozyme divergence and evolution in the genus *Lens*. Plant Syst Evol 151:131–140

Pitarelli GW, Stavely JR (1975) Direct hybridisation of *Nicotiana rependa* x *N. tabacum*. J Hered 66:281–284

Rai R, Singh RP (1999) Effect of salt stress on interaction between lentil (*Lens culinaris*) genotypes and *Rhizobium* sp. Strains: symbiotic *N*-2 fixation in normal and sodic soils. Biol Fertil Soils 29:187–195

Reddy MRK, Rathour R, Kumar N, Katoch P, Sharma TR (2010) Cross-genera legume SSR markers for analysis of genetic diversity in *Lens* species. Plant Breed 129:514–518

Riccioni L, Haegi A, Valvassori M (2008) First report of vascular wilt caused by *Fusarium redolens* on lentil in Italy. Plant Dis 92:1132

Robaglia C, Caranta C (2006) Translation initiation factors: a weak link in plant RNA virus infection. Trends Plant Sci 11:40–45

Rolston MP (1978) Water impermeable seed dormancy. Bot Rev 44:365–396

Rosa de la L, Jouve N (1992) Genetic variation for isozyme and proteins in Spanish primitive cultivars and wild species of *Lens*. Euphytica 59:181–187

Rubeena S, Taylor PW, Ades PK, Ford R (2006) QTL mapping of resistance in lentil (*Lens culinaris* ssp. *culinaris*) to Ascochyta blight (*Ascochyta lentis*). Plant Breed 125:506–512

Saha GC, Muehlbauer FJ (2011) Lentil: an essential high protein food in South Asia. Grain Legum 57:49–51

Saha GC, Sarker A, Chen W, Vandenmark GJ, Muehlbauer FJ (2010) Inheritance and linkage map positions of genes conferring resistance to Stemphylium blight in lentil. Crop Sci 50:1831–1839

Salam MA, Islam MT (1994) Growth, yield and leaf-water attributes of some advanced mutant lentil lines under different moisture regimes. Lens Newsl 21:32–35

Sarker A, Erskine W (2006) Recent progress in the ancient lentil. J Agric Sci 144:19–29

Sarker A, Kumar S (2011) Lentils in production and food systems in West Asia and Africa. Grain Legum 57:46–48

Sarker A, Erskine W, Sharma B, Tyagi MC (1999) Inheritance and linkage relationship of days to flower and morphological loci in lentil (*Lens culinaris* Medikus subsp. *culinaris*). J Hered 90:270–275

Sarker A, Bayaa B, El Hassan H, Erskine W (2004) New sources of resistance to *Fusaium* wilt in lentil (*Lens culinaris* Medikus. ssp. *culinaris*). J Lentil Res 1:30–33

Sarker A, Erskine W, Singh M (2005) Variation in shoot and root characteristics and their association with drought tolerance in lentil landraces. Genet Resour Crop Evol 52:87–95

Scippa GS, Trupiano D, Rocco M, Viscosi V, Di Micheli M, D'Andrea A, Chiatante D (2008) An integrated approach to the characterization of two autochthonous lentil (*Lens culinaris*) landraces of Molise (south-central Italy). Heredity 101:136–144

Scippa GS, Rocco M, Ialicicco M, Trupiano D, Viscosi V, Di Michele M, Arena S, Chiatante D, Scaloni A (2010) The proteome of lentil (*Lens culinaris* Medik.) seeds: discriminating between landraces. Electrophoresis 31:497–506

Sharma B (2009) Genetics of economic traits. In: Erskine W, Muehlbauer FJ, Sarker A, Sharma B (eds) The lentil: botany, production and uses. CABI, Wallingford

Sharma B (2011) Genes for traits of economic importance in lentil. Grain Legum 57:15–17

Sharma SK, Knox MR, Ellis THN (1996) AFLP analysis of the diversity and phylogeny of *Lens* and its comparison with RAPD analysis. Theor Appl Genet 93:751–758

Sharpe AG, Ramsay L, Sanderson LA, Fedoruk MJ, Clarke WE, Li R, Kagale S, Vijayan P, Vandenberg A, Bett KE (2013) Ancient orphan crop joins modern era: gene-based SNP discovery and mapping in lentil. BMC Genomics 14:192

Siddique KHM, Loss SP, Pritchard DL, Regan KL, Tennant D, Jettner RL, Wilkinson D (1998) Adaptation of lentil (*Lens culinaris* Medik.) to Mediterranean-type environments: effect of time of sowing on growth, yield, and water use. Aust J Agric Res 49:613–626

Sillero JC, Fondevilla S, Davidson J, Patto MCV, Warkentin TD, Thomas J, Rubiales D (2006) Screening techniques and sources of resistance to rusts and mildews in grain legumes. Euphytica 147:255–272

Singh RK, Raghuvanski SS, Prakash D (1992) Genetic improvement in advanced generations of colchitetraploid lentil (*Lens culinaris* Medikus). J Genet Breed 46:151–154

Singh D, Dikshit HK, Singh R (2013a) A new phenotyping technique for screening for drought tolerance in lentil (*Lens culinaris* Medik.). Plant Breed 132:185–190

Singh M, Rana MK, Kumar K, Bisht IS, Dutta M, Gautam NK, Sarker A, Bansal KC (2013b) Broadening the genetic base of lentil cultivars through inter-subspecific and interspecific crosses of *Lens taxa*. Plant Breed 132:667–675

Skibinski DOF, Warren ME (1984) Variation at soluble enzyme loci in four *Lens* species. Lens Newsl 11:10–13

Solaiman Z, Colmer TD, Loss SP, Thomson BD, Siddique KHM (2007) Growth responses of cool-season grain legumes to transient waterlogging. Aust J Agric Res 58:406–412

Solh MB, Paredes OM, Tiwari AS (1980) Crossing technique in lentil under field conditions. Lens Newsl 7:9–14

Sonnante G, Pignone D (2001) Assessment of genetic variation in a collection of lentil using molecular tools. Euphytica 120:301–307

Sonnante G, Pignone D (2007) The major Italian landraces of lentil (*Lens culinaris* Medik.): their molecular diversity and possible origin. Genet Resour Crop Evol 54:1023–1031

Sonnante G, Galasso I, Pignone D (2003) ITS sequence analysis and phylogenetic inference in the genus *Lens* Mill. Ann Bot 91:49–54

Stebbins GL (1958) The inviability, weakness, and sterility of interspecific hybrids. Adv Genet 9:14–215

Stevenson PC, Dhillon MK, Sharma HC, El-Bouhssini M (2007) Insect pest of lentil and their management. In: Yadav SS, McNeil D, Stevenson PC (eds) Lentil: an ancient crop for modern times. Springer, Dordrecht

Stoddard FL, Balko C, Erskine W, Khan HR, Link W, Sarker A (2006) Screening techniques and sources of resistance to abiotic stresses in cool-season food legumes. Euphytica 147:167–186

Tahir M, Muehlbauer FJ, Spaeth SC (1994) Association of isozyme markers with quantitative trait loci in random single seed descent derived lines of lentil (*Lens culinaris* Medik.). Euphytica 75:111–119

Tar'an B, Buchwaldt L, Tullu A, Banniza S, Warkentin TD, Vandenberg A (2003) Using molecular markers to pyramid genes for resistance to ascochyta blight and anthracnose in lentil (*Lens culinaris* Medik). Euphytica 134:223–230

Tay J, Slinkard AE (1989) Transgressive segregation for *Ascochyta* resistance in lentil. Can J Plant Sci 69:547

Taylor PWJ, Ford R (2007) Diagnostics, genetic diversity and pathogenic variation of ascochyta blight of cool season food and feed legumes. Eur J Plant Pathol 119:127–133

Taylor P, Lindbeck K, Chen W, Ford R (2007) Lentil diseases. In: Yadav SS, McNeil D, Stevenson PC (eds) Lentil: an ancient crop for modern times. Springer, Dordrecht

Thavarajah D, Thavarajah P (2011) Lentil (*Lens culinaris*) as a biofortified crop with essential micronutrients: a food-based solution to micronutrient malnutrition. Grain Legum 57:29–31

Tivoli B, Baranger A, Avila CM, Banniza S, Barbetti M, Chen WD, Davidson J, Lindeck K, Kharrat M, Rubiales D, Sadiki M, Sillero JC, Sweetingham M, Muehlbauer FJ (2006) Screening techniques and sources of resistance to foliar diseases caused by major necrotrophic fungi in grain legumes. Euphytica 147:223–253

Toker C, Yadav SS, Solanki IS (2007) Mutation breeding. In: Yadav SS, McNeil D, Stevenson PC (eds) Lentil: an ancient crop for modern times. Springer, Dordrecht

Toklu F, Karakoy T, Hakli E, Bicer T, Brandolini A, Kilian B, Özkan H (2009) Genetic variation among lentil (*Lens culinaris* Medik.) landraces from Southeast Turkey. Plant Breed 128:178–186

Tullu A, Buchwaldt L, Warketin T, Taran B, Vandenberg A (2003) Genetics of resistance to anthracnose and identification of AFLP and RAPD markers linked to the resistance gene in PI 320937 germplasm of lentil (*Lens culinaris* Medikus). Theor Appl Genet 106:428–434

Tullu A, Buchwaldt L, Lulsdorf M, Banniza S, Barlow B, Slinkard AE, Sarker A, Tar'an B, Warkentin T, Vandenberg A (2006) Sources of resistance to anthracnose (*Colletotrichum truncatum*) in wild *Lens* species. Genet Resour Crop Evol 53:111–119

Tullu A, Tar'an B, Warkentin T, Vandenberg A (2008) Construction of an intraspecific linkage map and QTL analysis for earliness and plant height in lentil. Crop Sci 48:2254–2264

Tullu A, Banniza S, Bett K, Vandenberg A (2011) A walk on the wild side: exploiting wild species for improving cultivated lentil. Grain Legum 57:13–14

Tyagi MC, Sharma B (1989) Transgressive segregation for early flowering through conventional breeding in lentil. Lens Newsl 16:3–6

Tyagi MC, Sharma B (1995) Protein content in lentil (*Lens culinaris* Medik.). In: Sharma B, Kulshreshtha VP, Gupta N, Mishra SK (eds) Genetic research and education: current trends and the next fifty years. Indian Society of Genetics and Breeding, New Delhi

Vail S, Vandenberg A (2010) Evaluation of a clonal propagation protocol to obtain replicated disease data on infection by *Colletotrichum truncatum* in *Lens culinaris*. Crop Sci 50:926–932

Vaillancourt R, Slinkard AE (1992) Inheritance of new genetic markers in lentil. Euphytica 64:227–236

van Oss H, Aron Y, Ladizinsky G (1997) Chloroplast DNA variation and evolution in the genus *Lens* Mill. Theor Appl Genet 94:452–457

Vandenberg B (2011) Lentils – the little seeds with the big impact on human health. Grain Legum 57:25–26

Vandenberg B, Slinkard AE (1989) Inheritance of four new quantitative genes in lentil. J Hered 80:320–322

Varshney RK, Close TJ, Singh NK, Hoisington DA, Cook DR (2009) Orphan legume crops enter the genomics era! Curr Opin Plant Biol 12:202–210

Verma P, Shah N, Bhatia S (2013) Development of an expressed gene catalogue and molecular markers from the de novo assembly of short sequence reads of the lentil (*Lens culinaris* Medik.) transcriptome. Plant Biotechnol J 11:894–905. doi:10.1111/pbi.12082

Wang H, Chang KF, Hwang SF, Gossen BD, Turnbull GD, Howard RJ, Strelkov SE (2006) Response of lentil cultivars to rhizoctonia seedling diseases in Canada. J Plant Dis Prot 113:219–223

Williams EG (1987) Interspecific hybridization in pasture legumes. In: Janick J (ed) AVI Publishing Co., Westport. Plant Breed Rev 5:237–305

Wilson VE (1972) Morphology and technique for crossing *Lens esculenta* Moench. Crop Sci 12:231–232

Wilson VE, Law AG (1972) Natural crossing in *Lens esculenta* Moench. J Am Soc Hortic Sci 97:142–143

Wojciechowski MF (2006) IRLC (Inverted Repeat-lacking clade). Version 11 July 2006: http://tolweb.org/IRLC_%28Inverted_Repeat-lacking_clade%29/60358/2006.07.11. In: The tree of life web project, http://tolweb.org

Wojciechowski MF, Sanderson MJ, Steele KP, Liston A (2000) Molecular phylogeny of the "temperate herbaceous tribes" of papilionoid legumes: a supertree approach. In: Herendeen P, Bruneau A (eds) Advances in legume systematics, part 9. Royal Botanic Garden Kew, London, pp 277–298

Ye G, McNeil DL, Hill GD (2002) Breeding for resistance to lentil Ascochyta blight. Plant Breed 121:185–191

Yeung EC, Thorpe TA, Jensen CJ (1981) *In vitro* fertilization and embryo culture. In: Thorpe TA (ed) Plant tissue culture, methods and applications in agriculture. Academic, New York

Yuzbasioglu E, Acik L, Ozcan S (2008) Seed protein diversity among lentil cultivars. Biol Plant 52:126–128

Zamir D, Ladizinsky G (1984) Genetics of allozyme variants and linkage groups in lentil. Euphytica 33: 329–336

Závodná M, Kraic J, Paglia G, Gregova E, Morgante M (2000) Differentiation between closely related lentil (*Lens culinaris* Medik.) cultivars using DNA markers. Seed Sci Technol 28:217–219

Zimniak-Przybylska Z, Przybylska J, Krajewski P (2001) Genetic resources, *Lens*, lentil, SDS-PAGE, seed globulins. J Appl Genet 42:435–447

Zohary D (1999) Monophyletic vs. polyphyletic origin of crops on which agriculture founded in the Near East. Genet Resour Crop Evol 46:133–142

Pigeonpea

7

Nalini Mallikarjuna, Sandhya Srikanth, C.V. Sameer Kumar, Rakesh K. Srivastava, Rachit K. Saxena, and Rajeev K. Varshney

Abstract

Pigeonpea was labeled as an orphan crop but is now a trendy and pacesetter, with ample genetic and genomic information becoming available in recent times. It is now possible to cross wild relatives not only from the *Cajanus* group placed in the secondary and tertiary gene pool but also the related genera placed in the quaternary gene pool. This is no small achievement for a legume which is an important crop of Asia and Africa and plays a major role in the diet of majority of the people of this region. The need of the hour is further committed research on wide crosses in pigeonpea.

7.1 Introduction

In a time of resource constraint, land pressure, environmental concerns, and population explosion, genetic improvement of crops becomes the most promising approach by which food production can meet the demand. For genetic improvement to succeed, plant breeders rely on a rich source of genetic diversity. Although there is no shortage of genetic variation in nature, a process of genetic erosion due to various reasons, including plant breeding, threatens modern varieties. All crop species were originally domesticated from wild plants by humans; the very process of domestication inherently reduced genetic variation (Ladizinsky 1985). The limited genetic variation among modern crop varieties not only made them more susceptible to disease epidemics, but it also reduced the chance for plant breeders to identify new and useful combinations of genes, thus causing a slower rate of crop improvement in the long term. There is renewed interest in the utilization of wild relatives for the improvement of crop plants worldwide including India. Wild ancestors of most crop plants can still be found in their natural habitats, but their numbers are slowly dwindling. There is an urgent need to conserve all the available germplasm under long-term storage. The germplasm of crop plants represents a precious source of genetic variation that can potentially insure more rapid and sustained crop

N. Mallikarjuna (✉) • S. Srikanth • C.V.S. Kumar
R.K. Srivastava • R.K. Saxena • R.K. Varshney
Grain Legumes Program, International Crops Research Institute for Semi Arid Tropics (ICRISAT), Patancheru 501 324, Andhra Pradesh, India
e-mail: mysorenalinicgiar@yahoo.com; sandhya.thudumu@gmail.com; c.sameerkumar@cgiar.org; r.k.srivastava@cgiar.org; r.saxena@cgiar.org; r.k.varsheny@cgiar.org

M. Singh et al. (eds.), *Broadening the Genetic Base of Grain Legumes*,
DOI 10.1007/978-81-322-2023-7_7, © Springer India 2014

plant improvement for many years to come (Tanksley and McCouch 1997).

7.2 Origin and Potential

Pigeonpea [*Cajanus cajan* (L.) Millspaugh] is an important crop providing protein in the human diet to a large population of Asia, Africa, the Caribbean, and Latin America. It has a special place in the diet of the vast majority of Indians, who consume pigeonpea in one form or the other daily. It also has wide applications in traditional medicine (van der Maesen 2006). Cultivation of the pigeonpea dates back to at least 3,500 years. The center of origin is the eastern part of peninsular India, including the state of Orissa, where the closest wild relatives (*Cajanus cajanifolius*) occur in tropical deciduous woodlands (van der Maeson 1995). Archaeological finds of pigeonpea include those from two Neolithic sites in Orissa, Gopalpur and Golbai Sassan, dating between 3,400 and 3,000 years ago, and sites in South India, Sanganakallu and Tuljapur Garhi, also dating back to 3,400 years ago. From India, it traveled to East Africa and West Africa (Fuller and Harvey 2006). There, it was first encountered by Europeans, so it obtained the name Congo pea. By means of the slave trade, it came to the American continent, probably in the seventeenth century. Nutritionally they contain high levels of protein and the important amino acids like methionine, lysine, and tryptophan (Saxena et al. 2010). In combination with cereals, pigeonpea makes a well-balanced human diet. Pigeonpea is grown in low-input and risk-prone marginal environments. There is a large gap between potential yield (2,500 Kg/ha) and actual yield on farmer's fields which is 866.2 kg/ha in Asia and 736.2 kg/ha in Africa (Mula and Saxena 2010). It is a drought-tolerant and hardy crop of India, assuring sustainable returns from marginal lands with minimal inputs. Hence it is considered as a very suitable crop for subsistence agriculture. Pigeonpea is a cross-pollinated (20–70 %) species with a diploid number of $2n=2x=22$ and genome size of 858 Mbp. Pigeonpea is the first seed legume plant to have its genome sequenced. The first draft was done by a group of more than 30 ICRISAT and other institution scientists belonging to different organizations across the world (Varshney et al. 2012a).

7.3 Diversity in Crop Cultivars

Ample morphological diversity is exhibited by pigeonpea as a crop but the same is not true at the molecular level (Yang et al. 2006; Odeny et al. 2007). A low level of genetic diversity has led to a narrow genetic base, and this is reflected in the absence of resistance to various biotic and abiotic constraints. The reason for the low level of genetic diversity is the evolutionary bottleneck. There is some evidence that pigeonpea has a monophyletic evolution and has evolved from wild relative *Cajanus cajanifolius* (Kassa et al. 2012). There are 5–6 traits that differentiate *C. cajanifolius* from pigeonpea such as flower morphology, pod color and morphology, pod constriction, seed color, and 100-seed weight (Mallikarjuna et al. 2012a). Molecular studies revealed that a genetic dissimilarity index value ranging from 0.81 to 0.94 exists between the two species (Mallikarjuna et al. 2012a, b). But the crop has a rich source of variability in the form of wild relatives in different gene pools, which have played a major role in the introduction of disease resistance and agronomic traits such as high protein content, identification and diversification of the cytoplasmic base of CMS system, and more recently introgression of pod borer (*Helicoverpa armigera*), pod fly, bruchid resistance, and other agronomic traits (Mallikarjuna et al. 2011a).

7.4 Production-Related Problems

Pigeonpea is an important crop of Asia and especially in India, where it forms an important source of protein in the vegetarian diet. Despite its immense importance in sustainable agriculture and continued breeding efforts directed toward genetic improvement, the global production per hectare of pigeonpea has remained static over the last three decades. The yield gap between the potential yield and on-farm yield is mainly due to the prevalence of various abiotic and biotic stresses in pigeonpea-growing areas together with its cultivation in marginal lands with low input supply and lack of efficient management practices (Varshney et al. 2012b). Among the various diseases, fusarium wilt, caused by *Fusarium udum* Butler, is the most important disease in the Indian subcontinent and East Africa (Saxena 2008). Occurrence of wilting during the pod-filling stage causes infection to pigeonpea seeds and yield losses up to 50–70 % (Marley and Hillocks 1996). Another disease severely affecting pigeonpea yield is sterility mosaic disease (SMD) caused by the pigeonpea sterility mosaic virus (PPSMV) causing losses up to 95–100 % with infection occurring early at <45-day-old plants (Kannaiyan et al. 1984). Apart from wilt and mosaic, *Phytophthora* blight caused by the fungus *Phytophthora drechsleri* Tucker f. sp. *cajani* is another important disease that has got the status of an economic concern. However, the disease is limited in distribution and witnesses more severity in short-duration cultivars as compared to long- or medium-duration genotypes (Ratnaparkhe and Gupta 2007). Among the variety of insects feeding on pigeonpea, the pod borer, *Helicoverpa armigera* (Hubner), is the most damaging pest worldwide, and its frequent occurrence often results in complete crop failure. Besides *Helicoverpa*, other pests like *Maruca* (*Maruca vitrata* Geyer), pod-sucking bugs (*Clavigralla horrida* Germar), and pod fly (*Melanagromyza chalcosoma* Spencer) pose a big threat to pigeonpea production. Moreover, infestations from storage pests like bruchids (*Callosobruchus chinensis*) intensify the situation and result in profound seed damage during storage. Among the abiotic constraints, salinity and waterlogging severely affect the pigeonpea production (Saxena 2008; Choudhary et al. 2011). The limited success achieved so far in addressing the problem of production constraints is mainly due to the complex mechanism underlying these stresses together with the lack of precise and efficient screening techniques. Genetic diversity and resistance to many of the constraints are not up to the desired level in the cultivated germplasm. Low use of germplasm which is 0.4 % of the total 13,771 accessions to develop advanced breeding lines may have also contributed to the lack of good sources of resistance. Factors responsible for the low use of germplasm might be the large size of germplasm collections and lack of presence of useful traits present in them, thus emphasizing the importance of wild relatives with extensive genetic diversity.

7.5 Crop Gene Pools and Alien Introgressions

The classification of Harlan and de Wet (1971) of grouping the germplasm of a crop is followed to group the genetic resources of pigeonpea including their wild relatives. They constitute three basic gene pools and are divided as primary gene pool (GP1), secondary gene pool (GP2), and tertiary gene pool (GP3) and the related genera in the quaternary gene pool (GP4). In the genus *Cajanus* with 32 species and 11 related genera, *Cajanus cajan* is the only species cultivated throughout Asia and Africa for its leguminous proteins. The gene bank at ICRISAT conserves over 13,632 accessions of *Cajanus* species from 74 countries. This includes 555 accessions of wild relatives representing six genera and 57 species (Upadhyaya et al. 2007). Pigeonpea belongs to the subtribe *Cajaninae* which contains 13 genera. Earlier the genera *Atylosia* and *Cajanus* were considered closely related; later genus *Atylosia* was merged with the genus *Cajanus* (van der Maesen 1980). Subsequently, the genus *Cajanus* has 32 species, 18 of which are endemic to Asia and 13 to Australia and one to West Africa (van der Maesen 1986). Apart from these, there are other related genera, namely, *Rhynchosia*, *Dunbaria*, *Flemingia*, *Paracalyx*, *Eriosema*, *Adenodolichos*, *Bolusafra*, *Carissoa*, *Chrysoscias*, and *Baukea*. For details of the gene pools, please refer to the publication of Mallikarjuna et al. (2011a).

7.5.1 Primary Gene Pool

Considerable progress has been made in pigeonpea improvement by using variability within the cultivated species, and globally pigeonpea is grown on

4.5 million ha with a production of 3.5 million metric tons and productivity of 863 kg ha^{-1} (FAO 2009). The gene bank at ICRISAT, Patancheru, India, conserves the largest collection of pigeonpea germplasm assembled in the world. The germplasm collection includes 8,215 landraces, 4,795 breeding lines, and 67 improved cultivars (Upadhyaya et al. 2012). In spite of the large germplasm collection in the primary gene pool, it is not widely used (Wright 1997) as information on the presence of useful traits is not easily available and necessitating an extended period of research whenever utilized (Goodman 1990). To overcome these issues, core and mini core collections have been developed (Upadhyaya et al. 2006). Variation with the primary gene pool is of importance as they are easy to use with quicker gains and can be directly released as cultivars. Progress has been made in the utilization of material from the primary gene pool (Saxena 2000). Varieties BDN-1 and Maruthi released in 1989 are selections from pure-line breeding which are popular even today (Bantilan and Joshi 1996). Development of high yielding varieties such as ICPL 87, ICPL 151, Prabhat, T 21, Pusa Ageti, CO 5, and JA 3 has also been reported (Singh et al. 2005). In spite of the above successes, a perusal of the utilization pattern of *Cajanus* germplasm indicates that so far a very small proportion of germplasm has been used in pigeonpea improvement programs globally. In pigeonpea, 57 ancestors were used to develop 47 varieties. The top 10 ancestors contributed 48 % to the genetic base of the released varieties (Kumar et al. 2003). One of the reasons for such poor utilization may be the vast number of lines available in the primary gene pool which lack detailed characterization, evaluation, and genetic diversity data. In spite of the above constraints, 10 pigeonpea germplasm have been released as cultivars with two in Nepal, one in the Philippines, two in India, one in China, one in Fiji, and one each in Malawi and Venezuela between 1985 and 2005.

7.5.2 Secondary Gene Pool

The pigeonpea wild species collection at the ICRISAT gene bank has not been characterized and evaluated systematically, although the data is beginning to emerge. Main reasons could be low seed quantity, lack of resources, difficulties in phenology and growth habit, and lower priority than the cultivated species (Upadhyaya et al. 2012). However, limited evaluation of different species by researchers across the world indicated that the wild gene pool of pigeonpea, particularly the secondary gene pool, is a promising source for various biotic and abiotic stresses (Bohra et al. 2010; Mallikarjuna et al. 2011a; Jadhav et al. 2012a, b). Compatible wild relatives of pigeonpea which are placed in the secondary gene pool do not need specialized techniques in the crossability experiments in majority of the cases with a few exceptions (Mallikarjuna et al. 2011a). Cytoplasmic male sterile systems were developed for pigeonpea exploiting the cross-pollination mechanism and utilizing wild *Cajanus* species. So far eight CMS systems have been reported utilizing wild relatives of pigeonpea (Mallikarjuna et al. 2012b). Of these, seven have been developed utilizing wild relatives from the secondary gene pool as the female parent. One system has cultivated pigeonpea cytoplasm with the utilization of wild species as the male parent (Mallikarjuna and Saxena 2005). More recently, CMS trait was observed utilizing *C. lanceolatus*, which is a wild relative in the secondary gene pool (Sandhya S and Mallikarjuna N, unpublished data). Efforts are underway to identify their restorers. Once developed, this will be named as A_9 CMS system.

The process of outcrossing is important in the development of CMS systems, but this can lead to genetic deterioration. A partially cleistogamous line, which showed less than 1 % cross-pollination, was purified from the cross *C. cajan* × *C. lineatus*, which was governed by a single recessive gene (Saxena et al. 1992). Partial cleistogamous lines developed from the above cross were found to be stable in India as well as in Sri Lanka. The cleistogamous trait can be utilized in pigeonpea to obtain pure seeds from genetic stocks. Cleistogamous lines were identified utilizing *C. platycarpus* (Cherian et al. 2006), a wild relative in the tertiary gene pool of pigeonpea. These lines had open flowers. When CMS is developed on such lines, it will help better cross-pollination and thus better yield. High-protein breeding lines were developed from *C. sericeus*, *C. albicans*, and *C. scarabaeoides*. A significant positive correlation between seed size and protein content was observed in lines derived from *C. scarabaeoides*. Lines HPL 2, HPL 7, HPL 40, and HPL 51 are some of the high-protein and high-seed-weight lines derived from wild species (Saxena et al. 1987). More recently crosses between pigeonpea and *C. acutifolius* yielded progeny with high seed weight. High seed weight accompanied by beige seed color is a desirable trait (Jadhav et al. 2012a, b). *C. acutifolius*, a wild relative from the secondary gene pool and native of Australia, can be crossed with pigeonpea as a one-way cross. The reciprocal cross using *C. acutifolius* as the female parent aborts to give rise to immature seeds. In vitro interventions are necessary to obtain hybrid plants (Mallikarjuna and Saxena 2002). Advanced generation population from the cross utilizing *C. acutifolius* as the pollen parent has shown resistance for pod-borer damage (Mallikarjuna et al. 2007; Jadhav et al. 2012a), variation for seed color, and high seed weight. Some of the lines showed high level of resistance to pod borers, pod fly, and bruchid under unprotected field conditions (Jadhav et al. 2012a). Bruchid resistance (Jadhav et al. 2012a) is an important trait for pigeonpea seeds under storage as resistance to the pest has not been observed in cultivated pigeonpea. These lines are available in pigeonpea breeding and legume cell biology units of the Grain Legumes Program (CGIAR project on grain legumes). Some of the advance generation lines derived from *C. acutifolius* were screened for waterlogging by germinating them and later growing them under waterlogged conditions. A few lines grew under waterlogged conditions and formation of lenticels was observed in the region above the water surface. The region gave rise to roots which entered the soil through the water surface. This shows that some of these lines may survive under waterlogged conditions (Aneesha Begum and Mallikarjuna N, unpublished data). Another species from the secondary gene pool, namely, *C. lanceolatus*, was crossed successfully with cultivated pigeonpea at ICRISAT and progeny lines developed (Srikanth et al. 2013). F_1 hybrids flowered, but some of the hybrids were pollen sterile, and in the rest of the hybrids, pollen fertility varied from 25 to 55 %. All the hybrids were female fertile. Progeny lines developed from the cross were screened for bruchid resistance. Bruchid growth and survival was inhibited in the lines derived from *C. lanceolatus*. Some of the lines showed delayed bruchid growth and delayed life cycle thus showing antibiosis mechanism of resistance to bruchids. Lines were screened for protein content and some of the lines showed higher protein content than both their parents. Further, biochemical analysis showed higher content of proteinase inhibitor activity in some of the lines (Sandhya Srikanth and Mallikarjuna, N, unpublished data). A new source of CMS was identified in the progeny lines and experiments are underway to identify maintainers and restorers (Sandhya Srikanth and Mallikarjuna N, unpublished data). Previously, Sateesh Kumar (1985) attempted crossing pigeonpea with *C. lanceolatus* but obtained sterile hybrids which did not flower and remained in the vegetative stage.

7.5.3 Tertiary Gene Pool

There are 20 wild species in the tertiary gene pool of pigeonpea (Mallikarjuna et al. 2011a). Until now, only one wild *Cajanus* species from this gene pool has been successfully crossed and traits of interest transferred (Mallikarjuna et al. 2011b). *Cajanus platycarpus* was successfully crossed utilizing hormone-aided pollinations and in vitro interventions (Mallikarjuna 1998) to obtain hybrids. Progeny lines showed variation for days to flower, growth habit, seed weight and number, seed color, and resistance to pod borer, pod fly, bruchids, *Fusarium*, and sterility mosaic disease (Mallikarjuna N unpublished data). Some cleistogamous lines were identified in CMS lines (Cherian et al. 2006), a trait favoring total cross-pollination. Hence utilizing *C. platycarpus* not only broadened the genetic base of pigeonpea, but it was possible to introgress useful traits. Diversity Arrays Technology (DArT), a genome-wide marker technology, was used to genotype the parents and advance generation hybrids after four backcrosses. A total of 1,225 markers were found polymorphic among the parents and the progeny. The results of the study showed that apart from DNA stretches from the female and male parent, there was some novel DNA polymorphism observed in the progeny not seen in both parental species. It was interesting to observe that as per theoretical calculations, there should be 3.12 % of *C. platycarpus* genome after four backcrosses with cultivated parent *C. cajan* (Mallikarjuna et al. 2011b). Diversity Arrays Technology (DArT) analysis showed the presence of *C. platycarpus* genome ranging from 2.0 to 4.8 %. The presence of nonparental DNA sequences was presumably because of recombination, ranging from 2.6 to 10.4 % (Mallikarjuna et al. 2011b).

More recently, another species from the tertiary gene pool, namely, *C. volubilis*, was crossed with pigeonpea. In F_2 generation, extra short duration lines were recovered in 2011. These lines flowered earlier than the short-duration cultivar ICPL 85010 which was the female parent of the cross. The lines were again screened for the short duration trait in Rabi Season of 2012 and were found to retain the trait. Dwarf growth habit coupled with determinate and semi-determinate plants was observed. In the determinate types, the number of pods per inflorescence and the number of inflorescence were more than that observed in the extra early and determinate cultivar MN5 (Sandhya Srikanth and Mallikarjuna N, unpublished data).

7.5.4 Quaternary Gene Pool

There are 11 related genera, namely, *Rhynchosia*, *Flemingia*, *Dunbaria*, *Eriosema*, *Paracalyx*, *Adenodolichos*, *Bolusafra*, *Carissoa*, *Chrysoscias*, and *Baukea* including *Cajanus* under the subtribe *Cajaninae*. Many of these genera are classified as underexploited legumes. *Rhynchosia* is one such example as it harbors important nutritional and therapeutic properties (Drabu et al. 2011), with the presence of phytochemicals such as alkaloids, glycosides, anthraquinones, carotenoids, coumarins, dihydrochalcones, fatty acids, flavonoids, steroids, and triterpenoids (Bakshu and Venkataraju 2001). Some species of *Rhynchosia* are used in human and animal diet (Oke et al. 1995). Many of the tribal communities in India soak the seeds in water and consume the seeds after boiling and decanting many times to get rid of unwanted constituents (Murthy and Emmannuel 2011). Many of the *Rhynchosia* species are known to exhibit antitumor and thus curative properties. Normally during cancer treatment, iron deficiency and anemia are major issues. It was observed that treatment with *Rhynchosia* seeds restored hemoglobin (Hb), RBC, and WBC counts to normal levels. With the interest in dietary flavonoids and suppression of cancer, *Rhynchosia* species are surely going to attract more attention in the coming days.

None of the genera in the quaternary gene pool have been successfully crossed with pigeonpea until now. Among the genera in the quaternary gene pool, *Rhynchosia* was selected to initiate crossing/introgression/gene transfer experiments as it had many desirable properties as listed above. It was possible to successfully cross *Rhynchosia* with pigeonpea through hormone-aided pollinations. The success rate of crossing *Rhynchosia* was low not exceeding 1–2 %, but it was possible to obtain hybrids. Screening the hybrids with molecular markers confirmed the hybridity (Mallikarjuna N and Varshney R unpublished data). Although the initial process of crossing *Rhynchosia* with pigeonpea was challenging, nevertheless, hybrids were obtained. They were fertile and it was possible to obtain self and backcross progenies. Experiments to screen and study the progeny lines for different traits/constraints are in progress.

7.6 AB-QTL Mapping Populations Utilizing Wild *Cajanus* Species

Advanced backcross quantitative trait locus detection method abbreviated as AB-QTL increases the efficiency of identifying and transferring beneficial alleles from exotic germplasm (Tanksley and Nelson 1996). In this method, instead of using traditional F_2 or RIL mapping populations, it involves two or three backcrosses to the recurrent parent during population development, thus reducing the amount of donor introgressions in each individual. This method is especially advantageous, where wild relatives are used to develop mapping populations. AB-QTL method is used for simultaneous discovery and transfer of valuable QTLs from wild relatives. Since QTL analysis is delayed till BC_2 or BC_3 generations, it may be possible to detect dominant, partially dominant, overdominant, additive, and epistatic QTLs to name a few. Introgression of useful genes/traits accompanied by undesirable genomic fractions harboring deleterious alleles, collectively called linkage drag, can be overcome to identify favorable exotic quantitative trait locus (QTL) alleles for the improvement of agronomic traits. Two wild relatives, namely, *C. cajanifolius* (the progenitor species) and *C. acutifolius*, a wild species from the secondary gene pool and with many desirable traits (Mallikarjuna et al. 2011a), were used to develop AB-QTL mapping populations. The populations are ready for phenotyping and genotyping after two backcrosses and selfing. It is envisaged that such mapping populations will identify useful alleles present in the wild species as observed in other crops such as rice (Septiningsih et al. 2003) and groundnut (Chap. 8, this volume). Initial screening efforts have shown that AB-QTL lines have resistance to pod borers, *Phytophthora* blight, high yield, and a few other desirable traits.

7.7 Diversity in Expression of Traits Due to Alien Introgression

7.7.1 Segregation of Traits Following Classical Inheritance Pattern

Wild relatives from the secondary gene pool cross with cultigen following the Mendelian pattern of inheritance. This meant the F_1 produced broadly showed the expression of both the parental species in the ratio of 1:1. When the F_1 was selfed, the progeny lines segregated to traits in the ratio of 3:1, again following the Mendelian pattern of segregation. Similar results were observed when a tertiary gene pool species *C. platycarpus* was crossed with the cultigen (Mallikarjuna et al. 2011b). DArT analysis of the BC_4F_1 lines, where the recurrent parent was the cultigen, showed that it followed the Mendelian pattern of inheritance having approximately 94 % percent C. *Cajan* genome and the remaining of C. *platycarpus* (Mallikarjuna et al. 2011b).

7.7.2 Segregation Pattern of Genomic Dominance

C. volubilis, a wild relative in the tertiary gene pool of pigeonpea, produced F_1 hybrids when crossed with the cultigen. The F_1 hybrid showed traits of both the parental species of both *C. volubilis* and the cultigen, and the hybridity was further confirmed by SSR analysis (Sandhya Srikanth and Mallikarjuna N, unpublished data). F_1 hybrid was selfed to produce F_2 progeny lines. All the plants from F_2 onward showed genomic disequilibrium by resembling the cultigen with respect to morphological traits. None of the plants had any morphological traits or characters of *C. volubilis*. Molecular analysis of F_2 plants confirmed that they resembled the cultigen. Such a phenomenon has not been observed in pigeonpea wide crosses using compatible wild relatives from the secondary gene pool (Mallikarjuna and Saxena 2002; Saxena et al. 2010). Genomic asymmetry was observed in the present cross as distantly related genomes of *Cajanus*, i.e., of cultigen and of *C. volubilis*, were brought together for the first time through hybridization. Bringing together distantly related genomes involves radical and rapid mode of speciation by means of interspecific hybridization. It has been observed in wheat that, when different genomes were brought together, there was predominance of one genome over the others (Flagel et al. 2009; Rapp et al. 2009). In wheat A and B genome hybrids, B genome exhibited more or higher marker polymorphism than the A genome (Chao et al. 1989). Genome-wide transcriptome analysis in synthetic *Arabidopsis* allotetraploid showed that expression patterns from one genome could be dominant over the other (Wang et al. 2006a, b). Pumphrey et al. (2009) observed that some of the genes were similar to one of the parental genomes in synthetic hexaploid wheat. In the present investigation too, predominant expression of one of the parental genomes over the other was observed, and this can be explained taking examples of wheat, *Arabidopsis*, etc., that when distantly related genomes are brought together, there is silencing of the expression of the other genome completely or partially, although the DNA from the other parent is present, at least for many of the morphological traits female parental genome expression is obvious. Genome asymmetry has been observed not only for the expression of morphological traits but for other important traits in a few crops when different genomes are brought together by hybridization. Genome asymmetry in the control of storage proteins has been observed in wheat (Levy et al. 1988). In *Gossypium hirsutum*, genome asymmetry was found in the accumulation of seed storage proteins (Hu et al. 2011). Genome asymmetry in the control of agronomic, disease, and pest resistance traits has also been observed in wheat (Feldman et al. 2012). There were unsuccessful attempts in the past to cross pigeonpea with *C. volubilis* pollen (Pundir and Singh 1985). This is the first report of successful interspecific hybridization between pigeonpea and *C. volubilis*. Morphological traits of the F_1 hybrid were more skewed toward the female parent, and such phenomenon is not new when distant genomes are made to come together through wide hybridization. *C. volubilis* is a wild relative of pigeonpea placed in its tertiary gene pool (Mallikarjuna et al. 2011a; Bohra et al. 2010). Genomic studies have also shown its distant relationship with cultivated pigeonpea (Punguluri et al. 2007). Genome-wide transcription analysis in synthetic *Arabidopsis* allotetraploid showed that expression patterns from one genome could be dominant over the other genome (Wang et al. 2006a, b). Pumphrey et al. (2009) found that a small percentage of hybrids between wheat and synthetic hexaploids were similar to one of the parents. We report for the first time in pigeonpea that such a phenomenon is taking place in the hybrid between *C. cajan* (cultivated pigeonpea) and *C. volubilis* with the morphology of the F_2 hybrids skewed toward the female parent. Genetic control in storage proteins has been observed in allopolyploid wheat. Galili and Feldman (1984) showed that inactivation of an endosperm protein is brought about by an intergenomic suppression. Wheat genome-driven control of some agronomic, pest, and disease resistance was observed in wheat. Peng et al. (2000) observed that the R-gene cluster in the B genome of wheat and high marker clustering in the B genome than the A genome are the result of the expression of genome asymmetry. The ability of one genome to suppress the activity of genes in another in newly formed hybrids with different genomes may be to prevent defective organ formation/phenotype.

7.8 Conclusions

The synthesis of the review article shows the state-of-the-art information on experiments to broaden the genetic base of pigeonpea, elevating the crop to a trendsetting legume. Firstly, its genome sequence became available and now with interspecific and intergeneric hybrids utilizing wild relatives from secondary, tertiary, and quaternary gene pools is a path-breaking research. Concerted efforts might yield many more intergeneric crosses utilizing many more not only *Rhynchosia* species but with other genera such as *Flemingia*, *Dunbaria*, etc. At present there are 230 *Rhynchosia* species reported, and many of them have useful traits of interest. The need of the time is the committed research to exploit this vast gene pool with a major effort to cross many more species of *Cajanus* and utilize other related genera too. The results of the present investigation open new vistas in wide crosses research in pigeonpea. No other food legume crop has been investigated for alien introgression and succeeded in crossing wild relatives from all the gene pools, namely, secondary, tertiary, and quaternary. Pigeonpea is one crop where tremendous progress has been made to cross wild *Cajanus* species from different gene pools and introgress genes/traits successfully.

With these successes, it can no more have a narrow genetic base. With the advances in pigeonpea genomics, a major effort in sequencing the crop, and success in wide crosses in pigeonpea, it has emerged from being labeled as an orphan crop to a trendsetter. Recent successes in wide crosses show that it is possible to introduce desirable traits such as pod-borer resistance, develop CMS systems, develop lines with multiple disease and pest resistance, change plant type, and increase seed weight and yield.

References

Bakshu L, Venkataraju RR (2001) Antimicrobial activity of *Rhynchosia beddomei*. Fitoterpia 72:579–582

Bantilan MCS, Joshi PK (1996) Returns to research and diffusion investments on wilt resistance in pigeonpea. Impact series (No. 1) Patancheru 502 324, Andhra Pradesh, India. International Crops Research Institute for the Semi-Arid Tropics, p 36

Bohra A, Mallikarjuna N, Saxena K, Upadhyaya HD, Vales I et al (2010) Harnessing the potential of crop wild relatives through genomics tools for pigeonpea improvement. J Appl Biol 37:83–98

Chao S, Sharp PJ, Worland AJ, Koebner RMD, Gale MD (1989) RFLP-based genetic maps of wheat homoeologous group 7 chromosomes. Theor Appl Genet 78:495–504

Cherian CA, Mallikarjuna N, Jadhav DR, Saxena KB (2006) Open flower segregants selected from *Cajanus platycarpus* crosses. J SAT Agric Res 2:1–2, ISSN 0973-3094

Choudhary P, Khanna SM, Jain PK, Bharadwaj C, Kumar J, Lakhera PC et al (2011) Genetic structure and diversity analysis of the primary gene pool of chickpea using SSR markers. Genet Mol Res 11:891–905

Drabu S, Chaturvedi S, Sharma M (2011) Analgesic activity of methanolic extract from aerial parts of *Rhynchosia capitata* DC. Int J Pharm Technol 3:3590–3600

FAO (2009) FAO stat databases. Retrieved from: http://faostat.fao.org

Feldman M, Levy AA, Fahima T, Korol A (2012) Genomic asymmetry in allopolyploid plants: wheat as a model. J Exp Bot 63:5045–5059. doi:10.1093/jxb/ers192

Flagel LE, Chen L, Chaudhary B, Wendel JF (2009) Coordinated and fine scale control of homoeologous gene expression in allotetraploid cotton. J Hered 100:487–490

Fuller DQ, Harvey E (2006) The archaeobotany of Indian pulses: identification, processing and evidence for cultivation. Environ Archaeol 11:219–246

Galili G, Feldman M (1984) Intergenomic suppression of endosperm protein genes in common wheat. Can J Genet Cytol 26:651–656

Goodman MM (1990) Genetic and germplasm stocks worth conserving. J Hered 81:11–16

Harlan JR, de Wet JMJ (1971) Towards a rational classification of cultivated plants. Taxon 20:509–517

Hu G, Housron NL, Pathak D, Schmidt L, Wendel JF (2011) Genomically biased accumulation of seed storage proteins in allopolyploid cotton. Genetics 189:1103–1115

Jadhav DR, Mallikarjuna N, Sharma HC, Saxena KB (2012a) Introgression of *Helicoverpa armigera* resistance from *Cajanus acutifolius*-a wild relative from secondary gene pool of pigeonpea (*Cajanus cajan*). Asian J Agric Sci 4(4):242–248

Jadhav DR, Mallikarjuna N, Rathore A, Pokle D (2012b) Effect of some flavonoids on survival and development of *Helicoverpa armigera* (Hübner) and *Spodoptera litura* (Fab) (Lepidoptera: Noctuidae). Asian J Agric Sci 4(4):298–307

Kannaiyan J, Nene YL, Reddy MV, Ryan JG, Raju TN (1984) Prevalence of pigeonpea disease and associated crop losses in Asia, Africa and America. Trop Pest Manag 30:62–71

Kassa MT, Penmetsa RV, Carrasquilla-Garcia N, Sarma BK, Datta S, Upadhyaya HD, Varshney RV, von Wettberg EJB, Douglas RC (2012) Genetic patterns of domestication in pigeonpea (*Cajanus cajan* (L.) Millsp.), and wild *Cajanus* relatives. PLoS ONE 7:e39563. doi:10.1371/journal.pone.0039563

Kumar SP (1985) Crossability, genome relationships and inheritance studies in intergeneric hybrids of pigeonpea. Ph.D. thesis, University of Hyderabad, Hyderabad

Kumar S, Gupta S, Chandra S, Singh BB (2003) How wide is the genetic base of pulse crops? In: Ali M, Singh BB, Kumar S, Dhar V (eds) Pulses in new perspectives. Proceedings of the national symposium on crop diversification and natural resource management, Indian Institute of Pulses Research, Kanpur, UP, India, pp 211–221

Ladizinsky G (1985) Founder effect in crop-plant evolution. Econ Bot 39(2):191–199

Levy AA, Galili G, Feldman M (1988) Polymorphism and genetic control of high molecular weight glutenin subunit in wild tetraploid wheat *Triticum turgidum* var. *dicoccoides*. Heredity 61:63–72

Mallikarjuna N (1998) Ovule culture to rescue aborting embryos from pigeonpea (*Cajanus Cajan* L. Millspaugh) wide crosses. Indian J Exp Biol 36:225–228

Mallikarjuna N, Saxena KB (2002) Production of hybrids between *Cajanus acutifolius* and *C. cajan*. Euphytica 124(1):107–110

Mallikarjuna N, Saxena KB (2005) A new cytoplasmic male-sterility system derived from cultivated pigeonpea cytoplasm. Euphytica 142(1–2):143–148

Mallikarjuna N, Sharma HC, Upadhyaya HD (2007) Exploitation of wild relatives of pigeonpea and chickpea for resistance to *Helicoverpa armigera*. SAT eJournal I ejournalicrisatorg 3(1):4–7

Mallikarjuna N, Saxena KB, Jadhav DR (2011a) *Cajanus*. In: Kole C (ed) Wild crop relatives: genomic and breeding resources. Springer, Berlin/Heidelberg, pp 21–33

Mallikarjuna N, Senthivel S, Jadhav DR, Saxena KB, Sharma HC, Upadhyaya HD, Rathore A, Varshney RK (2011b) Progress in the utilization of *Cajanus platycarpus* (Benth.) Maesen in pigeonpea improvement. Plant Breed 130(5):507–514

Mallikarjuna N, Saxena KB, Lakshmi J, Varshney RK, Srikanth S, Jadhav DR (2012a) Differences between *Cajanus cajan* (L.) Millspaugh and *C. cajanifolius* (Haines) van der Maesen, the progenitor species of pigeonpea. Genet Resour Crop Evol 59:411–417

Mallikarjuna N, Jadhav DR, Saxena KB, Srivastava RK (2012b) Cytoplasmic male sterile systems in pigeonpea with special reference to A_7 CMS. Electron J Plant Breed 3(4):983–986

Marley PS, Hillocks RJ (1996) Effect of root-knot nematodes (*Meloidogyne* spp.) on *Fusarium* wilt in pigeonpea (*Cajanus cajan* (L.) Millspaugh). Field Crop Res 46:15–20

Mula MG, Saxena KB (2010) Lifting the level of awareness on pigeonpea – a global perspective. Research report. International Crops Research Institute for the Semi-Arid Tropics, Patancheru

Murthy KSR, Emmannuel S (2011) Nutritional AND anti-nutritional properties of the unexploited wild legume *Rhynchosia bracteata* benth. Bangladesh J Sci Ind Res 46:141–146

Odeny DA, Jayashree B, Ferguson M, Hoisington D, Cry LJ, Gebhardt C (2007) Development, characterization and utilization of microsatellite markers in pigeonpea. Plant Breed 126:130–136

Oke DB, Tewe OO, Fetuga BL (1995) The nutrient composition of some cowpea varieties. Niger J Anim Prod 22:32–36

Peng J, Korol AB, Fahima T, Rodert MS, Ronin YI, Li YC, Nevo E (2000) Molecular genetic maps in wild emmer wheat, *Triticum dicoccoides*: genome-wide coverage, massive negative interference and putative quasi-linkage. Genome Res 10:1509–1531

Pumphrey M, Bai J, Laudencia-Chingcuanco D, Gill BS (2009) Non-additive expression of homoeologous genes is established upon polyploidization in hexaploid wheat. Genetics 181:1147–1157

Pundir RPS, Singh RB (1985) Crossability relationships among *Cajanus*, *Atylosia* and *Rhynchosia* species and detection of crossing barriers. Euphytica 34:303–308

Punguluri SK, Janaiah K, Govil JN, Kumar PA, Sharma PC (2007) AFLP finger printing in pigeonpea (*Cajanus cajan* (L.) Mill sp) and its wild relatives. Genet Resour Crop Evol 53:423–431

Rapp RA, Udall JA, Wendel JF (2009) Genomic expression dominance in allopolyploids. BMC Biol 7:18. doi:10.1186/1741-7007-18

Ratnaparkhe MB, Gupta VS (2007) Pigeonpea. In: Kole C (ed) Genome mapping and molecular breeding in plants: pulses, sugar and tuber crops. Springer, Berlin, pp 133–142

Saxena KB (2000) Pigeonpea. In: Gupta SK (ed) Plant breeding – theory and techniques. Agrobios, Jodhpur, pp 82–112

Saxena KB (2008) Genetic improvement of pigeonpea–a review. Trop Plant Biol 1:159–178

Saxena KB, Faris DG, Kumar RV (1987) Relationship between seed size and protein content in newly developed high protein lines of pigeonpea. Plant Food Hum Nutr 36:335–340

Saxena KB, Ariyanayagam RP, Reddy LJ (1992) Genetics of a high-selfing trait in pigeonpea. Euphytica 59:125–127

Saxena KB, Ravikoti VK, Sultana R (2010) Quality nutrition through pigeonpea – a review. Health 2(11):1335–1344

Septiningsih E, Prasetiyono J, Lubis E, Tai TH, Tjubaryat T, Moeljopawiro S, McCouch SR (2003) Identification of quantitative trait loci for yield and yield components in an advanced backcross population derived from the *Oryza sativa* variety IR64 and the wild relative *O. rufipogon*. Theor Appl Genet 107:1419–1432

Singh NB, Singh IP, Singh BB (2005) Pigeonpea breeding. In: Ali M, Kumar S (eds) Advances in pigeonpea research. Indian Institute of Pulses Research, Kanpur, pp 67–95

Srikanth S, Rao MV, Mallikarjuna N (2013) Interspecific hybridization between *Cajanus cajan* (L.) Millsp. and *C. lanceolatus* (WV Fitgz) van der Maesen. Plant Genet Resour: Charact Util. doi:10.1017/S1479262113000361

Tanksley SD, McCouch SR (1997) Seed banks and molecular maps: unlocking genetic potential from the wild. Science 277:1063–1066

Tanksley SD, Nelson JC (1996) Advanced backcross QTL analysis: a method for the simultaneous discovery and transfer of valuable QTLs from unadapted germplasm into elite breeding lines. Theor Appl Genet 92:191–203

Upadhyaya HD, Reddy LJ, Gowda CLL, Reddy KN, Singh S (2006) Development of a mini core subset for enhanced and diversified utilization of pigeonpea germplasm resources. Crop Sci 46:2127–2132

Upadhyaya HD, Reddy KN, Gowda CLL, Singh S (2007) Phenotypic diversity in the pigeonpea (*Cajanus cajan*) core collection. Genet Resour Crop Evol 54:1167–1184

Upadhyaya HD, Reddy KN, Singh S, Gowda CLL (2012) Phenotypic diversity in *Cajanus* species and identification of promising sources for agronomic traits and seed protein content. Genet Resour Crop Evol 60:639–659. doi:10.1007/s10722-012-9864-0

van der Maesen LJG (1980) India is the native home of pigeonpea. In: Arends JC, Boelema G, de Groot CT, Leeuwenberg AJM (eds) Ibergratulatorius in honorem H. C.D. de Wit landbouwhogeschool Miscellaneous paper no 19. Veenman H, Zonen BV, Wageningen, pp 257–262

van der Maesen LJG (1986) *Cajanus* DC and *Alylosia* W. &. A. (Leguminosae). Agricultural University Wageningen papers 85-4 (1985). Agricultural University Wageningen, the Netherlands, p 222

van der Maeson LJG (1995) Pigeonpea *Cajanus cajan*. In: Smartt J, Simmonds NW (eds) Evolution of crop plants. Longman, Essex, pp 251–255 van der Maesen LJG (2006) *Cajanus cajan* (L.) Mill sp. In: Brink M, Belay G (eds) Plant resources of Africa 1. Cereals and pulses. Backhuys Publishers, Wageningen

Varshney RK, Chen W, Li Y, Bharti AK, Saxena RK, Schlueter JA et al (2012a) Draft genome sequence of pigeonpea (*Cajanus cajan*), an orphan legume crop of resource-poor farmers. Nat Biotechnol 30:83–89

Varshney RK, Kudapa H, Roorkiwal M, Thudi M, Pandey MK, Saxena RK, Chamarthi SK, Murali Mohan S, Mallikarjuna N, Upadhyaya HD, Gaur PM, Krishnamurthy L, Saxena KB, Nigam SN, Pande S (2012b) Advances in genomics research and molecular breeding applications in SAT legume crops by using next generation sequencing and high-throughput genotyping technologies. J Biosci 37:811–820

Wang J, Tian L, Lee HS, Chen ZJ (2006a) Non-additive regulation of FRI and FLC loci mediates flowering time variation in *Arabidiopsis* allopolyploids. Genetics 173:965–974

Wang J, Tian L, Lee HS et al (2006b) Genome wide non-additive gene regulation in *Arabidiopsis* allotetraploids. Genetics 172:507–517

Wright B (1997) Crop genetic resource policy: the role of ex situ gene banks. Aust J Agric Resour Econ 41:87–115

Yang S, Pang W, Ash G, Harper J, Carling J, Wenzel P, Hutter E, Zong X, Kilian A (2006) Low level of genetic diversity in cultivated pigeonpea compared to its wild relatives is revealed by diversity arrays technology. Theor Appl Genet 113:585–595

Groundnut

8

Nalini Mallikarjuna, Krishna Shilpa, Manish Pandey,
P. Janila, and Rajeev K. Varshney

Abstract

Groundnut, a crop rich in nutrients, originated in South America and spread to the rest of the world. Cultivated groundnut contains a fraction of the genetic diversity present in their closely related wild relatives, which is not more than 13 %, due to domestication bottleneck. Closely related ones are placed in section *Arachis*, which have not been extensively utilized until now due to ploidy differences between the cultivated and wild relatives. In order to overcome *Arachis* species utilization bottleneck, a large number of tetraploid synthetics were developed at the Legume Cell Biology Unit of Grain Legumes Program, ICRISAT, India. Evaluation of synthetics for some of the constraints showed that these were good sources of multiple disease and pest resistances. Some of the synthetics were utilized by developing ABQTL mapping populations, which were screened for some biotic and abiotic constraints. Phenotyping experiments showed ABQTL progeny lines with traits of interest necessary for the improvement of groundnut.

8.1 Introduction

Groundnuts (*Arachis hypogaea* L.) are rich in nutrients, providing over 30 essential nutrients and phytochemicals. They are a good source of niacin (Whitley et al. 2011), which plays a role in brain health and blood flow, folate, fiber, magnesium, vitamin E, manganese, and phosphorus (Savage and Keenan 1994). Plumpy'Nut, a ready-to-use therapeutic food made from groundnut, is a popular source of nutrient used by UNESCO to treat acute malnourished kids in Africa. Groundnuts contain about 25 % protein, a higher proportion than in any true nut. Recent research on groundnuts and nuts in general has found antioxidants (Yu et al. 2005) and other chemicals, which may provide health benefits. Roasted groundnuts rival the antioxidant content of blackberries and strawberries and are far richer in antioxidants than carrots or beets.

N. Mallikarjuna (✉) • K. Shilpa • M. Pandey
P. Janila • R.K. Varshney
Grain Legumes Program, International Crops Research Institute for Semi Arid Tropics (ICRISAT), Patancheru 502 324, Andhra Pradesh, India
e-mail: mysorenalinicgiar@yahoo.com; krishnashilpa.11@gmail.com; m.pandey@cgiar.org; p.janila@cgiar.org; r.k.varsheny@cgiar.org

M. Singh et al. (eds.), *Broadening the Genetic Base of Grain Legumes*,
DOI 10.1007/978-81-322-2023-7_8, © Springer India 2014

Research shows that groundnuts contain high concentrations of antioxidant polyphenols (Craft et al. 2010). Groundnuts are a significant source of resveratrol, and the amount of resveratrol equivalent to that present in red grapes (Sanders et al. 2000), a chemical associated with reduction in the risk of cardiovascular disease (Fraser et al. 1992; Hu et al. 1998; Prineas et al. 1993) and cancer (Awad et al. 2000) and having antiaging properties, hence would have high impact in both health and cosmetic industries. Groundnuts are a source of coenzyme Q10 (Pravst and Zmitek 2010), as are oily fish, beef, soybeans, and spinach. It is believed that the crop has originated in South America. According to Vavilov (1951), it was first domesticated in the Brazilian-Paraguayan region. The area of the valleys of Paraguay and Parana rivers is the most likely center of origin of this legume. Excavation in coastal Peru dating back to 800 BC shows the cultivation of groundnut. From South America, groundnut spread to other parts of the world. It was commonly found in the West Indies, but not in the United States in pre-Columbian times. Groundnut was introduced to the Old World in the sixteenth century when the Portuguese took the seeds from America to Africa. The Spaniards introduced it into the Philippines. It then spread to China, India, Japan, Malaysia, and other parts of the world.

Human interaction through selection of most suited lines over the centuries resulted in loss of much of the genetic diversity/desirable alleles and genes whose importance is now being realized. Further, we are still unaware of future preference for the so-called "lost" genetic diversity which includes genes for resistance/tolerance to biotic and abiotic stresses as well as taste and nutritional composition along with yield. Although such concerns are raised at majority of the scientific gatherings, still not much initiative has been taken even for very important food crops. Hence, this is the prime time to retrieve desirable alleles not only to address existing problems but also for the future as well in order to sustain food production. Although pre-breeding utilizing wild gene pool does not produce new varieties, it does turn up intermediate products that breeders find easier to use. It throws up enough variation in the crops such as groundnut to sustain breeding activities especially with the assistance of molecular markers. Many public agricultural research institutions, such as the International Rice Research Institute (IRRI), International Maize and Wheat Improvement Center (CIMMYT), International Institute of Tropical Agriculture (IITA), and International Crops Research Institute for Semi-Arid Tropics (ICRISAT), have active pre-breeding programs in their mandate crops. Pre-breeding is the link between conservation and use of crop wild relatives. Out of all the raw materials at the breeder's disposal, the diversity of crop wild relatives has been relatively neglected. The conserved germplasm in the gene banks is for the present and future use. With urbanization, explosion in population growth, and dwindling water resources and being in a 2 °C warmer world, we may not have a choice but to bring in new sources of variation through pre-breeding.

8.2 Taxonomy

The cultivated groundnut (*Arachis hypogaea* L.), an annual herb belonging to the family Fabaceae (Leguminosae), is classified into two subspecies, subsp. *fastigiata* Waldron and subsp. *hypogaea* Krap. et. Rig. The subsp. *fastigiata* contains four botanical varieties, var. *vulgaris*, var. *fastigiata*, var. *peruviana*, and var. *aequatoriana*. The subsp. *hypogaea* contains two varieties, var. *hypogaea* and var. *hirsuta*. Each of these botanical types has different plant, pod, and seed characteristics (Krapovickas and Gregory 1994). Groundnut is an allotetraploid ($2n=2x=40$) with "AA" and "BB" genomes. All species, except the cultivated species (*A. hypogaea* and *A. monticola*) in section *Arachis* and certain species in section *Rhizomatosae*, are diploid ($2n=2x=20$). The diploid progenitors, *A. duranensis* and *A. ipaensis*, contributed "AA" and "BB" genomes, respectively, to the cultivated groundnut (Kochert et al. 1996). The phylogenetic analyses based on intron sequences and microsatellite markers also provide evidence for this hypothesis (Moretzsohn et al. 2012). A single hybridization event between the diploid progenitors followed by chromosome doubling (Kochert et al. 1996) about 3,500 years ago led to the origin of cultivated groundnut. Krapovickas and Gregory (1994) used taxonomy and crossability studies to classify 69 *Arachis* species into nine sections. The additional 11 species described by Valls and Simpson (2005) also come under these nine sections, making the total to 80 *Arachis* species. Sections *Trierectoides*, *Erectoides*, *Extranervosae*, *Triseminatae*, and *Heteranthae* are treated as older sections while *Procumbentes*, *Caulorrhizae*, *Rhizomatosae*, and *Arachis* are of more recent origin.

8.3 Production-Related Problems

There is a large gap between potential pod yield and the realized pod yield in most of the groundnut growing areas (Johansen and Nageswara Rao 1996). Abiotic stresses such as drought and high temperature are important yield-reducing factors. Groundnut is attacked by several fungal pathogens such as late leaf spot (LLS) caused by *Phaeoisariopsis personata* (Berk. & Curt.) von Arx, early leaf spot (ELS) caused by *Cercospora arachidicola* Hori, and rust caused by *Puccinia arachidis* Spegazzini which are among the important foliar fungal diseases worldwide. Aflatoxins are potent carcinogens produced by *Aspergillus* spp. *A. flavus* is the predominant species in Asia and Africa. Stem and pod rot, caused by *Sclerotium rolfsii*, is a potential threat to groundnut. Apart from this, groundnut is prone to several virus diseases such as groundnut rosette disease (GRD), peanut bud necrosis disease (PBND), peanut stripe potyvirus (PStV), peanut stem necrosis disease (PSND), and peanut clump virus disease (PCVD). Bacterial wilt, caused by *Ralstonia solanacearum*, is predominant among bacterial diseases of groundnut. Globally, nematodes cause 11.8 % yield loss in groundnut (Sharma and McDonald 1990). Aphids (*Aphis craccivora* Koch), several species of thrips (*Frankliniella schultzei*, *Thrips palmi*, and *F. fusca*), leaf miner (*Aproaerema modicella*), red hairy caterpillar (*Amsacta albistriga*), jassids (*Empoasca kerri* and *E. fabae*), and *Spodoptera* are the major insect pests in groundnut, causing serious damage (Wightman and Amin 1988). Groundnut borer or weevil (*Caryedon serratus*) and rust-red flour beetle (*Tribolium castaneum*) are the major storage insect pests in groundnut. Nutritional traits which include oil, protein, sugar, iron, and zinc content, fatty acid profile, and freedom from toxins are important. The major issue being the presence of sources of resistance in cultivated germplasm is low to moderate compared to high levels of resistance in wild *Arachis* species.

8.4 Evolution and Diversity in Cultivated Groundnut

Cultivated groundnut contains a fraction of the genetic diversity which is not more than 13 % (Varshney et al. 2009), found in their closest wild relatives in section *Arachis* (Kochert et al. 1991), a legacy of the "domestication bottleneck." Groundnut has an interesting evolutionary history. The domesticated groundnut is an amphidiploid or allotetraploid, meaning that it has two sets of chromosomes from two different species, thought to be *A. duranensis* and *A. ipaensis* (Kochert et al. 1991, 1996; Seijo et al. 2007). These species combined in the wild to form the tetraploid species which gave rise to the domesticated groundnut. The first domestication bottleneck was the combination of two species *A. duranensis* and *A. ipaensis* among 26 species from section *Arachis*. Crossing experiments between *A. duranensis* and *A. ipaensis* have shown that the diploid hybrid is highly sterile (Mallikarjuna et al. 2011a). Had the hybrid been fertile, there would not have been the need to double its chromosome number to form the fertile amphidiploid. It would probably have remained a diploid than a tetraploid as it is today. So the second bottleneck was in the formation of a diploid sterile hybrid. The third bottleneck was in process of chromosome duplication to form the allotetraploid as it is known in literature that polyploidy causes genetic bottlenecks (Sanford 1983). Ancient farmers would have selected relatively few plants from the progenitors of modern crops in a limited number of places, and a similar situation would have existed for groundnut in South America. This can be visualized as the fourth bottleneck. The early Portuguese and Spanish traders during their expeditions spread the crop to the rest of the world thus giving rise to yet another, the fifth, bottleneck, superimposed by the sixth bottleneck which is the self-pollinating nature of the crop. To conclude, groundnut is the product of evolution after a series of six bottlenecks.

8.5 Utilization of Wild Relatives

Arachis species from section *Arachis*, which are true diploid with $2n=20$, have been extensively utilized in crosses with cultivated groundnut ($2n=40$) utilizing various pathways of introgression (Simpson and Starr 2001), and traits of interest such as resistance to nematode, late leaf spot, rust, and *Spodoptera litura* transferred (Garcia et al. 1996; Burrow et al. 1996; Singh et al. 2003; Mallikarjuna et al. 2004a, b). GPBD4, a groundnut genotype resistant to rust and LLS, was released for commercial cultivation in India (Gowda et al. 2002). One of its parents, ICGV 86855, is an interspecific derivative between *A. hypogaea* and *A. cardenasii*, resistant to rust and LLS, and was developed at ICRISAT, India. Members of section *Procumbentes* have been successfully crossed and fertile progeny has been obtained using in vitro techniques. Progeny lines were screened for various traits of interest. Another member of section *Procumbentes*, *A. kretschmeri*, was also crossed successfully (Mallikarjuna and Hoisington 2009). *A. chiquitana* that was previously placed in section *Procumbentes* (Krapovickas and Gregory 1994) has now been moved to section *Arachis* (Robledo et al. 2009). F_2 seeds were screened for *A. flavus* infection, and many of the seeds did not have any *A. flavus* infection, whereas the control and some of the other F_2 seeds had *A. flavus* infection (Mallikarjuna 2005). *A. glabrata* belonging to section *Rhizomatosae* was successfully crossed with *A. duranensis* and *A. diogoi* (Mallikarjuna 2002) and *A. hypogaea* (Mallikarjuna and Sastri 2002) using embryo rescue techniques (Mallikarjuna and Sastri 1985). The interest at present is to exploit the genetic diversity present in section *Arachis* through the development of tetraploid amphidiploid and autotetraploid groundnuts which are also called synthetic groundnuts.

Inadequate levels of resistance in peanut germplasm are one of the important factors for not having resistance to *A. flavus* aflatoxin in peanut. This means sources of resistances have to be scouted beyond the cultivated primary gene pool. The report from Xue et al. (2005) showed that *Arachis* species *A. duranensis* (8 accessions) and *A. cardenasii* (2 accessions) from section *Arachis* had high levels of resistance to aflatoxin production and interspecific derivatives obtained from them continued to show the trait. ICRISAT screened advance generation lines derived from *A. cardenasii* (10,017 lines) in aflatoxin sick plot for three consecutive years and found many of the lines with low aflatoxin production. This opens up new avenues for aflatoxin resistance breeding in peanut (Mallikarjuna N and Sudini H, unpublished data). Sources of resistance to late leaf spot caused by *Cercosporidium personatum* (Berk. & M.A. Curtis) are higher in wild *Arachis* species (Subrahmanyam et al. 1985b) as compared to moderate levels of resistance in cultivated germplasm (Dwivedi et al. 2002). *Arachis cardenasii*-derived lines showed resistance to LLS, when screened under unprotected field conditions in different locations (Mallikarjuna N and Sudini H, unpublished data). Peanut bud necrosis disease (PBND) is an economically important virus disease of peanut in many Asian countries where peanut is grown. The disease causes crop losses exceeding 89 million US dollars in India alone (Anon 1992). Sources of resistance are absent in cultivated germplasm (Reddy 1998). Many of the *Arachis* species have been found to be resistant to the disease (Reddy et al. 2000). Stable lines derived from *Arachis* species were screened for PBND under disease hot spot locations, and a few resistant lines were identified (Sunkad G and Mallikarjuna N, unpublished data).

Among the soilborne fungal diseases of peanut, stem rot caused by *S. rolfsii* is a potential threat to peanut production throughout the world.

The disease causes severe damage during any stage of crop growth, and yield losses over 25 % have been reported by Mayee and Datar (1988). Sources of resistance to the constraint is not up to the desired level in cultivated gene pool. Stable lines derived from *Arachis* species were screened for *S. rolfsii* in the disease hot spot location at Dharwad, Karnataka State, India, and a few lines with durable resistance were obtained (Kenchanagowdar and Mallikarjuna N, unpublished data). *Spodoptera litura*, also called fall armyworm, a polyphagous insect, is becoming an important insect pest of groundnut with sources of resistance to the pest absent in the cultivated gene pool. Yield losses of peanuts have been directly associated with higher density of larvae of *S. litura* and the intensity of defoliation (Panchbhavi and Nethradani 1987). Currently, no cultivars of peanut are known to express reasonable resistance to *S. litura*. However, some wild relatives of peanut were found resistant to *S. litura*. Neonate larvae suffer high levels of mortality, and the development of older larvae on resistant wild species is severely inhibited (Stevenson et al. 1993b). Stevenson et al. (1993a) identified flavonoids chlorogenic acid, quercetin, and rutin present in *Arachis kempff-mercadoi* responsible for resistance to *S. litura*. Mallikarjuna et al. (2004a) developed lines utilizing *A. kempff-mercadoi* and screened lines for *S. litura* resistance. Resistant derivatives were found to have high levels of flavonoids, and antibiosis mechanism prevented larval growth. Susceptible derivatives and the female parent *A. hypogaea* had low levels of flavonoids (Mallikarjuna et al. 2004b).

8.6 Development of Synthetics for Groundnut Improvement

Tremendous progress has been made in wheat and brassica, two similar polyploid genera in the development and utilization of synthetics, by combining the putative genome donors of the cultivated species. This triggered the development of tetraploid amphidiploids (synthetics) in groundnut. Until recently, there were three sources of amphiploids or new sources of tetraploid *A. hypogaea* available in public domain. The first one originated from a cross between *A. cardenasii* Krapov. et W.C. Gregory, *A. diogoi* Hoehne, and *A. batizocoi* Krapov. et W.C. Gregory, and it was utilized to develop backcross progeny lines (Simpson et al. 1993). Two groundnut cultivars, namely, COAN (Simpson and Starr 2001) and NemaTAM, were released utilizing this source. More recently, an amphidiploid was constructed utilizing *A. ipaensis* and *A. duranensis* (Favero et al. 2006). This amphidiploid is being used by Brazil and Senegal to develop backcross population and chromosome substitution lines. Preliminary mapping data indicated low level of marker segregation distortion in F_2 population utilizing amphidiploid (Dwivedi et al. 2008). A successful effort for genome-wide segment introgressions from a synthetic amphidiploid (*A. ipaensis* × *A. duranensis*) to a cultivated variety (Fluer 11) using molecular markers has already been reported (Fonceka et al. 2009). The third one is a cross between *A. gregoryi* and *A. linearifolium* (GCP 2005), and there is no information of using this source for peanut improvement. With this background, the Legume Cell Biology Unit of ICRISAT under the stewardship of N. Mallikarjuna has generated many new sources of tetraploid groundnut utilizing many more *Arachis* A, B, and K genome species, not used until now to generate new sources of tetraploid/synthetic groundnut (Mallikarjuna et al. 2011a). Traditionally, wild relatives of peanut were directly used in crossing program producing triploids as *Arachis* species in the compatible gene pool are diploids and cultivated groundnut is a tetraploid. Triploids are cumbersome to use for groundnut improvement, as they demand an elaborate backcross program, but such efforts have not gone without dividends (Mallikarjuna et al. 2004a, b). Among the published 17 new sources of synthetics (Mallikarjuna et al. 2011a), one (ISATGR 1212) had putative genome donors and it comprised of *A. duranensis* and *A. ipaensis*, and another one (ISATGR 40A) had the reciprocal combination (*A. ipaensis* × *A. duranensis*). Three of them had at least one of the genome donors of *A. hypogaea*, i.e., either *A. duranensis* or

A. ipaensis. Since there is only one accession of *A. ipaensis* reported to date, it can be presumed that it is indeed the B genome donor. With respect to *A. duranensis*, there are many accessions available in different gene banks across the world, and molecular analysis has shown that there are some differences between the accessions (Husain and Mallikarjuna 2012) as well as with respect to traits. *A. duranensis* accession ICG 8139 is of the earliest flowering accessions in the ICRISAT gene bank (Mallikarjuna N, unpublished data). Five amphidiploids had K genome either with A or B genome; hence these would be totally new combinations available for the improvement of groundnut. Among the five autotetraploids synthesized with A genome species, there is diversity with respect to the groups they belong, for example, one autotetraploid ISATGR 90B comprised of two A genome species *A. kempff-mercadoi* and *A. stenosperma. A. kempff-mercadoi* belongs to group Chiquitano. Members of the group Chiquitano grow in the southern and western portion of the Chiquitania biogeographic region of the Santa Cruz Department of Bolivia (Robledo et al. 2009). Another autotetraploid ISATGR 99B is made up of *A. diogoi* and *A. cardenasii. A. diogoi* belongs to the Pantanal group and *A. cardenasii* belongs to Chiquitano group. Robledo et al. (2009) have shown variability in heterochromatin and 18S–26S rRNA loci between the members belonging to Chiquitano and Pantanal groups. Hence, the autotetraploids generated are also important sources of variation which can be exploited for broadening the genetic base of cultivated groundnut.

Apart from the published new sources of synthetics, a few more synthetics have been generated since then (Mallikarjuna N, unpublished data). They include ISATGR 47A comprising of *A. valida* × *A. duranensis*. Here, the accession of *A. valida* was different than that of ISATGR 168B reported by Mallikarjuna et al. (2011a). ISATGR 72B is made up of *A. duranensis* and *A. cardenasii*. ISATGR 163B is made up *A. kempff-mercadoi* and *A. stenosperma*. A different accession of *A. kempff-mercadoi* was used in this cross than the one used to develop ISATGR 80A (Mallikarjuna et al. 2011a). In attempts to broaden the genetic base and introduce variation into cultivated groundnut, double synthetics were generated by combining the genomes of two synthetics. Care was taken to see that at least three *Arachis* species, and a maximum of four, contributed their genomes. There were five such sources developed, ISATGR 1212 (*A. duranensis* × *A. ipaensis*) X ISATGR 9A (*A. batizocoi* × *A. cardenasii*), ISATGR 1212 (*A. duranensis* × *A. ipaensis*) X ISATGR 5B (*A. magna* × *A. batizocoi*), ISATGR 278-18 (*A. duranensis* × *A. batizocoi*) X ISATGR 5B (*A. magna* × *A. batizocoi*), and ISATGR 1212 (*A. duranensis* × *A. ipaensis*) X ISATGR 265-5A (*A. kempff-mercadoi* and *A. hoehnei*) (Shilpa et al. 2013), and it was observed that although synthetics selected to develop double synthetics were stable and fertile, not all double synthetics were fertile. One of them [ISATGR 278-18 (*A. duranensis* × *A. batizocoi*) X ISATGR 11A (*A. magna* × *A. valida*)] did not set any pods/seeds, hence was a genetic dead end. The rest of the double synthetics set pods/seeds and hence were considered valuable sources of variation. All the synthetics and double synthetics produced single-seeded pods being larger in size in majority of the cases but resembling the wild species pods in shape and reticulation. An interesting feature was observed in the pod traits from double synthetic ISATGR 278-18 (*A. duranensis* × *A. batizocoi*) X ISATGR 5B (*A. magna* × *A. batizocoi*). Many of the pods were double seeded which did not have peg, as seen in many wild *Arachis* pods, and resembled those of *A. hypogaea* pod shape and pod wall architecture. Such a feature was not observed in any *Arachis* species conserved in the ICRISAT gene bank, or in the combination of *Arachis* species during the formation of synthetics, or in any other double synthetics generated in this study. Double-seeded pods from this cross may be a case of evolution fast-forward, as they resemble those of *A. hypogaea*. It will be interesting to see if only double-seeded pods can be recovered from this cross a few generations later.

Cultivated groundnut is susceptible to late leaf spot (LLS) caused by *Phaeoisariopsis personata* [(Berk. & M.A. Curtis) Aex], and the resistance is low to moderate in the primary gene pool of groundnuts (Dwivedi et al. 2002). Closely related wild species in the secondary gene pool are highly resistant to the disease (Subrahmanyam et al. 1985a). Diploid *Arachis* species are difficult to use for the introgression of LLS resistance due to ploidy differences between cultivated and *Arachis* species. Synthetics were screened for LLS in a disease hot spot location at Raichur. All the tetraploids, except for ISATGR 155, showed resistance to the disease (Shilpa et al. 2013). Utilizing one of the synthetics, namely, ISATGR 265-5, and crossing it with cultivated groundnut yielded BC_2F_3 lines. These lines were screened for LLS. Screening results showed that some of the progenies had LLS resistance (Sudini H and Mallikarjuna N, unpublished data), thus showing that synthesized tetraploids are good sources of LLS resistance. Mallikarjuna et al. (2012a) studied the components of LLS resistance such as incubation period, leaf area damage, lesion number and diameter, latent period of infection, and infection frequency in some of the synthetics using detached leaf technique. The studies gave a clearer picture of LLS resistance in the synthetics. Synthetics were screened for the presence of resveratrol, and its presence was observed in the few lines (Padmashree and Mallikarjuna, unpublished data). Peanut bud necrosis disease is an economically important disease causing yield losses up to 89 million USD in India alone (Anon 1992), and the sources of high levels of resistance are lacking in the cultivated germplasm, but many wild *Arachis* species have reasonable resistance (Reddy et al. 2000). Some of the synthetics were screened for PBND in a disease hot spot at Raichur, India, and many of them were found to have immune to resistant reaction to the disease (Shilpa et al. 2013). Since tetraploids can be easily crossed with cultivated peanut, sources of resistance in tetraploids open up new opportunities to breed PBND-resistant groundnuts. Plant proteinase inhibitors are known for improving defense against insects and pathogens (Ryan 1990). Trypsin and chymotrypsin inhibitors have been described as potential cancer or chemo-protective agents (Clemente et al. 2005). Many of the synthetics used in the study by Shilpa et al. (2013) showed high PI activity compared to that present in the cultivated groundnut varieties. This may be one of the reasons for the presence of disease and pest resistance in the synthetics. PIs are known to prevent the target insect from digesting protein by competitively binding to the active site of protein, which is the actual binding site of proteinase. As the insect cannot digest protein, it is subjected to starvation and/or death. PIs are also known to cause increased levels of insect deformity, due to the potential inhibition of the proteinases involved in the metamorphosis of the larvae (Prasad et al. 2010). Screening the synthetics for the presence of proteinase inhibitory activity against bovine pancreatic trypsin and chymotrypsin showed activity against midgut trypsin-like proteinases of *Spodoptera litura* (Padmashree and Swathi unpublished data). A new role for PI in the modulation of apoptosis or programmed cell death has been identified in soybean (Koslak et al. 1997). Although PI was also present in the cultivated groundnut, S. Marri and K. Padmashree (unpublished data) observed that the molecular components of those present in the cultivated and the synthesized tetraploids were different, and these components have a differential role with respect to insect midgut trypsin-like proteinases of *Spodoptera litura*. Nutritional composition which is composed of oil, fatty acid composition, O/L ratio (oleic to linoleic fatty acid ratio), amount of protein, and iodine value were studied (Shilpa et al. 2013). O/L ratio is an important factor in deciding the stability of the oil. Fatty acid composition is made up of unsaturated fatty acids (TUSF) and saturated fatty acids (TSF), which make up the physical and chemical properties of the oil. Variation was observed for total protein concentration. In the cultivated peanut accessions, total protein concentration ranged from 65 to 85 mg/g compared to a range of protein concentration of 43–134 mg/g in the synthetics. High-protein concentration synthetics can be selected to breed for high-protein groundnut lines in those regions of the world, where groundnut is not only an oilseed crop but is used as a food crop too. Some difference was observed with respect to percent oil content between the cultivated groundnut varieties (approx. 40 %) and synthetics (45–57 %) (Shilpa et al. 2013). With respect to O/L ratio, it was 1.0 in cultivated varieties, and in the synthetics, it varied from 0.8 to 1.3 (Shilpa et al. 2013). Not much difference was observed between the cultivated varieties and synthetics with respect to fatty acid composition (excluding oleic and linoleic). Iodine content (IV), a measure of the degree of unsaturation which has been commonly used as an indicator of predicting shelf life (Mercer et al. 1990), was 88–105 in synthetics compared to 90–95 in groundnut varieties.

In order to study meiotic recombinations between the cultivated groundnut and the newly developed synthetics, a range of synthetics were crossed with a few cultivars of *A. hypogaea* (Mallikarjuna et al. 2012b). The study showed good recombination between cultivated and synthetics with high pollen fertility in the hybrids, thus showing that synthetics form a good source of variation, which can be successfully utilized for broadening the genetic base of the cultigen. More recently, double synthetics were used in the crossing program, and a high level of meiotic recombination was observed (Mallikarjuna N, unpublished data). Fonceka et al. (2012) used a synthetic amphidiploid (Favero et al. 2006) and applied a conventional breeding scheme to capture the genetic diversity in peanut wild relatives. In their study, a set of 122 introgression lines (IL) that offered an extensive coverage of the cultivated peanut genome with generally a unique fragment per line and overlapping fragments between contiguous lines were developed. Their findings opened new avenues for peanut improvement using new sources of tetraploid/synthetic groundnuts. Realizing the scope for advanced backcross (ABQTL) breeding in the improvement of groundnut, initiatives have been taken at ICRISAT to develop three ABQTL mapping populations utilizing three sources of synthetics. These populations were segregated for several biotic, abiotic, and agronomic traits. A subset of BC2F1 individuals was genotyped with DArT markers to construct genetic maps. Advance generation lines (BC2F3) from all populations were screened for a range of biotic and abiotic traits such as O/L ratio, LLS, rust, PBND, and other yield-related traits. These initial screening experiments revealed a range of useful traits present in the populations. Population one was developed by utilizing synthetic amphidiploid ISATGR 1212 (Mallikarjuna et al. 2011a), which is composed of putative genome donors of *A. hypogaea* (*A. duranensis* × *A. ipaensis*). Screening the synthetic ISATGR 1212 for LLS and rust showed that it is highly resistant to LLS and rust (Mallikarjuna et al. 2012b). In population one (pop I), 16 % of the lines had moderate levels of resistance to LLS (score of 4, on a scale of 1–9). Majority of the lines had a score of 2 for rust (on a scale of 1–9). O/L ratio in the population varied from 1 to 4 with 50 % of the lines having a ratio of 2–3 compared to a ratio of 1 in cultivated lines used in the study. Only one line had a ratio of 4. The population was screened for peanut bud necrosis disease (PBND) in Raichur, which is a disease hot spot in the Karnataka State of India. The cultivated check line showed 48 % disease incidence. Two lines were devoid of the disease. Many (40 %) of the test lines from pop I had 20 % or less disease incidence, and 5 % of the lines had 10 % or less disease damage. Stem rot of groundnut caused by *Sclerotium rolfsii* is an economically important disease with sources of resistance lacking in the cultivated gene pool. Pop I (97 lines) was screened for the disease in a hot spot location, and 36 lines (37 % of the screened lines) did not show any disease symptoms. Hence, these can be classified as immune to the disease. The progenies were evaluated in a replicated trail along with the recurrent parent and some popular checks for dry pod yield and other yield parameters. The dry pod yield per plant of the progenies was between 3 and 17 g per plant higher compared to the recurrent parent ICGV 91114 that produced about 13 g per plant. Similarly, shelling outturn (ranged from 54 to 74 %) and 100-seed weight (HSW) (23 to 42 g) were higher than the recurrent parent mean values in the trial (63 % shelling outturn and 32 g of HSW). The proportion of sound mature kernels (SMK) in the progenies ranged from 75 to 97 % that is slightly higher than ICGV 91114 (93 %). Population 2 (pop II) was developed utilizing synthetic amphidiploid ISATGR 265-5 (*A. kempff-mercadoi* × *A. hoehnei*), which does not have any putative genome donors of *A. hypogaea*

(Mallikarjuna et al. 2011a). Population two had better levels of resistance to LLS compared to pop I. Forty percent of the lines had a score of 3 for LLS, and 58 % of the lines had a score of 4 (on a scale of 1–9). Most of the lines had a score 2 for rust (on a scale of 1–9). O/L ratio in the lines was higher in pop II compared to pop I. In four lines, the ratio was 4 or above but less than 5. In 41 % percent of lines, the ratio was above 2 and 57 % percent of lines had a ratio of more than 3. Thirty-eight percent of lines from pop II did not show any *S. rolfsii* symptoms, when screened for the disease in the hot spot location. These lines can be classified as immune to the disease. The dry pod yield in the progenies ranged between 4 and 27 g, which was lower as compared to the recurrent parent ICGV 87846, which had a mean of 36 g. Nevertheless, the other yield parameters of the progenies were slightly better than the mean value of ICGV 87846. The SMK of progenies ranged from 58 to 95 %, shelling outturn ranged from 55 to 74 %, and HSW ranged from 26 to 51 g, as compared to the recurrent parent ICGV 87846 (SMK 86 %, shelling outturn 68 %, and HSW of 44 g).

Breeders often encounter comprised yield and/or pod and kernel features when they use wild species in groundnut breeding programs. However, the results from the evaluation studies of ABQTL populations showed that it is possible to circumvent this constraint. It is possible to combine high levels of resistance from the wild species with high yield potential of elite breeding lines as seen in the progenies. The progenies having high levels of resistance as well as desirable yield levels can be carried forward to the advance generations for use in groundnut breeding as potential sources of variability or breeding lines. Nevertheless, to draw valid conclusions, a more thorough evaluation in fixed lines (of advance generations) may be desirable given the quantitative nature of the inheritance of yield and other yield parameters.

Population 3 (pop III) was derived utilizing synthetic ISATGR 278-18 (*A. duranensis* × *A. batizocoi*) (Mallikarjuna et al. 2011a). *A. duranensis* is one of the genome donors of *A. hypogaea*, but *A. batizocoi* is not. Two lines had a LLS disease score of 2 and majority of the lines had a score of 3. All the lines had a score of 2 for rust. The population was studied for the presence of proteinase inhibitors. Many of the lines showed maximum TI activity of 8–9 TI units/mg protein, and some lines had a minimum of 2–3 TI units/mg proteins, and in CI activity maximum was 3 CI units/mg protein, whereas in the cultivated species, maximum TI activity was 2–3 TI units/mg protein and CI activity maximum was 2 CI units/mg protein. To examine the specificity of inhibitors on *Spodoptera litura* larvae feeding on groundnut, the gut extracts were also assayed for inhibition of trypsin and chymotrypsin specificities. Maximum of 2 gut units were observed in pop III, and 1 gut unit was observed in cultivated species. The presence of PI in pop III shows that they may play a major role in pest resistance (Shilpa and Mallikarjuna unpublished data).

8.7 Molecular Markers, Genome Mapping, and Genomics as an Adjunct to Breeding

The first groundnut variety developed through integrated marker-assisted selection (MAS) was NemaTAM, a root-knot nematode-resistant variety (Simpson et al. 2003). Identification of a major QTL (QTL rust 01) contributing up to 82.96 % phenotypic variation for rust resistance, it was introgressed through MABC to improve three popular groundnut varieties (ICGV 91114, JL 24, and TAG 24) for rust resistance using GPBD4 as a donor genotype. Several promising introgression lines with remarkable reduction in disease spread and other desirable agronomic traits have been selected for further multiplication and generation advancement (Pandey et al. 2012). Availability of a large number of markers in recent years has ensured limiting linkage drag through stringent background selection and tracking the presence of non-desirable genomic region from the wild relatives. Several reviews provided in-depth information on development, availability, and deployment of genomic resources in groundnut published most recently (Pandey et al. 2012; Janila et al. 2013) and suggested the use of molecular markers in routine breeding programs. Although simple sequence repeat (SSR) markers still rule the hearts of plant breeders for use in genetics and breeding applications, other genotyping systems such as diversity array technology (DArT) and single nucleotide polymorphisms (SNPs) hold the key role for future groundnut improvement. ICRISAT in collaboration with DArT Pvt. Ltd, Australia, developed DArT arrays with 15,360 features which showed low polymorphism among genotypes of primary gene pool. Despite low polymorphism, these markers are of great help in monitoring the alien genome introgression in the cultivated species as observed in the case of pigeon pea (Mallikarjuna et al. 2011b). Furthermore, realizing the great potential role of SNPs, thousands of SNPs were identified in groundnut by the University of Georgia (8486 SNPs) and the University of California-Davis (>2,000 SNPs). To deploy abovementioned marker resources, a range of cost-effective SNP genotyping platforms have become available such as Illumina GoldenGate assays for genotyping 768 SNPs by the University of California-Davis, USA, and 1536 SNPs for groundnut by the University of Georgia, USA. Similarly, an alternative genotyping assay (KASP assay) developed by LGC Genomics (www.lgcgenomics.com/genotyping/kasp-genotyping-reagents, Semagn et al. 2013) provides flexibility to genotype any number of samples with any number of SNPs. Thus, ICRISAT has developed KASP assays for 90 SNPs in groundnut (Khera et al. 2013).

Breeders have been continuously struggling to handle linkage drag using conventional approaches. On the other side, genomic approaches provided reliable and precise solution for monitoring genome-wide alien introgression in elite lines. Thus, integration of genomic tools with conventional breeding approaches promises to enrich cultivated gene pool which will help in harnessing available rich diversity of wild relatives possessing superior alleles. Several synthetics have been developed so far (Simpson et al. 1993; Favero et al. 2006; Mallikarjuna et al. 2011a), providing opportunity to diversify the primary gene pool and conduct ABQTL analysis. A subset of the abovementioned two populations (ABQTL pop I and ABQTL pop II) was genotyped with DArT markers to construct genetic maps and conduct ABQTL analysis. Already one study reported genome-wide segment introgressions using markers from a synthetic amphidiploid (*A. duranensis* × *A. ipaensis*) into the genetic background of the cultivated variety (Fluer 11) (Foncéka et al. 2009). Therefore, availability of cost-effective genotyping systems in recent years has accelerated the introgression of useful traits and thus improvement of groundnut.

8.8 Conclusion

There is ample genetic diversity in the wild gene pool which harbors several useful genes for groundnut improvement. Direct utilization of diploid *Arachis* species, which are closely related, is cumbersome due to ploidy differences between the cultivated and wild *Arachis* species germplasm. Utilization of distantly related *Arachis* species needs in vitro interventions (Mallikarjuna and Sastri 1985), and utilization of both closely and distantly related species in secondary and tertiary gene pools needs an elaborate backcross program for alien introgressions and obtaining stable tetraploid lines. Ample variation is now available in the form of synthetics and double synthetics (Mallikarjuna et al. 2012a, b; Shilpa et al. 2013). Research experience in the utilization of synthetics has shown that stable tetraploid lines with alien introgressions can be achieved in a shorter period of time and the utilization of suitable molecular markers further accelerates the research programs. The best option for alien introgressions in groundnut is through the utilization of already available tetraploid synthetics and double synthetics in the breeding programs. It is also necessary to develop new sources of tetraploid synthetics so that ample variation is available for groundnut improvement. A range of groundnut synthetics (Mallikarjuna et al. 2011a) are available at ICRISAT (Sharma et al. 2013) for groundnut researchers for utilization in their breeding programs for groundnut genetic improvement.

References

Anon (1992) The medium term plan, 1994-1998. Main report, vol 1. ICRISAT, Patancheru, p 80

Awad HM, Boersma MG, Vervoort J, Rietjens IM (2000) Peroxidase catalyzed formation of quercetin quinone methide-glutathione adducts. Arch Biochem Biophys 378:224–233

Burrow MD, Starr JL, Paterson AH, Simpson CE (1996) Identification of peanut (*Arachis hypogaea L*) RAPD markers diagnostic of root- knot nematode (*Meloidogyne arenaria*) resistance. Mol Breed 2:369–379

Clemente A, Gee JM, Johnson IT, Mackenzie DA, Domoney C (2005) Pea (*Pisum sativum* L.) protease inhibitors from the Bowman-Birk class influence the growth of human colorectal adenocarcinoma HT29 cells in vitro. J Agric Food Chem 16(23):8979–8986

Craft BD, Kosinska A, Amarowwicz R, Ronald BP (2010) Antioxidant properties of extracts obtained from raw and dry roasted and oil-roasted US peanuts of commercial importance. Plant Food Hum Nutr 65(3):311–318

Dwivedi SL, Gurtu S, Nigam SN (2002) ALFP diversity among selected foliar diseases resistant groundnut (*Arachis hypogaea* L.) germplasm. Indian J Plant Genet Resour 15:46–50

Dwivedi SL, Upadhyaya HD, Stalker HT, Blair MW, Bertioli D, Nielen S, Ortiz R (2008) Enhancing crop gene pools of cereals and legumes with beneficial traits using wild relatives. Plant Breed Rev 30:179–280

Favero AP, Simpson CE, Valls JFM, Vello NA (2006) Study of the evolution of cultivated peanut through crossability studies among *Arachis ipaënsis*, *A. duranensis*, and *A. hypogaea*. Crop Sci 46:1546–1552

Fonçéka D, Hodo-Abalo T, Rivallan R, Faye I, Sall MN (2009) Genetic mapping of wild introgressions into cultivated peanut: a way toward enlarging the genetic basis of a recent allotetraploid. BMC Plant Biol 9:103

Fonceka D, Tossim HA, Rivallan R, Vignes H, Faye I (2012) Fostered and left behind alleles in peanut: interspecific QTL mapping reveals footprints of domestication and useful natural variation for breeding. BMC Plant Biol 12:26

Fraser GE, Sabates J, Beeson WL, Strahan TM (1992) A possible protective effect of nut consumption on risk of coronary heart disease: The Adventist Health Study. Arch Intern Med 152:1416–1424

Garcia GM, Stalker HT, Shroeder E, Kochert G (1996) Identification of RAPD, SCAR, and RFLP markers tightly linked to nematode resistance genes introgressed from *Arachis cardenasii* into *Arachis hypogaea*. Genome 39:836–845

GCP (2005) Unlocking the genetic diversity in peanut's wild relatives with genomic and genetic tools targeted subprogram: SP3- trait capture for crop. In: Improvement proceedings of generation challenge program 2005. Annual report and year three (2006) Workplan, Mexico, 56pp

Gowda MVC, Motagi BN, Naidu GK, Diddimani SN, Sheshagiri R (2002) GPBD4: a Spanish bunch groundnut genotype resistant to rust and late leafspot. Int Arachis Newsl 22:29–32

Hu FB, Stampfer MJ, Manson JE, Rimm EB, Colditz GA, Rosner BA, Speizer FE, Hennekens CH, Willett WC (1998) Frequent nut consumption and risk of coronary heart diseases in women: prospective cohort study. Br Med J 317:1341–1345

Husain F, Mallikarjuna N (2012) Genetic diversity in Bolivian landrace lines of groundnut (*Arachis hypogaea* L.). Indian J Genet 72(3):384–389

Janila P, Nigam SN, Pandey MK, Nagesh P, Varshney RK (2013) Groundnut improvement: use of genetic and genomic tools. Front Plant Sci 4:23

Johansen C, Nageswara Rao RC (1996) Maximizing groundnut yields. In: Renard C, Gowda CLL, Nigam SN, Johansen C (eds) Achieving high groundnut yields. Proceedings of international workshop, 25–29 Aug 1995, Laixi City, Shandong. ICRISAT, Patancheru, pp 117–127

Khera P, Upadhyaya HD, Pandey MK, Roorkiwal M, Sriswathi M, Janila P, Yufang Guo Y, Michael-McKain MR, Ervin D, Nagy ED, Steven J, Knapp SJ, James Leebens-Mack J, Conner JA, Ozias-Akins P, Varshney RK (2013) The Plant Genome 6. Crop Science Society of America, Madison. doi:10.3835/plantgenome2013.06.0019

Kochert G, Halward T, Branch WD, Simpson CE (1991) RFLP variability in peanut (*Arachis hypogaea* L) cultivars and wild species. Theor Appl Genet 81:565–570

Kochert G, Stalker HT, Gimenes M, Galgar L, Lopes CR, Moore K (1996) RFLP and cytogenetic evidence on the origin and evolution of allotetraploid domesticated peanut, *Arachis hypogaea* (Leguminosae). Am J Bot 83:1282–1291

Koslak RM, Chamberlin MA, Palmer RG, Bowen BA (1997) Programmed cell death in the root cortex of soybean root necrosis mutants. Plant J 11:729–745

Krapovickas A, Gregory WC (1994) Taxonomia del genero *Arachis* (Leguminosae). Bonplandia 8:1–186

Mallikarjuna N (2002) Gene introgression from *A. glabrata* into *A. hypogaea*, *A. duranensis* and *A. diogoi*. Euphytica 124:99–105

Mallikarjuna N (2005) Hybrids between *Arachis hypogaea* and *A. chiquitana* (section *Procumbentes*). Peanut Sci 32:148–152

Mallikarjuna N, Hoisington D (2009) Peanut improvement: production of fertile hybrids and backcross progeny between *Arachis hypogaea* and *A. kretschmeri*. Food Sci 1:457–462

Mallikarjuna N, Sastri DC (1985) In vitro culture of ovules and embryos from some interspecific in the genus *Arachis*. In: Proceedings of an international workshop on the cytogenetics of arachis. ICRISAT, Patancheru, Andhra Pradesh, India, pp 153–158

Mallikarjuna N, Sastri DC (2002) Morphological, cytological and disease resistance studies of the intersectional hybrids between *Arachis hypogaea* L. and *A. glabrata Benth.* Euphytica 126(2):161–167

Mallikarjuna N, Pande S, Jadhav DR, Sastri DC, Narayan Rao J (2004a) Introgression of disease resistance genes from *Arachis kempff-mercadoi* into cultivated groundnut. Plant Breed 123(6):573–576

Mallikarjuna N, Jadhav DR, Kranthi KR, Kranthi S (2004b) Influence of foliar chemical compounds on the development of *Spodoptera litura* (Fab.) on interspecific derivatives of groundnut. J Appl Entomol 128(5):321–328

Mallikarjuna N, Senthilvel S, Hoisington D (2011a) Development of synthetic groundnuts (*Arachis hypogaea* L) to broaden the genetic base of cultivated groundnut. Genet Resour Crop Evol 58:889–907

Mallikarjuna N, Senthivel S, Jadhav DR, Saxena K, Sharma HC, Upadhyaya HD, Rathore A, Varshney R (2011b) Progress in the utilization of *Cajanus platycarpus* (Benth.) Maesen in pigeonpea improvement. Plant Breed 130(5):507–514

Mallikarjuna N, Srikanth S, Vellanki RK, Jadhav DR, Das K, Upadhyaya HD (2012a) Meiotic analysis of hybrids between cultivated and synthetic tetraploid groundnuts. Plant Breed 131:135–138

Mallikarjuna N, Jadhav DR, Reddy K, Husain F, Das K (2012b) Screening new *Arachis* amphidiploids, and autotetraploids for resistance to late leaf spot by detached leaf technique. Eur J Plant Pathol 132:17–21

Mayee CD, Datar VV (1988) Diseases of groundnut in the tropics. Rev Trop Plant Pathol 5:169–198

Mercer LC, Wynne JC, Young CT (1990) Inheritance of fatty acid content in peanut oil. Peanut Sci 17:17–21

Moretzsohn MC, Gouvea EG, Inglis PW, Leal-Bertioli SCM, Valls JFM, Bertioli DJ (2012) A study of the relationships of cultivated peanut (*Arachis hypogaea*) and its most closely related wild species using intron sequences and microsatellite markers. Ann Bot 111:113–126

Panchbhavi KS, Nethradani CR (1987) Yield of groundnut as affected by varying larval density of *Spodoptera litura* (Fabricius) (Lepidoptera: Noctuidae). Indian J Agric Sci 57:525–527

Pandey MK, Monyo ES, Ozias-Akins P, Liang X, Guimarães P, Nigam SN, Upadhyaya HD, Janila P, Zhang X, Guo B, Cook DR, Bertioli DJ, Michelmore R, Varshney RK (2012) Advances in *Arachis* genomics for peanut improvement. J Biotechnol Adv 30:639–651

Prasad ER, Dutta-Gupta A, Padmashree K (2010) Insecticidal potential of Bowman-Birk proteinase inhibitors from red gram (*Cajanus cajan*) and black gram (Vigna mungo) against Lepidopteran insect pests. Pestic Biochem Physiol 98:80–88

Pravst I, Zmitek K (2010) Coenzyme Q10 contents in foods and fortification strategies. Crit Rev Food Sci Nutr 50(4):269–280

Prineas RJ, Kushi LH, Folsom AR, Bostick RM, Wu Y (1993) Walnuts and serum lipids (letter). N Engl J Med 329:359

Reddy DVR (1998) Control measures for the economically important peanut viruses. In: Hadidi A, Khetarpal RK, Koganezawo A (eds) Plant virus disease control. American Phytopathological Society, APS Press, St. Paul, pp 541–546

Reddy AS, Reddy LJ, Mallikarjuna N, Abdurahman MD, Reddy YV, Bramel PJ, Reddy DVR (2000) Identification of resistance to Peanut bud necrosis virus (PBNV) in wild *Arachis* germplasm. Ann Appl Biol 37:135–139

Robledo G, Lavia GI, Seijo G (2009) Species relations among wild *Arachis* species with the A genome as revealed by FISH mapping of r-DNA loci and heterochromatin detection. Theor Appl Genet 118:1295–1307

Ryan CA (1990) Protease inhibitors in plants. Genes for improving defenses against insects and pathogens. Annu Rev Phytopathol 28:425–449

Sanders TH, McMichel RW, Hendria KW (2000) Occurrence of resveratrol in edible peanuts. J Agric Food Chem 48:1234–1246

Sanford JC (1983) Ploidy manipulation. In: Moore JN, Janick J (eds) Method in fruit breeding. Purdue University Press, West Lafayette, pp 100–123

Savage GP, Keenan JI (1994) The composition and nutritive value of groundnut kernels. In: Smartt J (ed) The groundnut crop. A scientific basis for improvement. Chapman & Hall, London/New York

Seijo JG, Lavia GI, Fernández A, Krapovickas A, Ducasse DA, Bertioli DJ, Moscone EA (2007) Genomic relationships between the cultivated peanut (*Arachis hypogaea*, Leguminosae) and its close relatives revealed by double GISH. Am J Bot 94:1963–1971

Semagn K, Yoseph B, Marilyn LW, Amsal T, Stephen M, Barbara M, Sehabiague P, Prasanna BM (2013) Meta-analyses of QTL for grain yield and anthesis silking interval in 18 maize populations evaluated under water-stressed and well-watered environments. BMC Genomics 14:313

Sharma SB, McDonald D (1990) Global status of nematode problems of groundnut, pigeonpea, chickpea, sorghum and pearl millet and suggestion for future work. Crop Prot 9:453–458

Sharma S, Upadhyaya HD, Varshney RK, Gowda CLL (2013) Pre-breeding for diversification of primary gene pool and genetic enhancement of grain legumes. Front Plant Sci 4:309

Shilpa K, Sunkad G, Kurella S, Marri S, Padmashree K, Jadhav DR, Sahrawat KL, Mallikarjuna N (2013) Biochemical composition and disease resistance in newly synthesized amphidiploid and autotetraploid peanuts. Food Nutr Sci 4(2):169–176

Simpson CE, Starr JL (2001) Registration of 'COAN'. Peanut Sci 41:918–918

Simpson CE, Starr JL, Nelson SC, Woodard KE, Smith OD (1993) Registration of TxAG-6 and TxAG-7 peanut germplasm. Crop Sci 33:1418

Simpson CE, Starr JL, Church GT, Burrow MD, Paterson AH (2003) Registration of NemaTAM peanut. Crop Sci 43:1561

Singh AK, Dwivedi SL, Pande S, Moss JP, Nigam SN, Sastri DC (2003) Registration of rust and late leaf spot resistant peanut germplasm lines. Crop Sci 43:440–441

Stevenson PC, Blaney WM, Simmonds MJS, Wightman JA (1993a) The identification and characterization of resistance in wild species of *Arachis* to Spodoptera litura (Lepidoptera: Noctuidae). Bull Entomol Res 83:421–429

Stevenson PC, Anderson JC, Blaney WM, Simmonds MSJ (1993b) Developmental inhibition of *Spodoptera litura* (Fab.) larvae by a novel caffeoylquinic acid from the wild groundnut *Arachis paraguariensis* (Chod et Hassl.). J Chem Ecol 19:2917–2933

Subrahmanyam P, Ghanekar AM, Nolt BL, Reddy DVR, McDonald D (1985a) Resistance to groundnut diseases in wild *Arachis* species. ICRISAT (International Crops Research Institute for the Semi-Arid Tropics) In: Moss JP, Feakin SD (eds) Proceedings of an international workshop on cytogenetics of *Arachis*. ICRISAT Center, Patancheru, India, 31 Oct–2 Nov 1983, pp 49–55

Subrahmanyam P, Moss JP, McDonald D, Rao PVS, Rao VR (1985b) Resistance to leaf spot caused by *Cercosporidium personatum* in wild *Arachis* species. Disease 69:951–954

Valls JFM, Simpson CE (2005) New species of *Arachis* L. (Leguminosae) from Brazil, Paraguay and Bolivia. Bonplandia 14:35–64

Varshney RK, Bertioli DJ, Moretzsohn MC, Vadez V, Krishnamurthy L, Aruna R, Nigam SN, Moss BJ, Seetha K, Ravi K, Knapp HSJ, Hoisington DA (2009) The first SSR- based genetic linkage map for cultivated groundnut (*Arachis hypogaea* L.). Theor Appl Genet 118:729–739

Vavilov NI (1951) Phytogeographic basis of plant breeding. The origin, variation, immunity and breeding of cultivated plants. Chron Bot 13:1–366

Whitley ML, Isleib TG, Hendrix KW, Sanders TH, Dean LO (2011) Environmental and varietal effects on niacin content of raw and roasted peanuts. Peanut Sci 38:20–25

Wightman JA, Amin PW (1988) Groundnut pests and their control in semi arid tropics. Trop Pest Manag 34:218–226

Xue HQ, Isleib TG, Stalker HT, Payne GA, Obrian G (2005) Evaluation of *Arachis* species and interspecific tetraploid lines for resistance to aflatoxin production by *Aspergillus flavus*. Peanut Sci 31:134–141

Yu J, Ahmedna M, Goktepe I (2005) Effects of processing methods and extraction solvents on concentration and antioxidant activity of peanut skin phenolics. Food Chem 90:199–206

9 Vigna Species

Norihiko Tomooka, Takehisa Isemura, Ken Naito, Akito Kaga, and Duncan Vaughan

Abstract

The scientific information about the genus *Vigna*, which contains nine important food legumes, has been accumulated in the past decade. In this chapter, progress of the genetics of domestication, important agronomic traits, ecological adaptations, and genomic information are summarized. Domestication genetics revealed by a detailed Quantitative trait locus (QTL) analysis for mung bean, black gram, azuki bean, rice bean, and yard-long bean have been described and compared. Amazing abilities of some wild *Vigna* species to adapt harsh environments were described. Some outstanding examples are adaptation to sandy and saline soils by *V. marina* and *V. trilobata*, alkaline limestone rock soils by *V. exilis*, exposed windy cliff top environments by *V. riukiuensis*, waterlogged riverside by *V. luteola*, and shady forests by *V. minima*. *Vigna* genome project which is under way and aims to sequence 16 *Vigna* species will provide a foundation of clarifying genes which are responsible for the abilities to survive under extreme environments.

9.1 Introduction

Vigna species play a particularly important role in diets of people in Asia. In East Asia, azuki bean has a cultural significance as it is mixed with glutinous rice (red rice) on days of celebration (Lumpkin and McClary 1994). In the hills of Southeast Asia, rice bean is a source of protein for poor people. Mung bean is used in a variety of forms including noodles, sprouts, flour, and whole seeds and is an ingredient of many Asian dishes. In South Asia, several different *Vigna* species form an essential component of a vegetarian's diet. Several of the wild *Vigna* species are harvested as an occasional food. In addition, cowpea introduced to Asia long ago from Africa is used for both its seeds and, in the form of yard-long bean, its pods as a vegetable.

The Asian *Vigna* generally refers to the *Vigna* species in the subgenus *Ceratotropis* (Tomooka

N. Tomooka (✉) • T. Isemura • K. Naito
A. Kaga • D. Vaughan
Genetic Resources Center, National Institute of Agrobiological Sciences, Tsukuba, Japan
e-mail: tomooka@affrc.go.jp; isemura@affrc.go.jp; knaito@affrc.go.jp; kaga@affrc.go.jp; duncan@affrc.go.jp

M. Singh et al. (eds.), *Broadening the Genetic Base of Grain Legumes*,
DOI 10.1007/978-81-322-2023-7_9, © Springer India 2014

et al. 2002a). However, there are *Vigna* species in other subgenera that occur in Asia; some of them are domesticated such as *V. unguiculata* and *V. vexillata*. In this chapter, we will discuss all the *Vigna* crops of Asia (Table 9.1). Despite the obvious importance of *Vigna* in Asia, where most of the world's poor and undernourished live, it is only recently that there has been increased attention to their understudied genetic resources and application of biotechnology tools to them.

Table 9.1 *Vigna* species in Asia

Subgenus	Section	Species	Crop name
Ceratotropis			
	Angulares	*V. angularis* (Willd.) Ohwi & Ohashi var. *angularis*	Azuki bean
		Vigna angularis (Willd.) Ohwi & Ohashi var. *nipponensis*	
		V. dalzelliana (O. Kuntze) Verdcourt	
		V. exilis Tateishi & Maxted	
		V. hirtella Ridley	
		V. minima (Roxb.) Ohwi & Ohashi	
		V. nakashimae (Ohwi) Ohwi & Ohashi	
		V. nepalensis Tateishi & Maxted	
		V. reflexo-pilosa Hayata var. *glabra*	Creole bean
		V. reflexo-pilosa Hayata var. *reflexo-pilosa*	
		V. riukiuensis (Ohwi) Ohwi & Ohashi	
		V. tenuicaulis N. Tomooka & Maxted	
		V. trinervia (Heyne ex Wall.) Tateishi & Maxted	
		V. umbellata (Thunb.) Ohwi & Ohashi	Rice bean
	Ceratotropis	*V. grandiflora* (Prain) Tateishi & Maxted	
		V. mungo (L.) Hepper	Black gram
		V. silvestris (Lukoki, Maréchal et Otoul, Aitawada, Bhat et Yadav)	
		V. sahyadriana Aitawade, Bhat et Yadav	
		V. radiata (L.) Wilczek var. *radiata*	Mung bean
		V. radiata (L.) Wilczek var. *sublobata*	
		V. subramaniana (Babu ex Raizada) M. Sharma	
	Aconitifolia	*V. aconitifolia* (Jacq.) Maréchal	Moth bean
		V. aridicola N. Tomooka & Maxted	
		V. khandalensis (Santapau) Raghavan & Wadhwa	
		V. stipulacea Kuntze	
		V. trilobata (L.) Verdc.	
		V. indica Dixit, Bhat & Yadav	
Vigna	*Catiang*	*V. unguiculata* (L.) Walpers cv.-gr. Unguiculata	Cowpea
		V. unguiculata (L.) Walpers cv.-gr. Sesquipedalis	Yard-long bean
	Vigna	*V. marina* (Burm.) Merrill	
	Vigna	*V. hosei* (Craib) Backer ex K. Heyne	
	Vigna	*V. luteola* (Jacq.) Benth.	
Plectotropis	*Plectotropis*	*V. vexillata* (L.) A. Rich.	Tuber cowpea

Surprisingly, a recent paper on "orphan grain legumes" does not mention the Asian *Vigna* (Varshney et al. 2009), so perhaps the Asian *Vigna* should be called the "forgotten grain legumes." The international agricultural research community of the Consultative Group on International Agriculture Research (CGIAR) system has not conducted research on the *Vigna* crops of Asia, although an affiliate organization, the World Vegetable Center (AVRDC), does conduct research on mung bean. Despite this, there have been two recent genetic resource monographs, one on the Asian and the other on the African *Vigna* (Maxted et al. 2004; Tomooka et al. 2002a). Laboratories in different countries have conducted various genomic and genetic studies on the *Vigna* crops of Asia, but this research community has, unfortunately, not benefitted from a globally coordinated effort. A summary of the current status of *Vigna* genome maps (updated from Kaga et al. 2005) and molecular mapping of agronomically important traits is presented (Tables 9.2 and 9.3). In this chapter, we will discuss first the recent progress that has been made in understanding the *Vigna* crops of Asia and their wild progenitor species. Then the other wild *Vigna* species of Asia that may have a role to play in future crop improvement are discussed.

9.2 The *Vigna* Crops of Asia and Their Wild Progenitor Species

9.2.1 Mung Bean [*Vigna radiata* (L.) Wilczek]

Wild mung bean is very widely distributed from West Africa to Northern Australia. Within its range, germplasm from Australia and the island of New Guinea represents a distinct gene pool (Sangiri et al. 2007). Based on the diversity of domesticated mung bean and archaeobotanical records, mung bean seems to have been domesticated in India (Sangiri et al. 2007; Fuller and Harvey 2006). Recently, a comprehensive genome map of mung bean was published (Isemura et al. 2012). In common with many crops, domestication traits in mung bean are controlled by a few major genes including some minor genes. However, compared to other crops, including domesticated *Vigna*, domestication-related traits in mung bean were more dispersed across the genome. Several genes, such as hilum and seed color, are found at a similar location in mung bean and other Asian *Vigna* species. Several QTLs for 100-seed weight in mung bean were located in a similar genomic position to other *Vigna* species, such as QTLs on linkage groups 1 and 2 that were at a similar location to those on the azuki bean and rice bean genome maps (Fig. 9.1). However, the QTL with the largest effect for 100-seed weight in mung bean was found on linkage group 8 and is specific to mung bean. Further studies of the differences among QTLs and their effect in relation to seed size in mung bean and other domesticated *Vigna* species will help to explain the marked differences for seed size among domesticated *Vigna*. Both the mitochondrial and chloroplast genomes of mung bean have been sequenced (Alverson et al. 2011; Tangphatsornruang et al. 2010). The mung bean mitochondrial genome sequence was the first for a legume and is of unexceptional size. The genome is unusually depauperate in repetitive DNA compared with the mitochondrial genomes of other species (Alverson et al. 2011). The chloroplast genome includes a pair of inverted repeats (IRs). In addition, compared to other plant chloroplast genome sequences, *V. radiata* has two distinct rearrangements of 50 and 78 kb (Tangphatsornruang et al. 2010).

Among domesticated pulses, mung bean has one of the smallest seed sizes, and seeds are usually green. It has a short growth duration and can readily be intercropped with cereals. Powdery mildew is a major disease of mung bean and breeding for resistance to this disease is a priority. However, breeding progress has been slow because there appear to be many races of the disease, and these races have not yet been described. Several sources of resistance or partial resistance to the disease have been identified. Results of different studies have identified sources of a major resistance QTL on both linkage groups 9 and 6

Table 9.2 Genome linkage maps for *Vigna* species

	Cross combination	Population (plants/lines) analyzed	Markers used	Linkage groups resolved	Map distance (cM)	Level of distortion (%)	Reference
Asian *Vigna*							
Azuki bean	*V. nepalensis* (JP107881) × *V. angularis* (JP81481)	BC_1F_1 (187)	205 SSRs, 187 AFLPs, and 94 RFLPs	11	832.1	3.9	Han et al. (2005)
	V. nepalensis (JP107881) × *V. angularis* (JP81481)	F_2 (141)	74 SSRs	11	649.7	28.4	Isemura et al. (2007)
	V. angularis var. *angularis* (JP109685 cv. 'Kyoto Dainagon') × *V. angularis* var. *nipponensis* (JP110658)	F_2 (188)	191 SSRs, 2 STSs, 1 CAPS, 2 SCARs, and 36 AFLPs	10	771.9	3.9	Kaga et al. (2008)
Black gram	*V. mungo* (JP219132) × *V. mungo* var. *silvestris* (JP107873)	BC_1F_1 (180)	61 SSRs, 59 RFLPs, 27 AFLPs, and 1 morphological trait	11	783	0.0	Chaitieng et al. (2006)
	V. mungo var. *mungo* (TU94-2) × *V. mungo* var. *silvestris*	F_9 RIL (104)	47 SSRs, 254 AFLPs, 86 RAPDs, and 41 ISSRs	11	865.1	44.6	Gupta et al. (2008)
Mung bean	*V. radiata* var. *radiata* (VC3890) × *V. radiata* var. *sublobata* (TC1966)	F_2 (58)	151 genomic and 20 cDNA RFLPs and 1 pest locus	14	1,570	12.0	Menacio-Hautea et al. (1992, 1993)
	V. radiata var. *radiata* (cv. 'Berken') × *V. radiata* var. *sublobata* (ACC41)	F_2 (67)	52 RFLPs, 56 RAPDs, and 2 morphological traits	12	758.3	14.5	Lambrides et al. (2000)
		F_7 RIL (67)	113 RAPDs and 2 morphological traits	12	691.7	24.3	
	V. radiata var. *sublobata* (TC1966) × *V. radiata* (cv. 'Pagasa 7')	F_9 RIL (76)	103 AFLPs	9	655.5	58.0	Sholihin and Hautea (2002)
	V. radiata (cv. 'Berken') × *V. radiata* ssp. *sublobata* (ACC41)	F_8 RIL (80)	255 RFLPs	13	737.9	30.8	Humphry et al. (2002)
	V. radiata (cv. 'Berken') × *V. radiata* ssp. *sublobata* (ATF 3640)	F_7 RIL (147)	52 RFLPs	10	350	15.4	Humphry et al. (2003)

	V. radiata (cv. 'Berken') × *V. radiata* ssp. *sublobata* (ACC41)	F_7 RIL (227)	78 RFLPs and 1 morphological trait	13	684.7	34.2	Humphry et al. (2005)
	V. radiata (cv. 'Berken') × *V. radiata* ssp. *sublobata* (ACC41)	F_9 RIL (202)	97 SSRs, 76 RFLPs, 4 RAPDs, and 2 STSs	12	831.8	30.7	Zhao et al. (2010)
	V. radiata var. *radiata* ('Kamphaeng Saen 1') × *V. radiata* var. *radiata* (V4718)	F_2 (155)	56 SSRs	11	995.6	–[a]	Chankaew et al. (2011)
	V. radiata (wild: JP211874) × *V. radiata* (cultivated: JP229096 cv. 'Sukhothai')	BC_1F_1 (250)	237 SSRs and 193 EST-SSRs	11	727.6	14.4	Isemura et al. (2012)
	V. radiata (susceptible cultivar: 'Kamphaeng Saen 2') × *V. radiata* (resistant line: NM10-12-1)	F_8 RIL (122)	47 SSRs and 7 AFLPs	12	991.98	–[a]	Prathet et al. (2012)
	V. radiata var. *radiata* (KUML29-1-3) × *V. radiata* var. *sublobata* (W021)	F_2 (186)	150 SSRs	12	1,174.2	–[a]	Kajonphol et al. (2012)
	V. radiata (cultivated line: NM92) × *V. radiata* ssp. *sublobata* (TC1966)	F_{12} RIL (200)	356 AFLPs, 113 RAPDs, 20 SSRs, 1 SCAR, and 1 CAPS	11	690.7	8.8	Chen et al. (2013)
Rice bean	*V. umbellata* (cultivated: JP217439) × *V. umbellata* (wild: JP210639)	BC_1F_1 (198)	223 SSRs and 103 AFLPs	11	796.1	18.7	Isemura et al. (2010)
Interspecific cross	*V. angularis* (cv. 'Erimoshouzu') × *V. nakashimae*	F_2 (80)	19 RFLPs, 108 RAPDs, and 5 morphological traits	14	1250	19.7	Kaga et al. (1996)
	V. angularis (cv. 'Erimoshouzu') × *V. umbellata* (cv. 'Kagoshima')	F_2 (86)	114 RFLPs, 74 RAPDs, and 1 morphological trait	14	1,701.9	29.8	Kaga et al. (2000)
	V. angularis var. *angularis* (cv. 'Tanba Dainagon') × *V. riukiuensis*	BC_1F_1 (77)	100 SSRs, 47 RFLPs, and 234 AFLPs	11	624	2.9	Kaga et al. (2003)
	V. umbellata (cultivated: JP100304) × *V. nakashimae* (JP107879)	F_2 (74)	101 SSRs and 74 RFLPs	11	652	24.0	Somta et al. (2006)

(continued)

Table 9.2 (continued)

	Cross combination	Population (plants/lines) analyzed	Markers used	Linkage groups resolved	Map distance (cM)	Level of distortion (%)	Reference
African *Vigna*							
Cowpea	*V. unguiculata* (IT2246-4) × *V. unguiculata* ssp. *dekindtiana* (TVNI 963)	F_2 (58)	87 genomic and 5 cDNA RFLPs, 5 RAPDs, and 2 morphological traits	10	684	22.0	Menacio-Hautea et al. (1993) Fatokun et al. (1997)
	V. unguiculata (IT84S-2049) × *V. unguiculata* (524B)	F_8 RIL(94)	133 RAPDs, 19 RFLPs, 25 AFLPs, 3 morphological traits, and 1 biochemical marker	12	972	18.0	Menéndez et al. (1997)
	V. unguiculata (IT84S-2246-4 improved line) × *V. unguiculata* ssp. *dekindtiana* var. *pubescens* (TVNu-110-3A)	F_8 RIL (94)	77 RAPDs and 3 morphological traits	12	669.8	21.7	Ubi et al. (2000)
	V. unguiculata (IT84S-2049) × *V. unguiculata* (524B)	F_9 RIL (94)	133 RAPDs, 36 RFLPs, 267 AFLPs, 3 morphological traits, and 1 biochemical marker	11	2,670	19.7	Ouédraogo et al. (2002a)
	V. unguiculata ('Sanzi') × *V. unguiculata* (VITA 7)	F_{10} RIL (145)	5 SSRs and 134 AFLPs	11	1,620.1	35.9	Omo-Ikerodah et al. (2008)
	V. unguiculata (524B) × *V. unguiculata* (IT84S-2049)	RIL (79)	436 SNPs	11	665	41.6 (a total of 986 were polymorphic in at least one population. 410 markers exhibited segregation distortion in one or more populations)	Muchero et al. (2009a)
	V. unguiculata ('California Blackeye No.27') × *V. unguiculata* (24-125B-1)	RIL (90)	299 SNPs	11	651		
	V. unguiculata ('California Blackeye No.46') × *V. unguiculata* (IT93K-503-1)	RIL (103)	388 SNPs	11	601		
	V. unguiculata ('Dan Ila') × *V. unguiculata* (TVu-7778)	RIL (109)	288 SNPs	11	665		
	V. unguiculata (TVu-14676) × *V. unguiculata* (IT84S-2246-4)	RIL (137)	349 SNPs	11	600		
	V. unguiculata ('Yacine') × *V. unguiculata* (58-77)	RIL (114)	415 SNPs	11	657		
	Consensus map	RIL (632)	928 SNPs	11	680		

V. unguiculata (IT93K503-1) × *V. unguiculata* ('California Blackeye No.46')	F_8 RIL (127)	306 AFLPs	11	643	0.0	Muchero et al. (2009b)
V. unguiculata (cultivated: 524B) × *V. unguiculata* (wild: 219-01)	F_7 RIL (159)	202 SSRs	11	677	10.7	Andargie et al. (2011)
V. unguiculata ('California Blackeye No.27') × *V. unguiculata* (IT97K-566-6)	RIL (92)[b]	438 SNPs	23	505.56	–[a]	Lucas et al. (2011)
V. unguiculata ('California Blackeye No.27') × *V. unguiculata* (IT82E-18)	RIL (160)[b]	430 SNPs	23	701.15	–[a]	
V. unguiculata ('California Blackeye No.27') × *V. unguiculata* (UCR 779)	RIL (56)[b]	560 SNPs	22	489.40	–[a]	
V. unguiculata ('California Blackeye No.46') × *V. unguiculata* (IT93K-503-1)	RIL (114)[b]	374 SNPs	17	639.59	–[a]	
V. unguiculata (524B) × *V. unguiculata* (IT84S-2049)	RIL (89)[b]	438 SNPs	22	710.09	–[a]	
V. unguiculata ('Dan Ila') × *V. unguiculata* (Tvu-7778)	RIL (79)[b]	288 SNPs	22	549.56	–[a]	
V. unguiculata ('Yacine') × *V. unguiculata* (58-77)	RIL (97)[b]	435 SNPs	22	650.98	–[a]	
V. unguiculata ('Sanzi') × *V. unguiculata* ('Vita 7')	RIL (182)[b]	413 SNPs	19	753.22	–[a]	
V. unguiculata (IT84S-2246) × *V. unguiculata* (IT93K-503)	$F_{3:4}$ (88)	155 SNPs	14	302.46	–[a]	
V. unguiculata (IT84S-2246) × *V. unguiculata* ('Mouride')	$F_{3:4}$ (87)	347 SNPs	15	595.33	–[a]	
V. unguiculata (TVu14676) × *V. unguiculata* (IT84S-2246-4)	RIL (136)[b]	345 SNPs	14	666.89	–[a]	
V. unguiculata ('California Blackeye No.27') × *V. unguiculata* (24-125B-1)	RIL (87)[b]	329 SNPs	23	526.75	–[a]	

(continued)

Table 9.2 (continued)

	Cross combination	Population (plants/lines) analyzed	Markers used	Linkage groups resolved	Map distance (cM)	Level of distortion (%)	Reference
	V. unguiculata (LB30#1) × *V. unguiculata* (LB1162#7)	RIL (90)[b]	180 SNPs	20	409.94	–[a]	
	Consensus map	RIL (1293)	1,107 SNPs	11	680		
Yard-long bean (asparagus bean)	*V. unguiculata* ssp. *sesquipedialis* (ZN016) × *V. unguiculata* ssp. *sesquipedialis* (Zhijiang 282)	$F_{7:8}$ RIL (114)	191 SNPs and 184 SSRs	11	745	13.8	Xu et al. (2011a)
	V. unguiculata ssp. *unguiculata* cv.-gr. sesquipedalis (JP81610) × *V. unguiculata* ssp. *unguiculata* var. *spontanea* (JP89083 = TVnu457)	BC_1F_1 (190)	226 SSRs	11	852.4	9.7	Kongjaimun et al. (2012a)
	V. unguiculata ssp. *unguiculata* cv.-gr. sesquipedalis (JP81610) × *V. unguiculata* ssp. *unguiculata* var. *spontanea* (JP89083 = TVnu457)	F_2 (188)	113 SSRs	11	977.1	48.7	Kongjaimun et al. (2012b)
Bambara groundnut	*V. subterranea* var. *subterranea* ('DipC') × *V. subterranea* var. *subterranea* ('Tiga necaru')	F_3 (73)	29 SSRs and 209 DArTs	21	608.6	31.8	Ahmad (2012)
	V. subterranea var. *subterranea* ('DipC') × *V. subterranea* var. *spontanea* (VSSP11)	F_2 (98)	12 SSRs, 106 DArTs, and 76 AFLPs	20	901.2	27.6	
Others							
V. vexillata	*V. vexillata* var. *angustifolia* (TVnu1443) × *V. vexillata* var. *vexillata* (TVnu73)	F_2 (94)	70 RAPDs, 47 AFLPs, 1 SSR, and 2 morphological traits	14	1,564.1	33.5	Ogundiwin et al. (2005)

[a]There is no description
[b]F_8 to F_{10} generations

Table 9.3 Molecular mapping of agronomically important traits in *Vigna*

Cross combination	Population (plants/lines) analyzed	Molecular markers used	Trait of interest	Gene/QTL (linkage group, LG)	References
Asian *Vigna*					
Azuki bean (*V. angularis*) *V. nepalensis* (JP107881) × *V. angularis* (JP81481)	BC_1F_1 (187)	SSR, RFLP, AFLP	Domestication-related traits (qualitative trait: seed coat color)	Gene (LG1)	Isemura et al. (2007)
			Domestication-related traits (qualitative trait: black mottle on seed coat)	Gene (LG4)	
			Domestication-related traits (three qualitative and 29 quantitative traits)	QTLs (for each trait, 1–6 QTLs are detected mainly on LG1, LG2, LG4, LG7, and LG9)	
	F_2 (141)	SSR	Domestication-related traits (qualitative trait: seed coat color)	Gene (LG1)	
			Domestication-related traits (qualitative trait: black mottle on seed coat)	Gene (LG4)	
			Domestication-related traits (qualitative trait: epicotyl color)	Gene (LG4)	
			Domestication-related traits (three qualitative and 30 quantitative traits)	QTLs (for each trait, 1–6 QTLs are detected mainly on LG1, LG2, LG4, LG7, and LG9)	
Azuki bean (*V. angularis*) *V. nepalensis* (JP107881) × *V. angularis* (JP81481)	BC_1F_1 (187)	SSR, RFLP, AFLP	Bruchid (*Callosobruchus chinensis*) resistance	QTLs (LG1, LG2, LG4, and LG10)	Kaga et al. 2008; Somta P et al. (2008)
			Bruchid (*Callosobruchus maculatus*) resistance	QTL (LG3)	
			Seed weight	QTLs (LG1, LG2, LG8, and LG9)	
	F_2 (141)	SSR	Bruchid (*Callosobruchus chinensis*) resistance	QTLs (LG1, LG2, LG4, and LG10)	
			Bruchid (*Callosobruchus maculatus*) resistance	QTL (LG3)	
			Seed weight	QTLs (LG1, LG2, LG8, and LG9)	

(continued)

Table 9.3 (continued)

	Cross combination	Population (plants/lines) analyzed	Molecular markers used	Trait of interest	Gene/QTL (linkage group, LG)	References
Black gram (*V. mungo*)	*V. mungo* (JP219132) × *V. mungo* var. *silvestris* (JP107873)	BC_1F_1 (180)	SSR, RFLP, and AFLP	Twisted and curly leaf	Gene (LG8)	Chaitieng et al. (2006)
Black gram (*V. mungo*)	*V. mungo* var. *mungo* (TU94-2) × *V. mungo* var. *silvestris*	F_9 RIL (104)	SSR, AFLP, RAPD, and ISSR	Bruchid resistance	QTLs (LG1, LG2, LG3, LG4, and LG10)	Souframanien et al. (2010)
Black gram (*V. mungo*)	*V. mungo* var. *mungo* (DPU 88-31) × *V. mungo* var. *mungo* (AKU 9904)	F_2 (168)	SSR	Resistance to mung bean yellow mosaic India virus	Gene (closely linked to SSR marker CEDG180)	Gupta et al. (2013)
Mung bean (*V. radiata*)	*V. radiata* var. *radiata* (VC3890) × *V. radiata* var. *sublobata* (TC1966)	F_2 (58)	RFLP	Seed weight	QTLs (LG: I, II, III, and VI)	Fatokun et al. (1992)
Mung bean (*V. radiata*)	*V. radiata* var. *radiata* (VC3890) × *V. radiata* var. *sublobata* (TC1966)	F_2 (58)	RFLP	Bruchid resistance	Gene (LG: VIII)	Young et al. (1992)
Mung bean (*V. radiata*)	*V. radiata* var. *radiata* (VC3890) × *V. radiata* var. *sublobata* (TC1966)	F_2 (58)	RFLP	Powdery mildew resistance	QTLs (LG3, LG7, and LG8)	Young et al. (1993)
Mung bean (*V. radiata*)	*V. radiata* var. *radiata* (Osaka-ryokuto) × *V. radiata* var. *sublobata* (TC1966)	$BC_{20}F_2$ (414)	RFLP and RAPD	Bruchid resistance	Gene (LG9)	Kaga and Ishimoto (1998)
Mung bean (*V. radiata*)	*V. radiata* var. *radiata* (cv. 'Berken') × *V. radiata* var. *sublobata* (ACC41)	F_2 (67)	RFLP and RAPD	Seed testa color	Gene (LG2b)	Lambrides et al. (2000)
				Pigmentation of the texture layer	Gene (LG2a)	
		F_7 RIL (67)	RFLP and RAPD	Seed testa color	Gene (LG2)	
				Pigmentation of the texture layer	Gene (LG2)	
Mung bean (*V. radiata*)	*V. radiata* var. *sublobata* (TC1966) × *V. radiata* (cv. 'Pagasa 7')	F_9 RIL (76)	AFLP	Drought resistance	QTLs (LG: A, B, D, E, and I)	Sholihin and Hautea (2002)

Mung bean (*V. radiata*)	*V. radiata* var. *radiata* (VC1210A) × *V. radiata* var. *sublobata* (TC1966)	F_2 (96)	RFLP and AFLP	Powdery mildew resistance	QTLs (two QTLs are detected. One is tightly linked to RFLP markers Mac71a, Mac71d, and Mac114. Another is linked to RFLP marker Bng065)	Chaitieng et al. (2002)
Mung bean (*V. radiata*)	*V. radiata* (cv. 'Berken') × *V. radiata* ssp. *sublobata* (ATF 3640)	F_7 RIL (147)	RFLP	Powdery mildew resistance	QTL (LG: K)	Humphry et al. (2003)
Mung bean (*V. radiata*)	*V. radiata* (cv. 'Berken') × *V. radiata* ssp. *sublobata* (ACC41)	F_7 RIL (227)	RFLP	Leaf lobing	Gene (LG: B)	Humphry et al. (2005)
				Seed weight	QTLs (LG: A, B, D, E, F, G, I, J, and K)	
				Hard seededness	QTLs (LG: A, B, C, and K)	
Mung bean (*V. radiata*)	*V. radiata* (cv. 'Berken') × *V. radiata* ssp. *sublobata* (ACC41)	F_8 RIL (80)	RFLP	Bruchid resistance	QTL (LG: I)	Mei et al. (2009)
				Seed mass	QTLs (LG: A, B, D, F, G, I, and K)	
		F_9 RIL (217)	RFLP	Bruchid resistance	QTL (LG: I)	
				Seed mass	QTLs (LG: A, B, D, E, G, I, J, and K)	
		F_9 RIL (210)	RFLP	Bruchid resistance	QTL (LG: I)	
				Seed mass	QTLs (LG: A, B, D, E, F, I, J, and K)	
		F_{10} RIL (191)	RFLP	Bruchid resistance	QTL (LG: I)	
				Seed mass	QTLs (LG: E, J, and K)	
Mung bean (*V. radiata*)	*V. radiata* var. *radiata* ('Kamphaeng Saen 1') × *V. radiata* var. *radiata* (VC6468-11-1A)	F_7 RIL (190)	SSR	Powdery mildew resistance	QTLs (LG: I and II)	Kasettranan et al. (2010)
Mung bean (*V. radiata*)	*V. radiata* var. *radiata* (Opaque leaf mutant) × *V. radiata* var. *radiata* ('Berken')	F_2 (85)	AFLP	Opaque leaf mutant	Gene (closely linked to AFLP marker AGG/ATA)	Rungnoi et al. (2010)
Mung bean (*V. radiata*)	*V. radiata* (Kamphaeng Saen 1) × *V. radiata* (NM10-12)	F_2 (160)	AFLP	Resistance to iron deficiency chlorosis	Gene (closely linked to AFLP markers E-ACT/M-CTA and E-ACC/M-CTG)	Srinives et al. (2010)

(continued)

Table 9.3 (continued)

	Cross combination	Population (plants/lines) analyzed	Molecular markers used	Trait of interest	Gene/QTL (linkage group, LG)	References
Mung bean (*V. radiata*)	*V. radiata* var. *radiata* ('Kamphaeng Saen 1') × *V. radiata* var. *radiata* (V4718)	F_2 (155)	SSR	Cercospora leaf spot resistance	QTL (LG3)	Chankaew et al. (2011)
		BC_1F_1 (760)		Cercospora leaf spot resistance	QTL (LG3)	
Mung bean (*V. radiata*)	*V. radiata* (wild: JP211874) × *V. radiata* (cultivated: JP229096 cv. 'Sukhothai')	BC_1F_1 (250)	SSR and EST-SSR	Domestication-related traits (qualitative traits: seed coat color)	Gene (LG5)	Isemura et al. (2012)
				Domestication-related traits (qualitative traits: black mottle on seed coat)	Gene (LG4)	
				Domestication-related traits (qualitative traits: hilum color)	Gene (LG5)	
				Domestication-related traits (qualitative traits: stem determinacy)	Gene (LG9)	
				Domestication-related traits (34 quantitative traits)	QTLs (for each trait, 1–7 QTLs are detected mainly on LG1, LG2, LG4, LG7, LG8, LG9, and LG10)	
Mung bean (*V. radiata*)	*V. radiata* var. *radiata* (KUML29-1-3) × *V. radiata* var. *sublobata* (W021)	F_2 (186)	SSR	Nine quantitative traits	QTLs (for each trait, 2–6 QTLs are detected mainly on LG 2 and LG4)	Kajonphol et al. (2012)
Mung bean (*V. radiata*)	*V. radiata* ('Kamphaeng Saen 2') × *V. radiata* (NM10-12-1)	F_8 RIL (122)	SSR and AFLP	Resistance to iron deficiency chlorosis	QTLs (LG2 and LG3)	Prathet et al. (2012)
Mung bean (*V. radiata*)	*V. radiata* ('Kamphaeng Saen 1') × *V. radiata* (V4718)	$F_{2:3}$ (134)	SSR	Powdery mildew resistance	QTLs (LG4 and LG9)	Chankaew et al. (2013)
	V. radiata ('Kamphaeng Saen 1') × *V. radiata* (V4718)	$F_{2:4}$ (134)	SSR	Powdery mildew resistance	QTLs (LG4 and LG9)	
	V. radiata ('Chainat 60') × *V. radiata* ('RUM5')	$F_{2:3}$ (190)	SSR	Powdery mildew resistance	QTLs (LG4, LG6, and LG9)	
	V. radiata ('Chainat 60') × *V. radiata* ('RUM5')	$BC_1F_{1:2}$ (74)	SSR	Powdery mildew resistance	QTLs (LG6 and LG9)	

Mung bean (*V. radiata*)	*V. radiata* (cultivated line NM92) × *V. radiata* ssp. *sublobata* (TC1966)	F_{12} RIL (200)	AFLP, RAPD, SSR, SCAR, and CAPS	Mung bean yellow mosaic India virus resistance	QTLs (LG7, LG8, and LG9)	Chen et al. (2013)
				Bruchid resistance	QTLs (LG7 and LG9)	
				Seed germination rate	QTLs (LG7 and LG9)	
				100-seed weight	QTLs (LG1, LG3, and LG9)	
Rice bean (*V. umbellata*)	*V. umbellata* (cultivated: JP217439) × *V. umbellata* (wild: JP210639)	BC_1F_1 (198)	SSR and AFLP	Domestication-related traits (qualitative traits: hilum color)	Gene (LG5)	Isemura et al. (2010)
				Domestication-related traits (30 quantitative traits)	QTLs (for each trait, 1–7 QTLs are detected mainly on LG1, LG2, LG4, and LG7)	
Interspecific cross						
V. angularis × *V. nakashimae*	*V. angularis* (cv. Erimoshouzu) × *V. nakashimae*	F_2 (80)	RFLP and RAPD	Epicotyl color	Gene (LG4)	Kaga et al. (1996)
				Intensity of epicotyl color	Gene (LG2)	
				Black mottle on seed testa	Gene (LG4)	
				Pod color	Gene (LG2)	
				Pod shattering	Gene (Unlinked)	
				Hilum cushion	Gene (LG7)	
V. angularis × *V. umbellata*	*V. angularis* (cv. 'Erimoshouzu') × *V. umbellata* (cv. 'Kagoshima')	F_2 (86)	RFLP and RAPD	Epicotyl color	Gene (LG1)	Kaga et al. (2000)
				Intensity of epicotyl color	Gene (LG5; mapped with a low LOD score)	
V. umbellata × *V. nakashimae*	*V. umbellata* (cultivated: JP100304) × *V. nakashimae* (JP107879)	F_2 (74)	SSR and RFLP	Bruchid (*Callosobruchus chinensis*) resistance	QTLs (LG1 and LG7)	Somta et al. (2006)
				Bruchid (*Callosobruchus maculatus*) resistance	QTLs (LG1 and LG4)	
				Seed weight	QTL (LG10)	
African *Vigna*						
Cowpea (*V. unguiculata*)	*V. unguiculata* (IT2246-4) × *V. unguiculata* ssp. *dekindtiana* (TVNI 963)	F_2 (58)	RFLP	Seed weight	QTLs (LG: II and VI)	Fatokun et al. (1992)
Cowpea (*V. unguiculata*)	*Vigna unguiculata* (IT84S-2246-4) × *Vigna unguiculata* ssp. *dekindtiana* (NI 963)	F_2 (58)	RFLP	Aphid resistance	Genes (LG1 and LG8)	Myers et al. (1996)

(continued)

Table 9.3 (continued)

	Cross combination	Population (plants/ lines) analyzed	Molecular markers used	Trait of interest	Gene/QTL (linkage group, LG)	References
Cowpea (*V. unguiculata*)	*V. unguiculata* (IT84S-2049) × *V. unguiculata* (524B)	F_8 RIL(94)	RFLP, RAPD, and AFLP	Flower color	Gene (LG1)	Menéndez et al. (1997)
				Pod color	Gene (LG1)	
				Petiole pigmentation	–[a]	
				Pod position	Gene (unlinked)	
				Internode length	–[a]	
				Dehydrin protein	Gene (LG7)	
				Chitinase	Gene (unlinked)	
				Nodal position of 1st flower	QTL (LG2)	
				Seed weight	QTL (LG5)	
Cowpea (*V. unguiculata*)	*V. unguiculata* (IT84S-2246-4 improved line) × *V. unguiculata* ssp. *dekindtiana* var. *pubescens* (TVNu-110-3A)	F_8 RIL (94)	RAPD	Mature pod color	Gene (LG: V)	Ubi et al. (2000)
				Inverted V mark on leaf	Gene (LG: I)	
				Pod dehiscence	Gene (LG: XII)	
				11 quantitative traits	QTLs (for each trait, 2–15 QTLs are detected mainly on LG I, II, IV, and V)	
Cowpea (*V. unguiculata*)	*V. unguiculata* (Tvx 3236) × *V. unguiculata* (IT82D-849)	F_2 (274)	AFLP	Resistance to *Striga gesnerioides* race 1	Gene (LG1 of map by Menéndez et al. (1997))	Ouédraogo et al. (2001)
	V. unguiculata (IT84S-2246–4) × *V. unguiculata* (Tvu 14676)	F_2 (150)	AFLP	Resistance to *Striga gesnerioides* race 3	Gene (LG1 of map by Menéndez et al. (1997))	
Cowpea (*V. unguiculata*)	*V. unguiculata* (IT84S-2049) × *V. unguiculata* (524B)	F_9 RIL (94)	RAPD, RFLP, and AFLP	Resistance to *Striga gesnerioides* race 1	Genes (LG1 and LG6)	Ouédraogo et al. (2002a)
				Resistance to *Striga gesnerioides* race 3	Gene (LG1)	
				Cowpea mosaic virus	Gene (LG3)	
				Cowpea severe mosaic virus	Gene (LG3)	
				Blackeye cowpea mosaic virus	Gene (LG8)	

				Southern bean mosaic virus	Gene (LG6)	
				Fusarium wilt	Gene (LG3)	
				Root-knot nematode	Gene (LG1)	
				Dehydrin protein	Gene (LG2)	
Cowpea (*V. unguiculata*)	*V. unguiculata* (Tvx3236) × *V. unguiculata* ('Gorom')	F_2 (150)	AFLP	Resistance to *Striga gesnerioides* race 1	Gene (LG6 of map by Ouédraogo et al. (2002a))	Ouédraogo et al. (2002b)
	V. unguiculata (Tvx3236) × *V. unguiculata* (IT81D-994)	F_2 (150)	AFLP	Resistance to *Striga gesnerioides* race 1	Gene (LG6 of map by Ouédraogo et al. (2002a))	
Cowpea (*V. unguiculata*)	*V. unguiculata* ('Sanzi') × *V. unguiculata* (VITA 7)	F_{10} RIL (145)	AFLP and SSR	Flower bud thrips	QTLs (LG1, LG2, LG3, LG6, and LG7)	Omo-Ikerodah et al. (2008)
Cowpea (*V. unguiculata*)	*V. unguiculata* (IT93K503-1) × *V. unguiculata* ('California Blackeye No.46')	F_8 RIL (127)	AFLP	Drought stress-induced senescence	QTLs (LG1, LG2, LG3, LG5, LG6, LG7, LG9, and LG10)	Muchero et al. (2009b)
				Maturity	QTLs (LG7 and LG8)	
Cowpea (*V. unguiculata*)	*V. unguiculata* ('Danila') × *V. unguiculata* (Tvu-7778)	RIL (113)	SNP	Bacterial blight resistance	QTLs (LG3, LG5, and LG9)	Agbicodo et al. (2010)
Cowpea (*V. unguiculata*)	*V. unguiculata* (IT93K503-1) × *V. unguiculata* ('California Blackeye No.46')	$F_{2:8}$ RIL (127)	AFLP	Foliar damage resistance	QTLs (LG5 and LG7)	Muchero et al. (2010a)
Cowpea (*V. unguiculata*)	*V. unguiculata* ('California Blackeye No.46') × *V. unguiculata* (IT93K-503-1)	$F_{2:8}$ RIL (113)	AFLP	Seedling drought tolerance	QTLs (LG1, LG2, LG3, and LG5)	Muchero et al. (2010b)
Cowpea (*V. unguiculata*)	*V. unguiculata* (cultivated: 524B) × *V. unguiculata* (wild: 219-01)	F_7 RIL (159)	SSR	Seed weight	QTLs (LG1, LG2, LG3, and LG10)	Andargie et al. (2011)
				Pod fiber layer thickness	QTLs (LG1, LG6, and LG10)	

(continued)

Table 9.3 (continued)

	Cross combination	Population (plants/lines) analyzed	Molecular markers used	Trait of interest	Gene/QTL (linkage group, LG)	References
Cowpea (*V. unguiculata*)	*V. unguiculata* (IT93K503-1) × *V. unguiculata* ('California Blackeye No.46')	$F_{2:8}$ RIL (108)	SNP and AFLP	Macrophomina phaseolina resistance	QTLs (LG2, LG3, LG5, LG6, and LG11)	Muchero et al. (2011)
				Maturity	QTL (LG5)	
Cowpea (*V. unguiculata*)	*V. unguiculata* ('California Blackeye No.27') × *V. unguiculata* (IT82E-18)	F_8 RIL (160)	SNP	Foliar thrips	QTLs (LG2 and LG10)	Lucas et al. (2012)
	V. unguiculata ('California Blackeye No.46') × *V. unguiculata* (IT93K-503-1)	F_8 RIL (114)	SNP	Foliar thrips	QTLs (LG2 and LG4)	
Cowpea (*V. unguiculata*)	*V. unguiculata* ('Sanzi') × *V. unguiculata* ('Vita 7')	F_{10} RIL (122)	SNP	Leaf shape morphology	QTL (LG15)	Pottorff et al. (2012)
Cowpea (*V. unguiculata*)	*V. unguiculata* ('California Blackeye No.27') × *V. unguiculata* (24-125B-1)	F_{10} RIL (90)	SNP	*Fusarium oxysporum* f. sp. *tracheiphilum* race 3	QTL (LG1)	Pottorff et al. (2013)
Cowpea (*V. unguiculata*)	*V. unguiculata* (IT97K-499-35) × *V. unguiculata* ('Canapu T16')	F_2 (286)	AFLP	Cowpea golden mosaic virus	Gene (tightly linked to AFLP markers E.AAC/M.CCC$_{515}$ and E.AGG/M.CTT$_{280}$)	Rodrigues et al. (2012)
Cowpea (*V. unguiculata*)	*V. unguiculata* var. *unguiculata* (524B) × *V. unguiculata* var. *spontanea* (219-01)	F_7 RIL (159)	SSR	Time of flower opening	QTLs (LG1, LG2, and LG10)	Andargie et al. (2013)
				Days to flower	QTLs (LG1, LG2, and LG7)	
Cowpea (*V. unguiculata*)	*V. unguiculata* ('California Blackeye No.27') × *V. unguiculata* (IT82E-18)	F_8 RIL (141)	SNP	Heat tolerance	QTLs (LG2, LG3, LG6, LG7, and LG10)	Lucas et al. (2013a)

Cowpea (*V. unguiculata*)	*V. unguiculata* ('California Blackeye No.27') × *V. unguiculata* (IT82E-18)	RIL (160)[b]	SNP	Seed weight	QTLs (LG5, LG6, and LG7)	Lucas et al. (2013b)
	V. unguiculata ('California Blackeye No.27') × *V. unguiculata* (UCR 779)	RIL (56)[b]	SNP	Seed weight	QTL (LG5)	
	V. unguiculata ('California Blackeye No.27') × *V. unguiculata* (24-125B-1)	RIL (87)[b]	SNP	Seed weight	QTLs (LG5 and LG6)	
	V. unguiculata ('California Blackeye No.46') × *V. unguiculata* (IT93K-503)	RIL (114)[b]	SNP	Seed weight	QTLs (LG2, LG7, and LG8)	
	V. unguiculata ('Dan Ila') × *V. unguiculata* (Tvu-7778)	RIL (79)[b]	SNP	Seed weight	QTL (LG10)	
	V. unguiculata (524B) × *V. unguiculata* (IT84S-2049)	RIL (89)[b]	SNP	Seed weight	QTL (LG2)	
	V. unguiculata (TVu14676) × *V. unguiculata* (IT84S-2246)	RIL (136)[b]	SNP	Seed weight	QTLs (LG2, LG5, and LG7)	
	V. unguiculata (IT84S-2246) × *V. unguiculata* ('Mouride')	F_4 (87)	SNP	Seed weight	QTLs (LG2, LG6, and LG10)	

(continued)

Table 9.3 (continued)

	Cross combination	Population (plants/lines) analyzed	Molecular markers used	Trait of interest	Gene/QTL (linkage group, LG)	References
Yard-long bean (asparagus bean) (*V. unguiculata* ssp. *sesquipedialis*)	*V. unguiculata* ssp. *sesquipedialis* (ZN016) × V.	$F_{7:8}$ RIL (209)	SSR and SNP	Flower color	Gene (LG8)	Xu et al. (2011b)
				Seed coat color	Gene (LG8)	
Yard-long bean (asparagus bean) (*V. unguiculata* ssp. *sesquipedialis*)	*V. unguiculata* ssp. *unguiculata* cv.-gr. sesquipedalis (JP81610) × *V. unguiculata* ssp. *unguiculata* var. *spontanea* (JP89083)	BC_1F_1 (190)	SSR	Pod length	QTLs (LG1, LG3, LG4, LG5, LG7, LG8, and LG11)	Kongjaimun et al. (2012a)
		F_2 (188)		Pod length	QTLs (LG1, LG3, LG4, LG5, LG7, LG8, and LG11)	
Yard-long bean (asparagus bean) (*V. unguiculata* ssp. *sesquipedialis*)	*V. unguiculata* ssp. *unguiculata* cv.-gr. sesquipedalis (JP81610) × *V. unguiculata* ssp. *unguiculata* var. *spontanea* (JP89083)	BC_1F_1 (190)	SSR	Domestication-related traits (qualitative traits: seed coat color)	QTLs (for each trait, 1–11 QTLs are detected mainly on LG3, LG7, LG8, and LG11)	Kongjaimun et al. (2012b)
		F_2 (188)	SSR	Domestication-related traits (qualitative traits: seed coat color)	Gene (LG7)	
				Domestication-related traits (qualitative traits: pod dehiscence)	Gene (not mapped; dehiscence:indehiscence = 9:7)	
				Domestication-related traits (18 quantitative traits)	QTLs (for each trait, 1–9 QTLs are detected mainly on LG3, LG7, LG8, and LG11)	
Yard-long bean (asparagus bean) (*V. unguiculata* ssp. *sesquipedialis*)	*V. unguiculata* ssp. *unguiculata* cv.-gr. sesquipedalis (JP81610) × *V. unguiculata* ssp. *unguiculata* var. *spontanea* (JP89083)	BC_1F_1 (190)	SSR	Pod tenderness	QTLs (LG7, LG8, and LG11)	Kongjaimun et al. (2013)
				Pod total soluble solid	QTL (not detected)	
		F_2 (188)		Pod tenderness	QTLs (LG7 and LG8)	
				Pod total soluble solid	QTLs (LG1 and LG3)	

Yard-long bean (asparagus bean) (*V. unguiculata* ssp. *sesquipedialis*)	*V. unguiculata* ssp. *sesquipedialis* (ZN016) × *V. unguiculata* ssp. *sesquipedialis* (ZJ282)	$F_{8:9}$ RIL (209)	SSR and SNP	Days to first flowering	QTLs (LG10 and LG11)	Xu et al. (2013)
				Nodes to first flower	QTLs (LG4 and LG11)	
				Leaf senescence	QTL (LG11)	
				Pod number per plant	QTLs (LG2 and LG3)	
Bambara groundnut (*V. subterranea* (L.) Verdc.)	*V. subterranea* var. *subterranea* ('DipC') × *V. subterranea* var. *subterranea* ('Tiga necaru')	F_3 (73)	SSR, DArT, and AFLP	26 quantitative traits	OTLs (for each trait, 1–3 QTLs are detected on LG1, LG7, LG9, LG10, LG13, LG14, and LG15)	Ahmad (2012)
	V. subterranea var. *subterranea* ('DipC') × *V. subterranea* var. *spontanea* ('VSSP11')	F_2 (98)	SSR, DArT, and AFLP	6 quantitative traits	QTLs (for each trait, 1–3 QTLs are detected on LG1, LG3, LG4, LG5, LG7, LG8, LG10, LG11, LG12, and LG18)	
Others						
V. vexillata	*V. vexillata* var. *angustifolia* (TVnu1443) × *V. vexillata* var. *vexillata* (TVnu73)	F_2 (94)	RAPD, AFLP, and SSR	Cowpea mottle carmovirus resistance	Gene (LG4)	Ogundiwin et al. (2005)
				Leaf shape	Gene (LG1)	
				Nine quantitative traits	QTLs (for each trait, 3–8 QTLs are detected mainly on LG1, LG2, and LG7)	

[a]There is no description
[b]These populations are at least F_8 generations

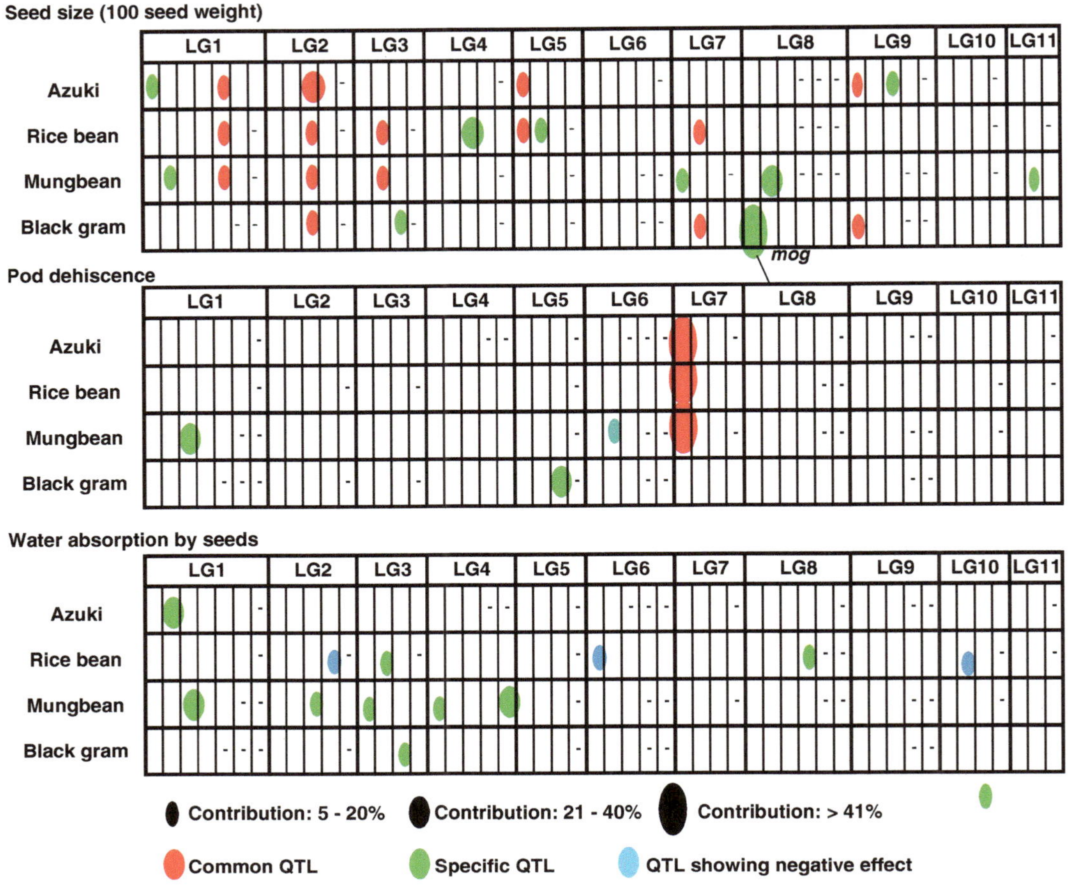

Fig. 9.1 Comparison of QTLs for three domestication-related traits among four *Vigna* crop species

(Chankaew et al. 2013; Kasettranan et al. 2010). Other resistance genes have been found on linkage group 4 (Chaitieng et al. 2002; Chankaew et al. 2013). Chaitieng et al. (2002) analyzed a mapping population using mung bean accession VC1210A as the resistant parent and found a QTL for field resistance to Thai races of powdery mildew that explained 65 % of the variation to powdery mildew resistance. Mung bean yellow mosaic viruses belong to the geminivirus group (Sunitha et al. 2013) and are highly destructive diseases of mung bean. Breeders have focused on finding resistance to MYMV (mung bean yellow mosaic virus) that is mainly found in south and west South Asia and MYMIV (mung bean yellow mosaic Indian virus) that is prevalent in central and northern South Asia. They are also a threat to a wide range of other legumes such as black gram and soybean.

There are a number of reports analyzing resistance to MYMV in different germplasm and both recessive and dominant genes. The resistant variety SML-668 has two recessive genes for resistance. Sudha et al. (2013) reported that the resistance of mung bean variety 'KMG189' is controlled by a single recessive gene. Two studies using different sources of MYMIV resistance, one using the wild mung bean (*V. radiata* var. *sublobata*) and the other a breeding line from Pakistan, have found a common major resistance QTL (variously named MYMIV'9_25, qMYMIV1, qMYMIV4) (Chen et al. 2013; Kitsanachandee et al. 2013). This locus was detected in different locations, years, sources of resistance, and scoring systems. The locus has been associated with specific markers; hence, these can be used in marker-assisted selection. Other QTLs have also been detected for MYMIV resistance, but these have not been consistently found in different tests.

Overcoming yellow mosaic virus may require various approaches, both genetic and agronomic, because this virus undergoes genetic recombination in its host whitefly (*Bemisia tabaci*) which is an efficient vector. Two iron deficiency chlorosis QTL (qIDC) for resistance to iron deficiency have been identified in a cross between susceptible 'Kamphaeng Saen 2' and a resistant line from Pakistan (NM10-12) (Prathet et al. 2012). The major QTL on linkage group 3 was the same as the dominant gene (IR) reported by Srinives et al. (2010) in a cross between 'Kamphaeng Saen 1' and the same resistant Pakistan line. Resistance to bruchid has been reported in mung bean cultivars (Somta P et al. 2006, 2008; Somta C et al. 2008); however, mung bean breeders are interested in new sources of resistance to this important pest from other Asian *Vigna* species such as *V. umbellata* and *V. nepalensis* (Pandiyan et al. 2010; Somta P et al. 2008; Somta C et al. 2008).

9.2.2 Azuki Bean [*Vigna angularis* (Willd.) Ohwi and Ohashi]

Wild azuki bean (*Vigna angularis* var. *nipponensis*) is widely distributed in the ecological and cultural zone associated with broad-leaved evergreen forests (Isemura et al. 2011) (Fig. 9.2).

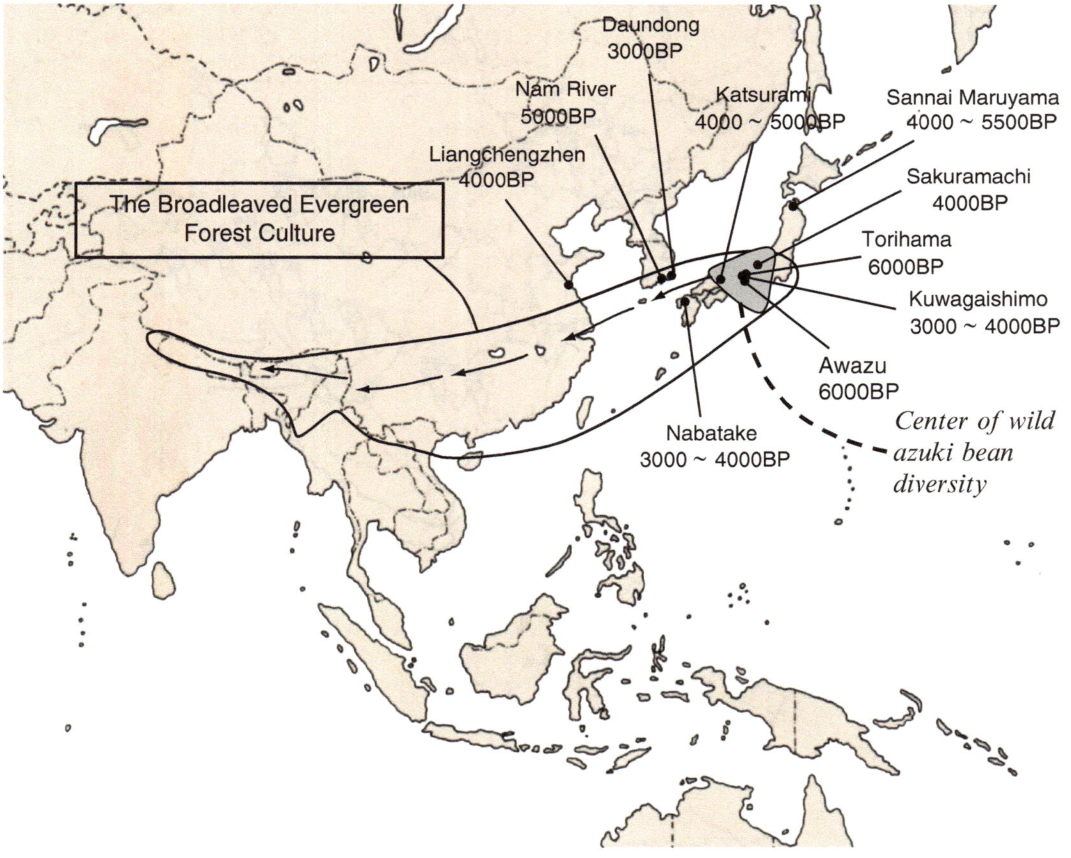

Fig. 9.2 Distribution of the "Broad-leaved Evergreen Forest Culture (Laurel Forest Culture)," center of wild azuki bean diversity and sites of the archaeological remains where putative azuki bean seeds were found (after Isemura et al. (2011))

Analysis of the present-day diversity of azuki beans and abundance of wild azuki beans suggests that Japan is the center of diversity of azuki beans (Yamaguchi 1992; Xu et al. 2000a, b, 2008). The earliest archaeobotanical remains of domesticated azuki bean have been found in the central regions of Japan, Shiga, and Fukui prefectures, dated at 5–6000 BP (Tomooka 2007). These dates are earlier than for archaeobotanical remains in China and Korea (Crawford 2005), suggesting azuki bean was domesticated in Japan. Two paper analyses domestication-related traits by the same method and with similar markers, but involving different parents (Isemura et al. 2007; Kaga et al. 2008). One paper results from a cross between *V. angularis* and its closely related species, *V. nepalensis* (female). The other mapping population was derived from wild and cultivated (female) azuki bean parents from Japan. While some results were similar, such as single QTL for pod dehiscence, about 60 % of the QTLs were considered to be different. This highlights the extent of genetic diversity that is available for *Vigna* breeders among the close relatives of *Vigna* crops. The useful diversity in species closely related to azuki bean was also shown when resistance to azuki brown stem rot was analyzed (Kondo and Tomooka 2012). Eight disease responses were observed, of which four were newly detected and 28 accessions (of 4 species) among 252 accessions from 26 Asian *Vigna* species were considered as potential sources of multiple resistance to azuki brown stem rot.

The chloroplast and mitochondrial genomes of azuki bean have been sequenced (Naito et al. 2013). The results were very similar to those for the chloroplast and mitochondrial genomes of mung bean (Alverson et al. 2011; Tangphatsornruang et al. 2010). The results have confirmed the relative stability of the chloroplast genome compared to the mitochondrial genome as has been shown in other species including mung bean. Since *Vigna* species in Asia are relatively recently evolved, the mitochondrial genome may be more suitable for understanding their evolutionary dynamics than the chloroplast genome (Naito et al. 2013).

9.2.3 Rice Bean [*Vigna umbellata* (Thunb.) Ohwi and Ohashi]

Wild rice bean (*V. umbellata*) is not taxonomically distinguished from its domesticated form. This reflects the lack of prominent differentiation between the two forms and possibly the recent domestication of rice bean (Tomooka et al. 2002a). The wild form is distributed in tropical monsoon climate areas from Nepal to East Timor but with most diversity centered on Southeast Asia (Isemura et al. 2011; Tian et al. 2013; Tomooka 2009). Rice bean is cultivated on small scale across a wide area of South, Southeast, and East Asia. It is most common in highlands, sometimes cultivated in the slash-and-burn agriculture of northeast India (Arora et al. 1980), Myanmar, Laos, Thailand, Vietnam, and southern China. It is occasionally grown in Japan, Korea, Indonesia, and East Timor. Rice bean is of particular interest, because it has the highest level of bruchid resistance among the species of *Vigna* subgenus *Ceratotropis* (Kashiwaba et al. 2002; Tomooka et al. 2000). Three chemicals, based on the structure of the flavonoid naringenin, were found to be associated with bruchid resistance in rice bean (Tomooka et al. 2006). There have been many efforts to introduce bruchid resistance from rice bean to other domesticated *Vigna* species. Somta et al. (2006) used a wild species *V. nakashimae* to develop rice bean linkage map and identify a source of resistance to bruchids. In India, direct crossing between mung bean and rice bean resulted in viable hybrids, and this may enable introduction of bruchid resistance to mung bean varieties (Pandiyan et al. 2010). In addition, resistance to mung bean yellow mosaic virus in rice bean has been analyzed in a cross with mung bean, and it was found to be conferred by a single recessive gene (Sudha et al. 2012, 2013). In Thailand, scientists have attempted to transfer useful traits from rice bean to mung bean by developing tetraploid interspecific hybrids from sterile F_1 hybrids (Chaisan et al. 2013). The fertile tetraploid produced artificially may have a potential for improving *V. reflexo-pilosa* var. *glabra* (*V. glabrescens*); the only domesticated Asian *Vigna* that is tetraploid. Rice bean is a highly prolific seed producer which may reflect its similarity to the wild form. In legumes, domestication has not always resulted in higher seed yield on plant basis (Kaga et al. 2008). A genome map developed for *V. umbellata* revealed many differences in the location of QTLs for domestication-related traits compared with *V. angularis*. Major QTLs for domestication-related traits in rice bean were reported to be on linkage group 4, whereas for azuki bean, they were mainly on linkage group 9 (Isemura et al. 2010). Among the QTLs detected in rice bean (69) and azuki bean (76) for domestication-related traits, only 15 were considered to be common. Of these, 15 QTLs were for seed size-related traits such as 100-seed weight, seed length, width, and thickness (Isemura et al. 2011). The numerous species-specific QTLs between these closely related species suggest that they can provide novel genes for breeding.

9.2.4 Black Gram (*Vigna mungo* (L.) Hepper)

Black gram is thought to have been domesticated in Gujarat and northern Peninsular India (Fuller 2007). Recently, the close relatives of black gram that still grow in northern Peninsular India have been studied and a new nomenclature has been published; *Vigna silvestris* is the presumed progenitor of black gram, and *V. sahyadriana* is a species closely related to black gram and other species in India of the section *Ceratotropis*, *V. hainiana*, *V. radiata*, and *V. subramaniana* (Aitawade et al. 2012). The largest production area for black gram is India. Black gram shows complete resistance to azuki bean weevil (*Callosobruchus chinensis*) but is susceptible to cowpea weevil (*C. maculatus*). In contrast, the wild progenitor of black gram shows complete resistance to both azuki bean weevil and cowpea weevil (Tomooka et al. 2000; Souframanien et al. 2010). Genome maps for black gram have been developed and compared to *V. angularis* (Chaitieng et al. 2006; Gupta et al. 2008). There have been a number of studies of a gigantism mutant of black gram that produces seeds, pods, and vegetative parts much larger than the parent from which it was derived, e.g., parent has 100-seed weight of 4.8 g compared to the mutant's 7.9 g (Tomooka et al. 2010). The mutant gene is located on linkage group 8 at a site different from seed size (100-seed weight) QTLs in other *Vigna* species. The potential of transferring this gene to other species is being explored to produce super-domesticated legumes (Vaughan et al. 2007).

Resistance to mung bean yellow mosaic India virus (MYMIV) has been found in an accession of black gram, and this resistance gene has been mapped using SSR markers (Gupta et al. 2013). An SSR marker closely linked to the resistant locus was found that can be used for marker-assisted selection.

9.2.5 Moth Bean (*Vigna aconitifolia* (Jacq.) Maréchal)

Moth bean is thought to have been domesticated in South Asia where its wild conspecific progenitor is reported to be widely distributed (Arora and Nayar 1984). The outstanding characteristic of this species is its drought and heat tolerance. Consequently, it is grown in arid and semiarid zones of northwest South Asia (Jain and Mehra 1980). Among 15 *Vigna* species, moth bean showed the highest heat tolerance, surviving conditions of 36 °C for 12 days followed by 40 °C for 11 days; all other *Vigna* species tested died at 40 °C (Tomooka et al. 2001).

9.2.6 Creole Bean (*Vigna reflexo-pilosa* Hayata var. *glabra* (*Vigna glabrescens*))

Vigna reflexo-pilosa var. *glabra* is a rare domesticated species, and very few accessions are to be found in the world's germplasm collections (Tomooka et al. 2002a). It seems, based on herbarium specimens and direct collection, to be mainly grown in Vietnam although there are reports of it being cultivated in Angola, Mauritius, West Bengal, and the Philippines.

Little is known about the agronomic aspects of this crop. On the other hand, the wild progenitor, var. *reflexo-pilosa*, is widely distributed across Southeast Asia into the Pacific (Tomooka et al. 2002a). There have been a number of studies to determine the origins of the two genomes that constitute this allotetraploid (Egawa and Tomooka 1994). The most recent and comprehensive survey confirms some earlier reports that the likely donor species are *V. trinervia* and *V. hirtella* (Chankaew et al. 2014b).

9.2.7 Cowpea Complex

Evidence suggests that cowpea (*V. unguiculata*) was domesticated in West Africa (Ng and Maréchal 1985). Early archaeological remains of domesticated cowpea in Ghana, West Africa, have been dated at between 3360 and 3840 BP (D'Andrea et al. 2007). It is presumed that sometime not long after domestication, the crop came to Asia by trading networks that developed from about 4,000 years ago (Fuller 2003; Fuller et al. 2011). In Africa, cowpea was selected for its seeds (*V. unguiculata* subsp. *unguiculata* cv.-gr. *Unguiculata*), but sometime after it arrived in Asia, divergent selection resulted in a new form of cowpea with long tender pods that could be eaten raw or cooked—yard-long bean (*V. unguiculata* subsp. *unguiculata* cv.-gr. Sesquipedalis). In some parts of Asia, such as Sri Lanka, cowpea is being planted instead of other *Vigna* species because of its higher yield and suitability to double cropping due to its short life cycle. Yard-long bean has the longest pod of any domesticated legume, and the pod can reach up to 90 cm long (35.4 in.) which is quite close to being a yard (36 in.) (http://www.gene.affrc.go.jp/databases-plant_images_detail_en.php?plno=5420610049). However, this is well short of the pod length of some wild legumes with *Cassia fistula* reported as having pods up to 238 cm (Jayasuriya 2012).

Three recent studies have analyzed the important agronomic traits related to the domestication of yard-long bean (Kongjaimun et al. 2012a, b, 2013). QTLs for domestication-related traits show co-localization on many linkage groups, but linkage groups 3, 7, 8, and 11 appear to be most important. Genomic dissection of pod length revealed 6 and 7 QTLs for this trait on different linkage groups in two populations, F_2 and BC_1F_1, respectively. The QTL with the main effect was on linkage group 7, explaining 30.5 % of the variation. A QTL with a large effect for pod length was also found on linkage group 7 in azuki bean. Linkage group 7 in yard-long bean appears to be important for a range of other domestication-related traits such as size of seed, stem, and leaf. This suggests that further study of genes for trait gigantism in yard-long bean should focus on linkage group 7.

9.3 Wild *Vigna* in Asia

9.3.1 *Vigna aridicola* N.Tomooka and Maxted

This species is recently found in the dry zone of Sri Lanka (Tomooka et al. 2002c). It is closely related to the wild form of *V. aconitifolia*. *V. aridicola* has been little studied, and there is an urgent need to understand the adaptive mechanism of this species and *V. aconitifolia* to harsh dry habitats.

9.3.2 *Vigna dalzelliana* (Kuntze) Verdc.

This species is poorly understood and previously was confused with *V. minima* (Tomooka et al. 2006). Recent germplasm collecting has found this species in Sri Lanka and India. There have also been collections of *Vigna* from southern Myanmar that appear to be *V. dalzelliana* (Tomooka et al. 2006).

9.3.3 *Vigna exilis* Tateishi & Maxted

Vigna exilis is only found growing in limestone outcrops (Tomooka et al. 2011b). It has been little studied but its distinctive habitat suggests that it may have useful genes for adaptation to specific high pH and dry soil conditions.

9.3.4 *Vigna grandiflora* (Prain) Tateishi & Maxted

This species should join *V. khandalensis* on the IUCN Red List of Threatened Species, since it is known currently from only a few rather vulnerable populations in Thailand.

9.3.5 *Vigna hirtella* Ridley and *V. tenuicaulis* N. Tomooka and Maxted

Based on AFLP analysis, *V. hirtella* appears to be a variable species with two eco-types, one of low altitude and another of high altitude (Seehalak et al. 2006; Tomooka et al. 2002b). *V. hirtella* grows sympatrically with a number of other species in section *Angulares* of the subgenus *Ceratotropis* in highland Southeast Asia. A natural hybrid between *V. hirtella* and *V. minima* has been found at one site (author's unpublished data). It is probable that this variable species may naturally hybridize with other species where it grows. *Vigna tenuicaulis* is relatively recently described using an accession collected in northern Thailand and is closely related to *V. hirtella* to which it is sometimes confused (Tomooka et al. 2002c). It has a low level of trypsin inhibitor activity that might be a useful characteristic to transfer to *Vigna* crops (Konarev et al. 2002).

9.3.6 *Vigna khandalensis* (Santapau) Raghavan & Wadhwa

V. khandalensis is poorly known; not even its chromosome number has been reported. It is considered "near threatened" in the IUCN Red List of Threatened Species (http://www.iucnredlist.org/details/19892969/0). Among the wild Asian *Vigna*, this species is an erect herb rather than a climbing vine. It is only found in India, mainly in the Western Ghats. It is reported that the seeds are sometimes used as a famine food (Babu et al. 1985).

9.3.7 *Vigna minima* (Roxb.) Ohwi & Ohashi, *V. nakashimae* (Ohwi) Ohwi and Ohashi, *V. riukiuensis* (Ohwi) Ohwi & Ohashi

Vigna minima is widely distributed in variable habitats across East Asia (China), Southeast Asia, and as far east as Papua New Guinea (Tomooka et al. 2002a). The closely related species *V. nakashimae* and *V. riukiuensis* have more restricted distributions. These three species are genetically closely related and are called *Vigna minima* complex (Yoon et al. 2000). *V. nakashimae* is widely distributed on the Korean peninsula and the Goto islands, Nagasaki, Japan (Yoon et al. 2007; Tomooka et al. 2013). Some populations of *V. nakashimae* from the Goto islands, living on a hill exposed to strong sea winds, are revealed to have a high level of salt tolerance. *V. riukiuensis* is found in the southern Okinawa Islands in Japan. It grows on a cliff near the sea and shows a high level of salt tolerance. *V. minima* is found in both shaded forest floor and open habitats such as around paddy field, suggesting a high level of diversity. Since it has been little studied, it may well consist of several varieties or subspecies. Recently, studies have shown that some accessions of *V. minima* and *V. nakashimae* have a high level of resistance to all races of soybean cyst nematode found in Japan (Kushida et al. 2013). These resistant sources are being used in azuki breeding since soybean cyst nematode is becoming an increasingly problematic pest on legumes in Hokkaido, Japan. *V. nakashimae* has been used to develop an interspecific linkage map with *V. umbellata* (Somta et al. 2006).

9.3.8 *Vigna nepalensis* Tateishi & Maxted

V. nepalensis is closely related to *V. angularis* with which it makes fertile hybrids. *V. nepalensis* was used as the female parent in a mapping population to analyze domestication-related traits in azuki bean (Isemura et al. 2007). It has a restricted distribution in Nepal and its adjacent localities in

India and Bhutan, and it was found growing between the altitude of 350 and 1,650 m (Tomooka et al. 2002a). QTLs for bruchid resistance have been reported from *V. nepalensis*, and some of these appear to be new sources of resistance including one QTL that is unrelated to seed size (Somta P et al. 2008).

9.3.9 *Vigna stipulacea* Kuntze

This species is characterized by extremely long peduncles 22–30 cm long that result in the flower and pod rising conspicuously above canopy. This characteristic is probably why this species has been harvested, and some harvested forms have been semidomesticated having increased vegetative organs and weak seed dormancy (Tomooka et al. 2011a). In Tamil Nadu, India, semidomesticated *V. stipulacea* is grown before or after rice in paddy fields and is considered good forage for cattle and green manure. Despite the labor required to harvest it because of shattering pods, this species is also sometimes used as a human food. Based on comments by farmers, *V. stipulacea* is resistant to many insects and diseases such as stinkbug and powdery mildew, grows faster (early flowering and maturing) than mung bean and black gram, and has high palatability (Tomooka et al. 2006, 2011b). Seeds are sold in a local seed shop in Tamil Nadu (Tomooka et al. 2011b). Given these positive attributes, the species warrants particular study of its agronomic potential and potential for furnishing new and useful genes to legume breeding.

9.3.10 *Vigna subramaniana* (Babu ex Raizada) M. Sharma and *Vigna hainiana* Babu, Gopinathan, and Sharma

V. subramaniana has the lowest level of trypsin inhibitor activity among the subgenus *Ceratotropis* species studied and absence of chymotrypsin inhibitor activity (Konarev et al. 2002). Taxonomic confusion surrounding this species and *V. hainiana* has been discussed by Tomooka et al. (2006). There remains a need to more thoroughly understand relationships in the mung bean complex to which these two species belong. There is now sufficient germplasm in various gene banks to permit these studies to be conducted.

9.3.11 *Vigna trinervia* (Heyne ex Wall.) Tateishi & Maxted

V. trinervia consists of two varieties. *Vigna trinervia* var. *trinervia* is widely distributed in South and Southeast Asia and also in the Indian Ocean Islands and Tanzania. *Vigna trinervia* var. *bourneae* (*Vigna bourneae*) is found only in southern India and is distinguished by being covered with white villose hairs. This species is an important cover crop in between plantations of coconut and rubber.

9.3.12 *Vigna trilobata* (L.) Verdc. and *V. indica* T.M. Dixit, K.V. Bhat & S.R. Yadav

Vigna trilobata characteristically is found in sandy soils; hence, it is found along the beaches in Sri Lanka. However, it is more commonly found inland where sandy soils predominate such as Tamil Nadu, India, and Sagaing, Myanmar. This species is characterized by its very long taproot and so is found along the coastline of the dry zone, whereas *V. marina* with shallower roots is found predominately along the coast of the wet zone in Sri Lanka (Liyanage 2013, personal communication to the authors). Previously, variation in the *V. trilobata* gene pool resulted in the species being divided into two subspecies, subsp. *trilobata* and subsp. *pusilla*. However, recently, subsp. *pusilla* has been raised to the rank of a species called *V. indica*, differing from *V. trilobata* in various characteristics of which cylindrical seeds with truncated ends are most obvious. The two subspecies in India are found in the western, central, and northern regions for *V. indica* and southern and eastern India for *V. trilobata*

(Dixit et al. 2011). *V. trilobata* is occasionally harvested as seeds or young pods in India (Tomooka et al. 2011a). It has been successfully hybridized with mung bean (female parent) (Pandiyan et al. 2012).

9.3.13 *Vigna vexillata* (L.) A. Rich. (Tuber Cowpea)

V. vexillata has a pantropical distribution and, like many legumes, produces tubers. In Bali and Timor, Indonesia, this species is cultivated. Since Bali's cultivated accession seeds increased their size remarkably and lost strong dormancy and pods are indehiscent, it may be considered as fully domesticated. In Bali and Timor, *V. vexillata* is used for its tubers and seeds as well as forage with estimated yields of 18–30 t ha^{-1} and 0.8–1.2 t ha^{-1} for tubers and seeds, respectively. Root protein content is ca. 15 % which is about 2.5 times higher than that of yam (6 %), 3 times higher than that of potato (5 %) and sweet potato (5 %), and 5 times higher than that of cassava (3 %) (Karuniawan et al. 2006). It is also reported to be cultivated in parts of India and harvested by aboriginals in Australia. Based on the genetic study of agronomic traits by hybridization, it was concluded that the wild African and Australian accessions could be used along with var. *macrosperma* (putative cultivated form) for breeding improved varieties of *V. vexillata* for forage, cover crop, and vegetable uses (Damayanti et al. 2010a). They also reported a high level of cross incompatibility between the wild and cultivated form in Bali (Damayanti et al. 2010b).

9.3.14 *Vigna hosei* (Craib) Backer (*V. parkeri* Backer)

V. hosei is widely distributed in Asia, Africa, and both North and South America. It is a useful legume in intensively grazed pastures and forms a good ground cover in lightly shaded areas. (http://www.tropicalforages.info/key/Forages/Media/Html/Vigna_parkeri.htm).

9.3.15 *Vigna luteola* (Jacq.) Benth.

V. luteola is native to Africa, Asia, and Australasia but has been used as a short-season and highly palatable pasture or green manure in Europe and the New World. It is particularly useful in wet or waterlogged conditions. It has a wide range of adaptation to soil types, light conditions, and temperatures. An Australian accession, CPI 60428, has been reported to have some jassid insect resistance in humid, subtropical Australia (http://www.tropicalforages.info/key/Forages/Media/Html/Vigna_luteola.htm). It has been used to develop a linkage map to analyze salt tolerance in *V. marina* (Chankaew et al. 2014a).

9.3.16 *Vigna marina* (Burm.) Merr.

This species consists of two subspecies that have different geographic distributions: subspecies *marina* is found in Asia, Australia, the coast of Indian Ocean, and countries of Africa, whereas subspecies *oblonga* is found on the coastal areas of the tropical Atlantic countries of Africa. The species is reportedly used by some African farmers as a cover crop and green manure and has a potential as a sand-binding agent (Padulosi and Ng 1993). It also produces edible tubers that are eaten by aboriginals in Australia, while in the Maldives seeds are harvested and eaten (Padulosi and Ng 1993). The seeds are used as a coffee substitute in Gabon (Burkill 1995). Diversity analysis has shown that *V. marina* subsp. *oblonga* is more closely related to *V. luteola* than subsp. *marina* (Sonnante et al. 1997). As a consequence, recent studies to identify genes for salt tolerance in *Vigna marina* have concentrated on an interspecific mapping population between subsp. *oblonga* and *V. luteola* (Chankaew et al. 2014a). The $F_{2:3}$ population was evaluated for salt tolerance under hydroponic conditions at the seedling and developmental stages. Segregation analysis indicated that salt tolerance in *V. marina* is controlled by a few genes. Multiple-interval mapping (MIM) consistently identified one major QTL for each trait—*Saltol1.1*, *Saltol1.2*, and *Saltol1.3* for the percentage of surviving plants in salt water at the seedling stage and leaf wilt and recovery score during vegetative stage, respectively. All of the QTLs detected were located on LG1. These QTLs explained 50.7 %, 41.4 %, and 20.0 % of variation in the evaluated trait, respectively. All of the detected salt-tolerant QTLs, alleles from *V. marina* subsp. *oblonga*, increased salt tolerance. The flanking markers of each QTL can facilitate transferring of the salt-tolerant allele from *V. marina* subsp. *oblonga* into related *Vigna* crops. Unlike other *Vigna* species, *Sinorhizobium* spp. not *Bradyrhizobium* spp. forms nodules on *V. marina* roots (Akatsu, personal communication). The isolated *Sinorhizobium* strains showed an extremely high level of salt tolerance. The isolates could grow even in a nutrient solution with 5 % NaCl (Tomooka et al. 2011b).

9.3.17 *Vigna adenantha* (G.F. Meyer) Maréchal, Mascherpa & Stainier

V. adenantha has a pantropical distribution and is occasionally cultivated (Brink and Belay 2006). The green pods and ripe seeds are used as an emergency food. In India, its tuberous roots are eaten in times of food scarcity. It grows in humid swampy locations, along seashores and rivers. The seed has a large cavity between the cotyledons that enables it to float, and distribution patterns suggest it can be dispersed by sea (Brink and Belay 2006).

9.3.18 Interspecific Hybridization

Cross compatibility studies have been reviewed by Tomooka et al. (2002a). Generally, there is no barrier to gene flow between domesticated forms and their closest relatives. Species in the same section of the subgenus *Ceratotropis* can usually cross with little difficulty. Natural interspecific hybrids have been found between *V. hirtella* and *V. minima* in northern Thailand (author's unpublished observations). Recently, Pandiyan et al. (2010) reported a number of cross-sectional and cross subgenus hybrids. Among these hybrids, the cross between *V. radiata* and *V. umbellata* is particularly significant as *V. umbellata* possesses a high level of resistance to bruchid beetles, one of the most serious pests of *Vigna*.

9.4 Domestication

The *Vigna* of Asia offer an unusually broad spectrum of species in various stages of domestication. Plant parts used by humans vary from seeds, green pods, to swollen roots. Several are useful forage species, such as *V. stipulacea*, *V. luteola*, and *V. marina*. Among them, *V. marina* may be de-domesticated since it has large seeds with non-dehiscent pods. Also in many parts of southern Japan and Southeast Asia, nonnative populations of escaped (de-domesticated or feral) cowpea are common with dehiscent pods (Bervillé et al. 2005).

The *Vigna* of Asia exhibit extremes in certain agronomic characteristics; among domesticated legumes, the longest pod is found in yard-long bean (*V. unguiculata* subsp. *unguiculata* cv.-gr. Sesquipedalis), and among the smallest seeds for a domesticated pulse used for its seeds are mung bean (*V. radiata*), black gram (*V. mungo*), and moth bean (*V. aconitifolia*).

9.5 Ecological Adaptation

The *Vigna* of Asia are adapted to a range of ecological conditions, among them harsh environments such as arid conditions (*V. aconitifolia*), sandy and saline soils (*V. marina* and *V. trilobata*), alkaline limestone rock soils (*V. exilis*), exposed windy cliff top environments (*V. riukiuensis*), waterlogged riverside (*V. luteola*), and shady forests (*V. minima*).

9.6 Genomic Studies

Currently, a large sequencing project is under way that aims to sequence 16 *Vigna* species of Asia (Naito et al. 2013). This information coupled with complete sequence information for the chloroplast and mitochondria of azuki bean and mung bean will provide a foundation of genome information resources for *Vigna* improvement in the future. Although the *Vigna* of Asia have been "forgotten" by much of the international agricultural research community, considerable progress has been made by national programs. However, the effort devoted to research on Asian *Vigna* does not do justice to their importance and potential contribution to agriculture in the future. The very wide ecological adaptation of *Vigna* in Asia, highlighted in this chapter, suggests that *Vigna* have much to offer other crops. Gigantism exhibited by the pods of yard-long bean and very high levels of salt tolerance in *Vigna marina* are two outstanding examples.

References

Agbicodo EM, Fatokun CA, Bandyopadhyay R, Wydra K, Diop N-N, Muchero W, Ehlers JD, Roberts PA, Close TJ, Visser RGF, van der Linden CG (2010) Identification of markers associated with bacterial blight resistance loci in cowpea [*Vigna unguiculata* (L.) Walp.]. Euphytica 175:215–226

Ahmad NS (2012) Genetic analysis of plant morphology in bambara groundnut [*Vigna subterranea* (L.) Verdc.]. PhD thesis, University of Nottingham. Available from http://etheses.nottingham.ac.uk/3150/1/Thesis_for_Nariman_Ahmad.pdf. Accessed July 2014

Aitawade MM, Sutar SP, Rao SR, Malik SK, Yadav SR, Bhat KV (2012) Section *Ceratotropis* of subgenus *Ceratotropis* of *Vigna* (Leguminosae – Papilionoideae) in India with a new species from the northern Western Ghats. Rheedea 22(1):20–27

Alverson AJ, Zhuo S, Rice DW, Sloan DB, Palmer JD (2011) The mitochondrial genome of the legume *Vigna radiata* and the analysis of recombination across short mitochondrial repeats. PLoS ONE 6(1):e16404. doi:10.1371/journal.pone.0016404

Andargie M, Remy S, Pasquet RS, Gowda BS, Muluvi GM, Timko MP (2011) Construction of a SSR-based genetic map and identification of QTL for domestication traits using recombinant inbred lines from a cross between wild and cultivated cowpea (*V. unguiculata* (L.) Walp.). Mol Breed 28:413–420

Andargie M, Pasquet RS, Muluvi GM, Timko MP (2013) Quantitative trait loci analysis of flowering time related traits identified in recombinant inbred lines of cowpea (*Vigna unguiculata*). Genome 56:289–294

Arora RK, Nayar ER (1984) Wild relatives of crops in India. NBPGR science monograph no. 7, Published by the National Bureau of Plant Genetic Resources, India

Arora RK, Chandel KPS, Joshi BS, Pant KC (1980) Rice bean: tribal pulse of eastern India. Econ Bot 34:260–263

Babu CR, Sharma SK, Johri BM (1985) Leguminosae-Papilionoideae: tribe – Phaseoleae. Bull Bot Surv India 27:1–28

Bervillé A, Breton C, Cunliffe K, Darmency H, Good AG, Gressel J, Hall LM, McPherson MA, Medail F, Pinatel C, Vaughan DA, Warwick SI (2005) Issues of ferality in oats, olives, the *Vigna* group, ryegrass species, safflower and sugarcane. In: Gressel J (ed) Crop ferality and volunteerism, vol 15. CRC Press, Boca Raton, pp 231–255

Brink M, Belay G (2006) Plant resources of tropical Africa. Cereals and Pulses Prota Foundation/Backhuy Publishers, Wageningen, p 298

Burkill HM (1995) The useful plants of West tropical Africa, Families J–L, vol 3, 2nd edn. Royal Botanic Gardens, Kew, p 857

Chaisan T, Somta P, Srinives P, Chanprame S, Kaveeta R, Dumrongkittikule S (2013) Development of tetraploid plants from an interspecific hybrid between mungbean (*Vigna radiata*) and rice bean (*Vigna umbellata*). J Crop Sci Biotechnol 16(1):45–51

Chaitieng B, Kaga A, Han OK, Wang XW, Wongkaew S, Laosuwan P, Tomooka N, Vaughan DA (2002) Mapping a new source of resistance to powdery mildew (*Erysiphe polygoni* DC.) in mungbean [*Vigna radiata* (L.) Wilczek]. Plant Breed 121:521–525

Chaitieng B, Kaga A, Tomooka N, Isemura T, Kuroda Y, Vaughan DA (2006) Development of a black gram [*Vigna mungo* (L.) Hepper] linkage map and its comparison with an azuki bean [*Vigna angularis* (Willd.) Ohwi and Ohashi] linkage map. Theor Appl Genet 113:1261–1269

Chankaew S, Somta P, Sorajjapinun W, Srinives P (2011) Quantitative trait loci mapping of Cercospora leaf spot resistance in mungbean, *Vigna radiata* (L.) Wilczek. Mol Breed 28:255–264

Chankaew S, Somta P, Isemura T, Tomooka N, Kaga A, Vaughan DA, Srinives P (2013) QTL mapping reveals conservation of major and minor loci for powdery mildew resistance in different sources of resistance in mungbean [*Vigna radiata* (L.) Wilczek]. Mol Breed 213:121–130

Chankaew S, Isemura T, Naito K, Ogiso-Tanaka E, Tomooka N, Somta P, Kaga A, Vaughan DA, Srinives P (2014a) QTL mapping for salt tolerance and domestication-related traits in *Vigna marina* subsp. *oblonga*; a halophytic species related to legume crops. Theor Appl Genet 127(3):691–702

Chankaew S, Isemura T, Isobe S, Kaga A, Tomooka N, Somta P, Hirakawa H, Shirasawa K, Vaughan DA, Srinives P (2014b) Detection of genome donor species of neglected tetraploid crop *Vigna reflexo-pilosa* (créole bean), and genetic structure of diploid species based on newly developed EST-SSR markers from azuki bean (*Vigna angularis*) (Accepted in PloS ONE)

Chen HM, Ku HS, Schafleitner R, Bains TS, Kuo CG, Liu CA, Nair RM (2013) The major quantitative trait locus for mungbean yellow mosaic Indian virus is tightly linked in repulsion phase to the major bruchid resistance locus in a cross between mungbean [*Vigna radiata* (L.) Wilczek] and its wild relative *Vigna radiata* ssp. *sublobata*. Euphytica 192:205–216

Crawford GW (2005) East Asian plant domestication. In: Stark MT (ed) The archaeology of Asia. Blackwell, Oxford, pp 78–95

D'Andrea AC, Kahlheber S, Logan AL, Watson DJ (2007) Early domesticated cowpea (*Vigna unguiculata*) from Central Ghana. Antiquity 81(2007):686–698

Damayanti F, Lawn CDRJ, Bielig LM (2010a) Expression of qualitative and quantitative traits in hybrids between domesticated and wild accessions of the tropical tuberous legume *Vigna vexillata* (L.) A. Rich. Crop Pasture Sci 61(10):798–811

Damayanti F, Lawn CDRJ, Bielig LM (2010b) Genetic compatibility among domesticated and wild accessions of the 2 tropical tuberous legume *Vigna vexillata* (L.) A. Rich. Crop Pasture Sci 61(10):785–797

Dixit TM, Sutar SP, Yadav SR, Bhat KV, Rao SR (2011) *Vigna indica*, a new name for *Vigna trilobata* var. *pusilla* and a note on section Aconitifoliae in India. Rheedea 21(1):1–7

Egawa Y, Tomooka N (1994) Phylogenetic differentiation of *Vigna* species in Asia. JIRCAS Int Symp Ser 2:112–120

Fatokun CA, Menancio-Hautea DI, Dariush D, Young ND (1992) Evidence for orthologous seed weight genes in cowpea and mung bean based on RFLP mapping. Genetics 132:841–846

Fatokun CA, Young ND, Myers GO (1997) Molecular markers and genome mapping in cowpea. In: Singh BB, Mohan Raj DR, Dashiell KE, Jackai LEN (eds) Advances in cowpea research. International Institute of Tropical Agriculture (IITA)/Japan International Research Center for Agricultural Sciences (JIRCAS), Ibadan

Fuller DQ (2003) African crops in prehistoric South Asia: a critical review. In: Neumann K, Butler A, Kahlheber S (eds) Food, fuel and fields. Progress in African archaeobotany, vol 15, Africa praehistorica. Heinrich-Barth Institute, Cologne, pp 239–271

Fuller D (2007) Contrasting patterns in crop domestication and domestication rates: recent archaeobotanical insights from the Old World. Ann Bot 100:903–924

Fuller DQ, Harvey E (2006) The archaeobotany of Indian pulses: identification, processing and evidence for cultivation. Environ Archaeol 11:219–246

Fuller DQ, Boivin N, Hoogervorst T, Allaby R (2011) Across the Indian Ocean: the prehistoric movement of plants and animals. Antiquity 85:544–558

Gupta SK, Souframanien J, Gopalakrishna T (2008) Construction of a genetic linkage map of black gram, *Vigna mungo* (L.) Hepper, based on molecular markers and comparative studies. Genome 51:628–637

Gupta S, Gupta DS, Anjum TK, Pratap A, Kumar J (2013) Inheritance and molecular tagging of MYMIV resistance gene in blackgram (*Vigna mungo* L. Hepper). Euphytica 193:27–37

Han OK, Kaga A, Isemura T, Wang XW, Tomooka N, Vaughan DA (2005) A genetic linkage map for azuki bean [*Vigna angularis* (Willd.) Ohwi & Ohashi]. Theor Appl Genet 111:1278–1287

Humphry ME, Konduri V, Lambrides CJ, Magner T, McIntyre CL, Aitken EAB, Liu CJ (2002) Development of a mungbean (*Vigna radiata*) RFLP linkage map and its comparison with lablab (*Lablab purpureus*) reveals a high level of colinearity between the two genomes. Theor Appl Genet 105:160–166

Humphry ME, Magner T, McIntyre CL, Aitken EAB, Liu CJ (2003) Identification of a major locus conferring resistance to powdery mildew (*Erysiphe polygoni* DC) in mungbean (*Vigna radiata* L. Wilczek) by QTL analysis. Genome 46:738–744

Humphry ME, Lambrides CJ, Chapman SC, Aitken EAB, Imrie BC, Lawn RJ, McIntyre CL, Liu CJ (2005) Relationships between hard-seededness and seed weight in mungbean (*Vigna vadiata*) assessed by QTL analysis. Plant Breed 124:292–298

Isemura T, Kaga A, Konishi S, Ando T, Tomooka N, Han OK, Vaughan DA (2007) Genome dissection of traits related to domestication in azuki bean (*Vigna angularis*) and comparison with other warm-season legumes. Ann Bot 100:1053–1071

Isemura T, Kaga A, Tomooka N, Shimizu T, Vaughan DA (2010) Construction of a genetic linkage map and genetic analysis of the domestication related traits in rice bean (*Vigna umbellata*). Ann Bot 106:927–944

Isemura T, Tomooka N, Kaga A, Vaughan DA (2011) Comparison of the pattern of crop evolution between two Asian beans, azuki bean (*Vigna angularis*) and rice bean (*V. umbellata*). Jpn Agric Res Q 45(1):23–30

Isemura T, Kaga A, Tabata S, Somta P, Srinives P, Jo U, Vaughan DA, Tomooka N (2012) Construction of a genetic linkage map and genetic analysis of domestication related traits in mungbean (*Vigna radiata*). PLoS ONE 7(8):e41304. doi:10.1371/journal.pone.0041304

Jain HK, Mehra KL (1980) Evolution, adaptation, relationships and uses of the species of *Vigna* cultivated in India. In: Summerfield RJ, Bunting AH (eds) Advances in legume science. Volume 1 of the proceedings of the international legume conference, Published by the Royal Botanic Gardens Kew, UK, pp 459–468

Jayasuriya AHM (2012) The longest legume inflorescence in Sri Lanka. Ceylon J Sci (Biol Sci) 41:79–82

Kaga A, Ishimoto M (1998) Genetic localization of a bruchid resistance gene and its relationship to insecticidal cyclopeptide alkaloids, the vignatic acids, in mungbean (*Vigna radiata* L. Wilczek). Mol Genet Genomics 258:378–384

Kaga A, Ohnishi M, Ishii T, Kamijima O (1996) A genetic linkage map of azuki bean constructed with molecular and morphological markers using an interspecific population (*Vigna angularis* x *V. nakashimae*). Theor Appl Genet 93:658–663

Kaga A, Ishii T, Tsukimoto K, Tokoro E, Kamijima O (2000) Comparative molecular mapping in *Ceratotropis* species using an interspecific cross between *V. umbellata* and *V. angularis*. Theor Appl Genet 100:207–213

Kaga A, Han OK, Wang XW, Egawa Y, Tomooka N, Vaughan DA (2003) *Vigna angularis* as a model for legume research. In: Jayasuriya AHM, Vaughan DA (eds) Conservation and use of wild relatives of crops. Proceedings of the Joint Department of Agriculture, Sri Lanka and National Institute of Agrobiological Sciences, Japan. Workshop. Department of Agriculture, Peradeniya, Sri Lanka, pp 51–74

Kaga A, Vaughan DA, Tomooka N (2005) Molecular markers in plant breeding and crop improvement of *Vigna*. In: Lörz H, Wenzel G (eds) Molecular markers in plant breeding and crop improvement, Biotechnology in agriculture and forestry. Springer, Heidelberg, pp 171–187

Kaga A, Isemura T, Tomooka N, Vaughan DA (2008) The domestication of the azuki bean (*Vigna angularis*). Genetics 178:1013–1036

Kajonphol T, Sangsiri C, Somta P, Toojinda T, Srinives P (2012) SSR map construction and quantitative trait loci (locus) identification of major agronomic traits in mungbean (*Vigna radiata* (L.) Wilczek). SABRAO J Breed Genet 44:71–86

Karuniawan A, Isawandi A, Kale PR, Heinzemann J, Grüneberg WJ (2006) *Vigna vexillata* (L.) A. Rich. Cultivated as a root crop in Bali and Timor. Genet Resour Crop Evol 53:213–217

Kasettranan W, Somta P, Srinives P (2010) Mapping of quantitative trait loci controlling powdery mildew resistance in mungbean (*Vigna radiata* (L.) Wilczek). J Crop Sci Biotechnol 13(3):155–161

Kashiwaba K, Tomooka N, Kaga A, Han OK, Vaughan DA (2002) Characterization of resistance to three bruchid species (*Callosobruchus* spp., *Coleoptera*, *Bruchidae*) in cultivated rice bean (*Vigna umbellata*) factors in rice bean (*Vigna umbellata*). J Econ Entomol 96:207–213

Kitsanachandee R, Somta P, Chatchawankanphanich O, Akhtar KP, Shah TM, Nair RM, Bains TS, Sirari A, Kaur L, Srinives P (2013) Detection of quantitative trait loci for mungbean yellow mosaic India virus (MYMIV) resistance in mungbean (*Vigna radiata* (L.) Wilczek) in India and Pakistan. Breed Sci 63:367–373

Konarev AV, Tomooka N, Vaughan DA (2002) Proteinase inhibitor polymorphism in the genus *Vigna* subgenus *Ceratotropis* and its biosystematic implications. Euphytica 123:165–177

Kondo N, Tomooka N (2012) New sources of resistance to *Cadophora gregata* f. sp. *Adzukicola* and *Fusarium oxysporum* f. sp. *Adzukicola* in *Vigna* spp. Plant Dis 96(4):562–568

Kongjaimun A, Kaga A, Tomooka N, Somta P, Shimizu T, Shu Y, Isemura T, Vaughan DA, Srinives P (2012a) An SSR-based linkage map of yardlong bean [*Vigna unguiculata* (L.) Walp. ssp. *unguiculata* cv.-gr. *sesquipedalis* (L.) Verdc.] and QTL analysis for pod length. Genome 55(2):81–92

Kongjaimun A, Kaga A, Tomooka N, Somta P, Vaughan DA, Srinives P (2012b) The genetics of domestication of yardlong bean *Vigna unguiculata* (L.) Walp. ssp. *unguiculata* cv.-gr. *sesquipedalis*. Ann Bot 109:1185–1200

Kongjaimun A, Somta P, Tomooka N, Kaga A, Vaughan DA, Srinives P (2013) QTL mapping of pod tenderness and total soluble solid in yardlong bean [*Vigna unguiculata* (L.) Walp. subsp. *unguiculata* cv.-gr. *sesquipedalis*]. Euphytica 189:217–223. doi:10.1007/s10681-012-0781-2

Kushida A, Tazawa A, Aoyama S, Tomooka N (2013) Novel sources of resistance to the soybean cyst nematode (*Heterodera glycines*) found in the wild relatives of azuki bean (*Vigna angularis*) and their characteristics of resistance. Genet Resour Crop Evol 60:985–994

Lambrides CJ, Lawn RJ, Godwin ID, Manners J, Imrie BC (2000) Two genetic linkage maps of mungbean using RFLP and RAPD markers. Aust J Agr Res 51:415–425

Lucas MR, Diop N-N, Wanamaker S, Ehlers JD, Roberts PA, Close TJ (2011) Cowpea–soybean synteny clarified through an improved genetic map. Plant Genome 4:218–225

Lucas MR, Ehlers JD, Roberts PA, Close TJ (2012) Markers for quantitative inheritance of resistance to foliar thrips in cowpea. Crop Sci 52:2075–2081

Lucas MR, Ehlers JD, Huynh B-L, Diop N-N, Roberts PA, Close TJ (2013a) Markers for breeding heat-tolerant cowpea. Mol Breed 31:529–536

Lucas MR, Huynh B-L, da SilvaVinholes P, Cisse N, Drabo I, Ehlers JD, Roberts PA, Close TJ (2013b) Association studies and legume synteny reveal haplotypes determining seed size in *Vigna unguiculata*. Front Plant Sci Plant Genet Genomics 4:95

Lumpkin TA, McClary DC (1994) Azuki bean: botany, production, and uses. CAB International, Wallingford, p 268

Maxted N, Mabuza-Dlamini P, Moss H, Padulosi S, Jarvis A, Guarino L (2004) An ecogeographic study. African *Vigna*. International Plant Genetic Resources Institute, Rome

Mei L, Cheng XZ, Wang SH, Wang LX, Liu CY, Sun L, Xu N, Humphry ME, Lambrides CJ, Li HB, Liu CJ (2009) Relationship between bruchid resistance and seed mass in mungbean based on QTL analysis. Genome 52:589–596

Menacio-Hautea D, Kumar L, Danesh D, Young ND (1992) A genome map for mungbean [*Vigna radiata* (L.) Wilczek] based on DNA genetic markers (2n = 2x = 22). In: O'Brien JS (ed) Genetic maps. A compilation of linkage and restriction maps of genetically studied organisms. Cold Spring Harbor, New York, pp 6259–6261

Menacio-Hautea D, Fatokun CA, Kumar L, Danesh D, Young ND (1993) Comparative genome analysis of mungbean (*Vigna radiata* L. Wilczek) and cowpea (*V. unguiculata* L.Walpers) using RFLP mapping data. Theor Appl Genet 86:797–810

Menéndez CM, Hall AE, Gepts P (1997) A genetic linkage map of cowpea (*Vigna unguiculata*) developed from a cross between two inbred, domesticated lines. Theor Appl Genet 95:1210–1217

Muchero W, Ehlers JD, Close TJ, Roberts PA (2009a) Mapping QTL for drought stress-induced premature senescence and maturity in cowpea [*Vigna unguiculata* (L.) Walp.]. Theor Appl Genet 118:849–863

Muchero W, Diop N-N, Bhatb PR, Fenton RD, Wanamaker S, Pottorff M, Hearne S, Cisse N, Fatokun C, Ehlers JD, Roberts PA, Close TJ (2009b) A consensus genetic map of cowpea [*Vigna unguiculata* (L) Walp.] and synteny based on EST-derived SNPs. Proc Natl Acad Sci U S A 106:18159–18164

Muchero W, Ehlers JD, Roberts PA (2010a) QTL analysis for resistance to foliar damage caused by *Thrips tabaci* and *Frankliniella schultzei* (Thysanoptera: Thripidae) feeding in cowpea [*Vigna unguiculata* (L.) Walp.]. Mol Breed 25:47–56

Muchero W, Ehlers JD, Roberts PA (2010b) Restriction site polymorphism-based candidate gene mapping for seedling drought tolerance in cowpea [*Vigna unguiculata* (L.) Walp.]. Theor Appl Genet 120:509–518

Muchero W, Ehlers JD, Close TJ, Roberts PA (2011) Genic SNP markers and legume synteny reveal candidate genes underlying QTL for *Macrophomina phaseolina* resistance and maturity in cowpea [*Vigna unguiculata* (L) Walp.]. BMC Genomics 12:8

Myers GO, Fatokun CA, Young ND (1996) RFLP mapping of an aphid resistance gene in cowpea (*Vigna unguiculata* L.Walp.). Euphytica 91:181–187

Naito K, Kaga A, Tomooka N, Kawase M (2013) *De novo* assembly of the complete organelle genome sequence of azuki bean (*Vigna angularis*) using next generation sequencers. Breed Sci 63(2):176–182

Ng NQ, Maréchal R (1985) Cowpea taxonomy, origin and germplasm. In: Singh SR, Rachie KO (eds) Cowpea research, production and utilization. Wiley, Chichester, pp 11–21

Ogundiwin EA, Thottappilly G, Aken'Ova ME, Pillay M, Fatokun CA (2005) A genetic linkage map of *Vigna vexillata*. Plant Breed 124:392–398

Omo-Ikerodah EE, Fawole I, Fatokun CA (2008) Genetic mapping of quantitative trait loci (QTLs) with effects on resistance to flower bud thrips (*Megalurothrips sjostedti*) identified in recombinant inbred lines of cowpea (*Vigna unguiculata* (L.) Walp). Afr J Biotechnol 7:263–270

Ouédraogo JT, Maheshwari V, Berner DK, St-Pierre CA, Belzile F, Timko MP (2001) Identification of AFLP markers linked to resistance of cowpea (*Vigna unguiculata* L.) to parasitism by Striga gesnerioides. Theor Appl Genet 102:1029–1036

Ouédraogo JT, Tignegre JB, Timko MP, Belzile FJ (2002a) AFLP markers linked to resistance against *Striga gesnerioides* race 1 in cowpea (*Vigna unguiculata*). Genome 45:787–793

Ouédraogo JT, Gowda BS, Jean M, Close TJ, Ehlers JD, Hall AE, Gillaspie RPA, Ismail AM, Bruening G, Gepts P, Timko MP, Belzile FJ (2002b) An improved genetic linkage maps for cowpea (*Vigna unguiculata* L.) combining AFLP, RFLP, RAPD, biochemical markers and resistance traits. Genome 45:175–188

Padulosi S, Ng NQ (1993) A useful and unexploited herb, *Vigna marina* (Leguminosae-Papilionoideae) and the taxonomic revision of its genetic diversity. Bull Jardin Bot Natl Belg 62:119–126

Pandiyan M, Senthil N, Ramamoorthi N, Muthiah AR, Tomooka N, Vaughan DA, Jayaraj T (2010) Interspecific hybridization of *Vigna radiata* X 13 wild *Vigna* species for developing MYMV donor. Electron J Plant Breed 1(4):600–610

Pandiyan M, Senthil N, Suersh R, Chakravarthy N, Packiaraj D, Jagadeesh S (2012) Interspecific hybridization of *Vigna radiata* × *Vigna trilobata*. Wudpecker J Agric Res 16:233–234

Pottorff M, Wanamaker S, Ma YQ, Ehlers JD, Roberts PA, Close TJ (2012) Genetic and physical mapping of candidate genes for resistance to *Fusarium oxysporum* f.sp. *tracheiphilum* Race 3 in Cowpea [*Vigna unguiculata* (L.) Walp]. PLoS One 7:e41600

Pottorff M, Ehlers JD, Fatokun C, Roberts PA, Close TJ (2013) Leaf morphology in cowpea [*Vigna unguiculata* (L.) Walp]: QTL analysis, physical mapping and identifying a candidate gene using synteny with model legume species. BMC Genomics 13:234

Prathet P, Somta P, Srinives P (2012) Mapping QTL conferring resistance to iron deficiency chlorosis in mungbean [*Vigna radiata* (L.) Wilczek]. Field Crop Res 137:230–236

Rodrigues MA, Santos CAF, Santana JRF (2012) Mapping of AFLP loci linked to tolerance to cowpea golden mosaic virus. Genet Mol Res 11:3789–3797

Rungnoi O, Chanprem S, Toojinda T, Godwin I, Lambrides C, Srinives P (2010) Characterization, inheritance, and molecular study of opaque leaf mutant in mungbean (*Vigna radiata* (L.) Wilczek). J Crop Sci Biotechnol 13:219–226

Sangiri C, Kaga A, Tomooka N, Vaughan DA, Srinives P (2007) Genetic diversity of the mungbean (*Vigna radiata*, Leguminosae) genepool based on microsatellite analysis. Aust J Bot 55:837–847

Seehalak W, Tomooka N, Waranyuwat A, Thipyapong P, Laosuwan P, Kaga A, Vaughan DA (2006) Genetic diversity of the *Vigna* germplasm from Thailand and neighboring regions revealed by AFLP analysis. Genet Resour Crop Evol 53:1043–1059

Sholihin, Hautea MD (2002) Molecular mapping of drought resistance in mungbean (*Vigna radiata*): 2. QTL linked drought resistance. Jurnal Bioteknologi Pertanian 7:55–61

Somta P, Kaga A, Tomooka N, Kashiwaba K, Isemura T, Chaitieng B, Srinives P, Vaughan DA (2006) Development of an interspecific *Vigna* linkage map between *Vigna umbellata* (Thunb.) Ohwi & Ohashi and *V. nakashimae* (Ohwi) Ohwi & Ohashi and its use in analysis of bruchid resistance and comparative genomics. Plant Breed 125:77–84

Somta P, Kaga A, Tomooka N, Isemura T, Vaughan DA, Srinives P (2008a) Mapping of quantitative trait loci for a new source of resistance to bruchids in the wild species *Vigna nepalensis* Tateishi & Maxted (*Vigna* subgenus *Ceratotropis*). Theor Appl Genet 117:621–628

Somta C, Somta P, Tomooka N, Ooi PAC, Vaughan DA, Srinives P (2008b) Characterization of new sources of mungbean (*Vigna radiata* (L.) Wilczek) resistance to bruchids, *Callosobruchus* spp. (Coleoptera: Bruchidae). J Stored Prod Res 44:316–321

Sonnante G, Spinosa A, Marangi A, Pignone D (1997) Isozyme and RAPD analysis of genetic diversity within and between *Vigna luteola* and *V. marina*. Ann Bot 80:741–746

Souframanien J, Gupta SK, Gopalakrishna T (2010) Identification of quantitative trait loci for bruchid (*Callosobruchus maculatus*) resistance in black gram [*Vigna mungo* (L.) Hepper]. Euphytica 176:349–356

Srinives P, Kitsanachandee R, Chalee T, Sommanas W, Chanprame S (2010) Inheritance of resistance to iron deficiency and identification of AFLP markers associated with the resistance in mungbean (*Vigna radiata* (L.) Wilczek). Plant Soil 335:423–437

Sudha M, Anusuya P, Ganesh NM, Karthikeyan A, Nagarajan P, Raveendran M, Senthil N, Pandiyan M, Angappan K, Balasubramanian P (2012) Molecular studies on mungbean (*Vigna radiata* (L.) Wilczek) and ricebean (*Vigna umbellata* (Thunb.)) interspecific hybridization for Mungbean yellow mosaic virus resistance and development of species-specific SCAR marker for ricebean. Phytopathol Plant Prot 46(5):503–517

Sudha M, Karthikeyan A, Anusuya P, Ganesh NM, Pandiyan M, Senthil N, Raveendran M, Nagarajan P, Angappan K (2013) Inheritance of resistance to Mungbean yellow mosaic virus (MYMV) in inter and intra specific crosses of mungbean (*Vigna radiata*). Am J Plant Sci 4:1924–1927

Sunitha S, Shanmugapriya G, Balamani V, Veluthambi K (2013) Mungbean yellow mosaic virus (MYMV) AC4 suppresses post-transcriptional gene silencing and an AC4 hairpin RNA gene reduces MYMV DNA accumulation in transgenic tobacco. Virus Genes 46:496–504

Tangphatsornruang S, Sangsrakru D, Chanprasert J, Uthaipaisanwong P, Yoocha T, Jomchai N, Tragoonrung S (2010) The chloroplast genome sequence of mungbean (*Vigna radiata*) determined by high-throughput pyrosequencing: structural organization and phylogenetic relationships. DNA Res 17:11–22

Tian T, Isemura T, Kaga A, Vaughan DA, Tomooka N (2013) The genetic diversity of the rice bean [*Vigna umbellata* (Thunb.) Ohwi & Ohashi] genepool as assessed by SSR markers. Genome 56: 717–727

Tomooka N (2007) Archaeological remains of azuki bean (*Vigna angularis*). Google map http://maps.google.co.jp/maps/ms?hl=ja&ie=UTF8&msa=0&msid=101458218381458921300.000001124500bf8f41d2b&z=5&om=1

Tomooka N (2009) The origins of rice bean (*Vigna umbellata*) and azuki bean (*V. angularis*). The evolution of two lesser-known Asian beans. In: Akimachi T (ed) An illustrated eco-history of the Mekong River Basin. Lotus Co. Ltd, Bangkok, pp 33–35

Tomooka N, Kashiwaba K, Vaughan DA, Ishimoto M, Egawa Y (2000) The effectiveness of evaluation using a species level core collection: a case study of searching for sources of resistance to bruchid beetles in species of the genus *Vigna* subgenus *Ceratotropis*. Euphytica 115:27–41

Tomooka N, Kaga A, Egawa Y, Vaughan DA, Kashiwaba K, Doi K (2001) Searching for sources of vegetative stage high temperature tolerance in the genus *Vigna* subgenus *Ceratotropis*. Jpn J Trop Agric 45:47–48. (in Japanese with English summary)

Tomooka N, Vaughan DA, Moss H, Maxted N (2002a) The Asian *Vigna*: the genus *Vigna* subgenus *Ceratotropis* genetic resources. Kluwer Academic, Dordrecht, p 270

Tomooka N, Yoon MS, Doi K, Kaga A, Vaughan DA (2002b) AFLP analysis of a *Vigna* subgenus *Ceratotropis* core collection. Genet Resour Crop Evol 49:521–530

Tomooka N, Maxted N, Thavarasook C, Jayasuriya AHM (2002c) Two new species, new species combinations and sectional designations in *Vigna* subgenus *Ceratotropis* (Piper) Verdcourt (Leguminosae, Phaseoleae). Kew Bull 57(3):613–624

Tomooka N, Kaga A, Vaughan DA (2006) The Asian *Vigna* (*Vigna* subgenus *Ceratotropis*) biodiversity and evolution. In: Sharma AK, Sharma A (eds) Plant genome: biodiversity and evolution. Part C phanerogams (angiosperms – dicotyledons), vol 1. Science Publishers, Enfield, pp 87–126

Tomooka N, Kaga A, Isemura T, Naito K, Vaughan DA (2010) Neo-domestication of *Vigna* (Leguminosae) and identification of *Multiple Organ Gigantism* (*mog*) mutant of cultivated black gram (*Vigna mungo*). Gamma Field Symp (NIAS) 49:41–49. (in English with Japanese summary)

Tomooka N, Pandiyan M, Senthil N (2011a) Conservation of leguminous crops and their wild relatives in Tamil Nadu, India. Annual report on exploration and introduction of plant genetic resources. NIAS 27:111–127. http://www.gene.affrc.go.jp/pdf/publications/plant-exp_2010(27)_p111.pdf

Tomooka N, Kaga A, Isemura T, Kuroda Y, Vaughan DA, Srinives P, Somta P, Thadavong S, Bounphanousay C, Kanyavong K, Inthapanya P, Pandiyan M, Sentil N, Ramamoorthi N, Jaiwal PK, Jing T, Umezawa K, Yokoyama T (2011b) *Vigna* genetic resources. Genetic resources and comparative genomics of legumes (*Glycine* and *Vigna*). In: 14th NIAS international workshop on genetic resources, NIAS, Tsukuba, pp 11–21 fullstop. http://www.gene.affrc.go.jp/pdf/misc/international-WS_14_11.pdf

Tomooka N, Fukui K, Chankaew S, Iizumi T, Hirashima S (2013) Collection and conservation of wild leguminous crop relatives on Goto islands, Nagasaki, Japan, 2012. Annual report on exploration and introduction of plant genetic resources, NIAS 29:19–43. http://www.gene.affrc.go.jp/pdf/publications/plant-exp_2012(29)_p19.pdf

Ubi BE, Mignouna H, Thottappilly G (2000) Construction of a genetic linkage map and QTL analysis using a recombinant inbred population derived from an intersubspecific cross of cowpea (*Vigna unguiculata* (L.) Walp.). Breed Sci 50:161–172

Varshney RK, Close TJ, Singh NK, Hoisington DA, Cook DR (2009) Orphan legume crops enter the genomics era! Curr Opin Plant Biol 12:202–210

Vaughan DA, Balázs E, Heslop-Harrison JS (2007) From crop domestication to super-domestication. Ann Bot 100(6):893–901

Xu RQ, Tomooka N, Vaughan DA, Doi K (2000a) The *Vigna angularis* complex: genetic variation and relationships revealed by RAPD analysis, and their implications for *in situ* conservation and domestication. Genet Resour Crop Evol 47:123–134

Xu RQ, Tomooka N, Vaughan DA (2000b) AFLP markers for characterizing the azuki bean (*Vigna angularis*) complex. Crop Sci 40:808–815

Xu HX, Jing T, Tomooka N, Kaga A, Isemura T, Vaughan DA (2008) Genetic diversity of the azuki bean [*Vigna angularis* (Willd.) Ohwi & Ohashi] genepool as assessed by SSR markers. Genome 51:728–738

Xu P, Wu X, Wang B, Liu Y, Ehlers JD, Close TJ, Roberts PA, Diop N-N, Qin D, Hu T, Lu Z, Li G (2011a) A SNP and SSR based genetic map of asparagus bean (*Vigna unguiculata* ssp. *sesquipedialis*) and comparison with the broader species. PLoS One 6(1):e15952

Xu P, Hu T, Yang Y, Wu X, Wang B, Liu Y, Qin D, Ehlers JD, Close TC, Lu Z, Li G (2011b) Mapping genes governing flower and seed coat color in asparagus bean (*Vigna unguiculata*. ssp. *sesquipedalis*) based on SNP and SSR markers. HortScience 46:1102–1104

Xu P, Wu X, Wang B, Hu T, Lu Z, Liu Y, Qin D, Wang S, Li G (2013) QTL mapping and epistatic interaction analysis in asparagus bean for several characterized and novel horticulturally important traits. BMC Genet 14:4

Yamaguchi H (1992) Wild and weed azuki beans in Japan. Econ Bot 46:384–394

Yoon MS, Doi K, Kaga A, Tomooka N, Vaughan DA (2000) Analysis of the genetic diversity in the *Vigna minima* complex and related species in East Asia. J Plant Res 113(4):375–386

Yoon MS, Lee J, Lim CY, Baek HJ (2007) Genetic relationships among cultivated and wild *Vigna angularis* (willd.) Ohwi et Ohashi and relatives from Korea based on AFLP markers. Genet Resour Crop Evol 54:575–883

Young ND, Kumar L, Menancio-Hautea D, Danesh D, Talekar NS, Shanmhgassundarum S, Kim DH (1992) RFLP mapping of a major bruchid resistance gene in mungbean (*Vigna radiata* L., Wilczek). Theor Appl Genet 84:839–844

Young ND, Danesh D, Menancio-Hautea D, Kumar L (1993) Mapping oligogenic resistance to powdery mildew in mungbean with RFLPs. Theor Appl Genet 87:243–249

Zhao D, Cheng X-Z, Wang L-X, Wang S-H, Ma Y-L (2010) Integration of mungbean (*Vigna radiata*) genetic linkage map. Acta Agron Sin 36:932–939 (in Chinese with English abstract)

Horse Gram 10

Nilamani Dikshit, Gopal Katna,
Chandra Sekhar Mohanty, Anath Bandhu Das,
and Natarajan Sivaraj

Abstract

The chapter discusses detailed account on germplasm enhancement-related information with regard to collection, systematic evaluation, and development of high yielding cultivars, which are necessary in consortium concept for its full exploitations. Key traits for broadening the genetic base in horse gram through interspecific hybridization would be disease and pest resistance, improving grain quality and protein percentage, and also reducing antinutritional factors. Utilization of *Macrotyloma axillare, a* wild species, which has many desirable traits like high number of pods per plant, high seed yield, and tolerance to cold and drought, has been highlighted.

N. Dikshit (✉)
National Bureau of Plant Genetic Resources, Regional Station, Akola 444 104, Maharashtra, India
e-mail: dikshitn@gmail.com; dikshitn@yahoo.com

G. Katna
Department of Crop Improvement, CSK Himachal Pradesh Agricultural University, Palampur 176 062, Himachal Pradesh, India
e-mail: gkatna@rediffmail.com

C.S. Mohanty
CSIR – National Botanical Research Institute, Rana Pratap Marg, Lucknow 226001, Uttar Pradesh, India
e-mail: cs.mohanti@nbri.res.in

A.B. Das
Department of Agricultural Biotechnology, Orissa University of Agriculture and Technology, Bhubaneswar 751 003, Odisha, India
e-mail: a_b_das@hotmail.com

N. Sivaraj
National Bureau of Plant Genetic Resources, Regional Station, Hyderabad 500 030, Andhra Pradesh, India
e-mail: sivarajn@gmail.com

10.1 Introduction

Macrotyloma uniflorum (Lam.) Verd. commonly known as “horse gram” is an arid food legume grown in diverse environmental conditions (Duke and Reed 1981). The species is native to Southeast Asia and tropical Africa, but the center of origin of cultivated species is considered to be southern India (Vavilov 1951; Zohary 1970). Horse gram has number of desirable traits like drought tolerance (Reddy et al. 1990), heavy metal stress tolerance (Sudhakar et al. 1992), high protein content, antioxidant activity (Reddy et al. 2005), and antimicrobial activity and various medicinal properties make it a crop of interest and potential food source of the future. It is widely grown as a pulse, fodder crop, green manure and medicinal purposes. Horse gram is cultivated in India, Myanmar, Nepal, Malaysia, Mauritius and Sri Lanka for food purposes,

M. Singh et al. (eds.), *Broadening the Genetic Base of Grain Legumes*,
DOI 10.1007/978-81-322-2023-7_10, © Springer India 2014

whereas in Australia and Africa it is grown for fodder purpose (Asha et al. 2006). Seed color ranges from black to chocolate brown, and protein content in cultivated horse gram is reported to be 16.9–30.4 % (Patel et al. 1995). It has also high lysine content, an essential amino acid, (Gopalan et al. 1989), rich in phosphorus, iron, and vitamins, viz., carotene, thiamine, riboflavin, niacin and vitamin C (Sodani et al. 2004). The identification and development of high-yielding varieties are key for the full exploitation of this dry land crop. The limited use of dry seeds of horse gram is due to its poor cooking quality although with a high protein percentage. As a result, it is consumed as soup in many parts of India (Sudha et al. 1995). The US National Academy of Sciences has identified this legume as a potential food source for the future (USNAS 1978) for its medicinal importance and its drought resistance ability. The crop is highly suitable for rainfed and marginal agriculture, but does not tolerate frost and water logging conditions. Sprouted seeds of horse gram also contain polyphenols that have high antioxidant properties than that of dry seeds (Ramesh et al. 2011).

10.2 Crop Gene Pool, Evolutionary Relationships and Systematics

The Genus *Macrotyloma* (Wight & Arn.) Verdc. belongs to the family *Fabaceae* and consist of about 25 wild species having chromosome numbers $2n=2x=20$ and $2n=2x=22$ (Allen and Allen 1981; Lackey 1981). Linnaeus (1753) included horse gram in the genus *Dolichos*, but Verdcourt (1970) on the basis of style and pollen characteristics distinguish *Macrotyloma* from *Dolichos*. The center of origin of *M. uniflorum* var. *uniflorum* is mentioned in India, though wild members of the species occur both in Africa and India (Vavilov 1951; Zohary 1970; Purseglove 1974; Smartt 1985). Arora and Chandel (1972) considered south-west India as the primary center of origin of *M. uniflorum* var. *uniflorum*, whereas Mehra and Magoon (1974) considered both Africa and India as gene rich centers. All the species of the genus *Macrotyloma* are distributed in African continent and *Macrotyloma uniflorum* is the only cultivated species grown in Indian subcontinent. The species occur exclusively in Africa includes *Macrotyloma africanum* (Wilczek) Verdc., *M. bieense (Torre) Verdc., M. bifloru., M. brevicaule, M. coddi., M. daltonii, M. decipiens, M. densiflorum, M. dewildemanianum (Wilczek) Verdc., M. ellipticum, M. fimbriatum, M. geocarpum, M. hockii, M. kasaiense, M. maranguense, M. oliganthum, M. prostratum, M. repestre, M. schweinfurthii, M. stenophyllum, M. stipulosum,* and *M. tenuiflorum. Macrotyloma axillare* is distributed both in Australia and Africa; *M. cilliatum* occurs in Asia and Africa, whereas *M. uniflorum* (Lam.) Verdc. is distributed in Asia, Africa and Australia. Zeven and de Wet (1982) considered maximum genetic diversity of this species from Old World Tropics mainly, southern India and Himalayas. Based on pod characteristics, four varieties have been recognized within *Macrotyloma uniflorum, i.e., Macrotyloma uniflorum* var. *uniflorum, Macrotyloma uniflorum* var. *stenocarpum* (Brenan) Verdc, *Macrityloma uniflorum* var. *verrucosum* Verdc., and *Macrotyloma uniflorum* var. *benadirianum* (Chiov.) Verdc. Bailey (1900) reported that var. *stenocarpum* and var. *verrucosum* are of African origin.

10.3 Assessment of Gene Flow for Crop Improvement

Gene flow is actually the transfer of genes from one population to another of the same species. Within a population, it can introduce or reintroduce genes to a population and changes in the frequency of alleles and thus increase the genetic variation of that population. Gene flow can occur both from domesticated to wild and in the opposite direction (Ellstrand et al. 1999; Jarvis and Hodgkin 1999). Effects of gene flow on fitness appear to be as complex in domesticated plants and their relatives as they are in wild form (Arnold 1992). It is an ongoing process affecting the genetic diversity of crop populations and may be a source of new genetic combinations in traditional farming systems (Jarvis and Hodgkin 1999; Elias et al. 2001).

10.4 Level of Diversity in Crop Germplasm

The Indian collection of horse gram exhibited tremendous variability for agro-morphological traits such as early plant vigor, plant growth habit and growth pattern, leaf color, leaf surface, stem color, number of primary branches, days to 50 % flowering, flower color, number of pods per plant, pod length, pod shape, pod surface, plant height, number of seeds per pod, days to 80 % maturity, seed yield per plant, test seed weight, and seed color (Patel et al. 1995). Latha (2006) observed variations in 22 accessions for seed related traits. Morpho-agronomical studies were also undertaken by several other workers like Savithriamma et al. (1990), Jayan and Maya (2001), Chahota et al. (2005), Mahajan et al. (2007), Kulkarni and Mogle (2011), and Kulkarni (2010).

10.5 Traits of Importance for Base Broadening

Horse gram has narrow genetic base due to its self-reproduction system. In the past, improvement work in horse gram has mainly been confined to single plant selection and limited hybridization. Past breeding efforts in the development of varieties have utilized only a fraction of the vast available diversity. There is a need to undertake an extensive hybridization program to incorporate useful variability from diverse parents and related species. Broadening genetic base necessitates collection of indigenous and exotic germplasm. The breeding work needs to be organized to make use of the variability present at the varietal, species, and generic levels. Induced mutations can create new genetic variability and can enlarge the genetic base of horse gram. Photo and thermo-insensitive, development of determinate, early maturing (90–110 days) varieties, resistant to yellow mosaic, and suitable for sole cropping under high plant population density for increasing production are important parameters. Interspecific hybridization also needs to be resorted for disease and pest resistance. Efforts should be made for improving the grain quality, protein percentage, and sulfur-containing amino acid and also to reduce anti-nutritional factors.

10.6 Interspecific Hybridization in Crop Species

Interspecific hybridization is a potential way of increasing genetic variation and introgression of resistance genes from wild species (Hawkes 1977). Concerted and systematic efforts are, however, required to harness the potential of this vast variability into the background of cultivated genepool. Though genus *Macrotyloma* consists of more than 25 species, there is no report regarding the evaluation of these wild species for desirable traits. Morris (2008) compared *Macrotyloma uniflorum* with *Macrotyloma axillare* and described a set of descriptors to differentiate these species. Evaluation of few wild species of *Macrotyloma* has been undertaken at the CSK, Himachal Pradesh Agricultural University Palampur India, to initiate systematic hybridization program among cultivated and wild species to transfer desirable traits from *Macrotyloma axillare* and *Macrotyloma* sar-*garhwalensis* into cultivated background (Chahota et al. 2013). *Macrotyloma axillare* has many desirable traits such as high number of pods per plant, high number of seeds per plant and tolerance to cold and drought (Staples 1966, 1982). The cultivated species, *Macrotyloma uniflorum*, is infected by a number of diseases, particularly in high rain fall areas, such as *Anthracnose, Cercospora* leaf spot, *Fusarium* wilt, rust, *Pellicularia* root rot and *Ascochyta* blight. Though the *Macrotyloma axillare* is reported to be resistant to many diseases, hybridization between *M. uniflorum* and *M. axillare* resulted in juvenile flowering in the first year of F_1 plant, hence prolonging the breeding process. *Macrotyloma axillare* and *M. africanum*, the two other species of this genus, have also shown potential as forage plants.

10.7 Molecular Markers, Genome Mapping, and Genomics as an Adjunct to Breeding

Horse gram (*Macrotyloma uniflorum*) can tolerate severe adverse environmental conditions such as drought (Reddy et al. 1990), salinity (Sudhakar et al. 1997; Reddy et al. 1998), and heavy metal contamination (Sudhakar et al. 1992; Reddy et al. 2005). Horse gram is thus an ideal candidate plant for mining genes for abiotic stress tolerance. The estimated genome-size of horse gram is 400 Mb. However, genomic-level question could be addressed through developing informative transcriptome (Mardis 2008; Ellegren 2008). When answering fundamental questions like mechanisms involved or related to a particular trait like drought tolerance, it becomes imperative to draw a conclusion based on a comparative study (Ashraf 2010). Comparison of plant genotypes differing in their sensitivities towards drought is an inevitable approach to discover natural drought tolerance mechanisms (Inan et al. 2004). Further, identification of stress-induced genes in this crop would provide a logical approach for improving drought tolerance in other related crop plants (Reddy et al. 2008). But, till date there is no much data related to global genomic, transcriptome or protein biology studies on horse gram relating to various attributes especially to drought tolerance. However, drought tolerance is an attribute maintained in plants by cross-talk between multiple and cascading metabolic pathways. Without a sequenced genome available for horse gram, it is difficult to comprehend such complex networks and intercalated genes associated with drought tolerance. Therefore, de novo transcriptome discovery and associated analysis were done for this highly drought tolerant yet underexploited legume to decipher its genetic makeup by Bhardwaj et al. (2013). As a first step towards characterization of genes that contribute to combating abiotic stresses, construction and analysis of subtracted cDNA library were carried out on Illumina GAIIx sequencing platform. On analysis, a total of 1,050 ESTs were isolated and sequenced. Out of the discovered ESTs, 959 high quality ESTs were clustered for further analysis. It revealed that, out of the detected ESTs, 531 sequences were unique and 30 % of these sequences had no homology to known proteins in the database. Further, to validate the identified differentially expressed genes, expression profiles of selected clones were analyzed using reverse-northern, northern blot analysis and showed that these clones were differentially expressed under various abiotic stress conditions (Reddy et al. 2008).

MicroRNAs (miRNAs) are a class of naturally occurring and small noncoding RNA molecules of about 21–25 nucleotides in length. The main function of these molecules is to downregulate gene expression in different manners like translational repression, mRNA cleavage, and epigenetic modification. To predict new miRNAs in plants, different computational approaches have been developed. The study carried out by Bhardwaj et al. (2010), an EST-based approach has been used to identify novel miRNAs in horse gram. Identification of miRNAs was initiated by mining the EST database available at NCBI. These ESTs were subjected to CAP3 assembly to remove the redundancy. This resulted in an output of 72 contigs and 606 singletons as nonredundant datasets. The miRNAs were then predicted by using miRNA-finder. A total of eight potential miRNAs were predicted and named as hor-miR1 to hor-miR8, unique and novel to the plant species. These miRNAs were inputted to miRU2 program to predict their targets. The target mRNAs for these miRNAs mainly belong to zinc finger, chromosome condensation, protein kinase, abscisic acid-responsive, calcineurin-like phosphoesterase, disease resistance and transcriptional factor family proteins. These targets appeared to be involved in plant growth and development and environmental stress responses. In some legumes, seed-coat color is quite heterogeneous within a sample and such seeds often comprise a collection of germplasm in a genebank. Correlation of different seed-coat colors with seed quality parameters usually provides a nondestructive tool (especially useful in a genebank) for sorting good quality seeds. Relationship between seed-coat color and quality parameters in horse gram (*Macrotyloma uniflorum*) was studied in seeds with different colors like pale brown, medium brown, and blackish brown. The colored seeds of 20 accessions of horse gram were carried out by Singh et al. (2009). Seed quality was assessed in terms of germination, seedling vigor, and seed storability. Random amplified polymorphic DNA (RAPD) markers were used to assess the genetic basis of heterogeneity in seed coat color. The light-colored fractions consistently showed highest germination percentage and seedling vigor followed by medium- and dark-colored fractions in all the accessions. Percentage seed moisture content and electrical conductivity were observed to be highest in the dark-colored seeds. Light-colored seeds showed better storability after 3 years of ambient storage, whereas the medium- and dark-colored seeds were poor in germination and showed significant decline. Out of the 18 random primers screened on three different colored seed subsamples (cream, red and black) from a single accession, five showed polymorphic pattern on template DNA. The results showed that color of seeds could be used as a visual indicator of seed quality and storability in horse gram.

The genetic diversity of 49 genotypes of *M. uniflorum* and one each of related wild species *M. axillare* and *M. sar-garhwalensis* was assessed using 25 random (RAPD and ISSR) and 76 newly designed sequence-based (EST-SSR and ILP) markers by Sharma et al. (2013). A considerable high polymorphism was detected in few genotypes which were proved to be promising lines for further genetic analyses. However, the study conducted by Varma et al. (2013) provided the estimates of genetic variability present in *M. uniflorum* germplasm countrywide, which could be harnessed for various breeding and improvement programs in this crop. Information on root nodule endosymbionts of this legume in India is limited. The study carried out on 69 isolates of naturally occurring root nodules of horse gram collected from two agro-eco-climatic regions of South India was analyzed by generation rate, acid/alkali reaction on YMA medium, restriction fragment length polymorphism analysis of 16S-23S rDNA intergenic spacer region (IGS), and sequence analyses of IGS and housekeeping genes glnII and recA. Based on the rDNA IGS RFLP with three restriction enzymes, the three studied rhizobia were grouped in five clusters (I–V). By sequence analysis of 16S–23S rDNA IGS, identified genotypes of horse gram rhizobia were distributed into five divergent lineages of *Bradyrhizobium* genus. All the isolated rhizobia nodulated *Macrotyloma* sp. and *Vigna* spp., and only some of them formed nodules on *Arachis hypogaea*. The isolates within each IGS type varied in their ability to fix nitrogen. Selection for high symbiotic effective strains could reward horse gram production in poor soils of South India where this legume is largely cultivated (Appunu et al. 2011).

10.8 Conclusions

Horse gram is a minor legume and said to be the poor man's crop. It provides high quality inexpensive protein, eaten both as boiled and fried. Nutritionally horse gram contains about 23 % protein and rich in lysine content when compared to pigeon pea and chickpea. It is also used as an ayurvedic medicine for curing piles, renal stones etc. besides its number of medicinal and therapeutic applications. It is indigenous to the Indian subcontinent and archaeological investigations have revealed the use of horse gram as food especially in India as origin around 2000 BC (Mehra and Magoon 1974). It is ubiquitous among Neolithic human dwellings found in South India (Bolivin et al. 2007). According to Mehra (2000) southern Indian plains and Eastern Ghats seems to be the limits for its domestication. Horse gram cultivation diffused into western and northern parts of Indian subcontinent during Neolithic period through the counter-migration of human beings (Fuller et al. 2004; Mehra 2000; Mittre 1961). Several wild species of *Macrotyloma* are traceable in African woods and savannas, but none seems to have contributed to its gene pool to the evolution of domestication of South Indian horse gram (Fuller and Harvey 2006). Horse gram is not a well-established pulse crop of Indian subcontinent and the germplasm information with regard to traits are scanty; therefore, collection and systematic evaluation work of germplasm is necessary in consortium concept. An urgent need of genetic database (phenotypic, biochemical, and DNA level diversity) of this crop is very essential. Development of DNA markers through DNA fingerprinting in a systematic manner can lead to establish a genetic map of potential germplasm for future breeding and crop improvement program. Of course genetic drifts could be immediately measured and phylogenetic analysis can be done applying Genomic in situ hybridization (GISH) or fluorescent in situ hybridization (FISH) program besides the chromosome number and meiotic behavior study of the germplasm.

References

Allen ON, Allen EK (1981) The leguminosae. A source book of characteristics, uses and nodulation. The University of Wisconsin Press, Madison

Appunu C, Ganesan G, Kalita M, Kaushik R, Saranya B, Prabavathy VR, Sudha N (2011) Phylogenetic diversity of rhizobia associated with horsegram (*Macrotyloma uniflorum* (Lam.) Verdc.) grown in South India based on gln II, rec A and 16S-23S intergenic sequence analysis. Curr Microbiol 62:1230–1238

Arnold ML (1992) Natural hybridization as an evolutionary process. Annu Rev Ecol Syst 23:237–261

Arora RK, Chandel KPS (1972) Botanical source areas of wild herbage legumes in India. Trop Grassl 6:213–221

Asha KI, Latha M, Abraham Z, Jayan PK, Nair MC, Mishra SK (2006) Genetic resources. In: Kumar D (ed) Horsegram in India. Scientific Publisher, Jodhpur, pp 11–28

Ashraf M (2010) Inducing drought tolerance in plants: recent advances. Biotechnol Adv 28:169–183

Bailey FM (1900) The Queensland flora. Part II. Government Printer, Brisbane

Bhardwaj J, Mohammad H, Yadav SK (2010) Computational identification of micro RNAs and their targets from the expressed sequence tags of horsegram (*Macrotyloma uniflorum* (Lam.) Verdc.). J Struct Funct Genomics 11:233–240

Bhardwaj J, Chauhan R, Swarnkar MK, Chahota RK, Singh AK, Shankar R, Yadav SK (2013) Comprehensive transcriptomic study on horse gram (*Macrotyloma uniflorum*): *De novo* assembly, functional characterization and comparative analysis in relation to drought stress. BMC Genomics 14:647

Bolivin N, Fuller D, Korisettar R, Petragalia M (2007) First farmers in South India: the role of internal processes and external influences in the emergence and transformation of South India's earliest settled societies. In: Tiwari RK (ed) First farmers in global perspective. Uttar Pradesh State Archaeology Department, Lucknow, pp 1–26

Chahota RK, Sharma TR, Dhiman KC, Kishore N (2005) Characterization and evaluation of horsegram (*Macrotyoma uniflorum* Roxb.) germplasm from Himachal Pradesh. Indian J Plant Genet Resour 18:221–223

Chahota RK, Sharma SK, Sharma TR, Kumar N, Kapoor C (2013) Induction and characterization of agronomically useful mutants in horse gram (*Macrotyoma uniflorum*). Indian J Agric Sci 83(10):1105–1109

Duke JA, Reed CF (1981) *Macrotyloma uniflorum* (Lam.) Verdc. In: Duke JA (ed) Handbook of legumes of world economic importance. Plenum Press, New York

Elias M, Penet L, Vindry P et al (2001) Unmanaged sexual reproduction and the dynamics of genetic diversity of a vegetatively propagated crop plant, Cassava (manihot esculenta Crantz), in a traditional farming system. Mol Ecol 10:1895–1907

Ellegren H (2008) Sequencing goes 454 and takes large-scale genomics into the wild. Mol Ecol 17:1629–1631

Ellstrand NC, Prentice HC, Hancock JF (1999) Gene flow and introgression from domesticated plants into their wild relatives. Annu Rev Ecol Syst 30:539–563

Fuller DO, Harvey EL (2006) The archaeobotany of Indian pulses: identification, processing and evidence for cultivation. Environ Archaeol 11:218–246

Fuller DO, Korisettar R, Venkatasubbaiah PC, Jones MK (2004) Early plant domestication in southern India: some preliminary archaeobotanical results. Veg Hist Archaeobot 13:115–129

Gopalan C, Ramashastri BV, Balasubramanyan SC (1989) Nutritive value of Indian foods. National Institute of Nutrition, ICMR, Hyderabad

Hawkes JG (1977) Importance of wild germplasm in plant breeding. Euphytica 26:615–621

Inan G, Zhang Q, Li P, Wang Z, Cao Z, Zhang H, Zhang C et al (2004) Salt cress- a halophyte and cryophyte *Arabidopsis* relative model system and its applicability to molecular genetic analyses of growth and development of extremophiles. Plant Physiol 135:1718–1737

Jarvis DI, Hodgkin T (1999) Wild relatives and crop cultivars: detecting natural introgression and farmer selection of new genetic combinations in agro-ecosystems. Mol Ecol 8:5159–5173

Jayan PK, Maya CN (2001) Studies on germplasm collections of horsegram. Indian J Plant Genet Resour 14:443–447

Kulkarni GB (2010) Evaluation of genetic diversity of horse gram (*Macrotyloma uniflorum*) germplasm through phenotypic trait analysis. Green Farm 1(6):563–565

Kulkarni GB, Mogle UP (2011) Characterization and evaluation of horse gram *Macrotyloma* (Lam) Verdcourt genotypes. Sci Res Rep 1(3):122–125

Lackey JA (1981) *Phaseolus*. In: Polhill RM, Raven PH (eds) Advances in legume systematic part I. Royal Botanic Garden, Kew, pp 301–327

Latha M (2006) Germplasm characterization and evaluation of horticultural crops. Annual report of NBPGR, New Delhi

Linnaeus C (1753) Species plantarum. Published by Laurentius Salvius, Stockholm, p Xi, 1200+XXXi

Mahajan V, Shukla SK, Gupta NS, Majumdera D, Tiwari V, Piiasad SVS (2007) Identifying phenotypically stable genotypes and developing strategy far yield improvement in horsegram (*Macrotyloma uniflorum*) for mid-altitudes of Northwestern Himalayas. Indian J Agric Sci 3:19–22

Mardis ER (2008) The impact of next-generation sequencing technology on genetics. Trends Genet 24:133–141

Mehra KL (2000) History of crop cultivation in prehistoric India. In: Choudhary SL, Nene YL (eds) Ancient and medieval history of India; agriculture and relevance to sustainable agriculture in 21st century. Rajasthan College of Agriculture, Udaipur, pp 11–16

Mehra KL, Magoon ML (1974) Gene centres of tropical and sub-tropical pastures legumes and their significance in plant introduction. In: Proceedings of the XII international grassland congress, Moscow, pp 251–256

Mittre V (1961) Plant economy in ancient Navdatoli-Maheshwar in technical report on archaeological remains. Department of Archaeology and Ancient Indian History, Deccan College, Pune, pp 13–52

Morris JB (2008) *Macrotyloma axillare* and *M uniflorum:* descriptor analysis, anthocyanin indexes, and potential uses. Genet Resour Crop Evol 55:5–8

National Academy of Sciences (1978) Mothbean in tropical legumes: resources for the future. National Academy of Sciences, Washington, DC

Patel DP, Dabas BS, Sapra RS, Mandal S (1995) Evaluation of horsegram (*Macrotyloma uniflorum*) (Lam.) germplasm. National Bureau of Plant Genetic Resources, Publication, New Delhi

Purseglove JW (1974) *Dolichos uniflorus*. Tropical crops: Dicotyledons. Longman, London/New York, pp 263–264

Ramesh CK, Rehman A, Prabhakar BT, Vijay Avin BR, Aditya SJ (2011) Antioxidant potential in sprouts vs. seeds of *Vigna radiata* and *Macrotyloma uniflorum*. J Appl Pharm Sci 1:99–103

Reddy PS, Sudhakar C, Veeranjaneyulu K (1990) Water stressed induced changes in enzymes of nitrogen metabolism in horsegram, (*Macrotyloma uniflorum* (Lam.)) seedlings. Indian J Exp Biol 28:273–276

Reddy PS, Ramanjulu S, Sudhakar C, Veeranjaneyulu K (1998) Differential sensitivity of stomatal and non-stomatal components of NaCl o r Na_2SO_4 salinity in horsegram (*Macrotyloma uniflorum* (Lam.)). Photosynthetica 35:99–105

Reddy AM, Kumar SG, Kumari GJ, Thimmanaik S, Sudhakar C (2005) Lead induced changes in antioxidant metabolism of horsegram (*Macrotyloma uniflorum* (Lam.) Verdc.) and bengalgram (*Cicer arietinum* L.). Chemosphere 60:97–104

Reddy PC, Sairanganayakulu G, Thippeswamy M, Reddy PS, Reddy MK, Sudhakar C (2008) Identification of stress-induced genes from the drought tolerant semi-arid legume crop horsegram (*Macrotyloma uniflorum* (Lam.) Verdc.) through analysis of subtracted expressed sequence tags. Plant Sci 175:372–384

Savithriamma DL, Shambulingappa KG, Rao MG (1990) Evaluation of horsegram germplasm. Curr Res 19(9):150–153

Sharma V, Ghani M, Verma B, Chahota RK, Sharma TR (2013) Assessment of genetic diversity in *Macrotyloma uniflorum* – a potential pulse crop of future. In: Plant and animal genome XXI, held at San Diego, California, 12–16 Jan 2013, p 0772. http://www.intlpag.com

Singh N, Devi C, Kak A, Singh G, Kumari A, Mahajan RK (2009) Influence of seed coat colour associated heterogeneity on quality and storability in horse gram (*Macrotyloma uniflorum*). Seed Sci Technol 37(1):232–240

Smartt J (1985) Evolution of grain legumes. Old and new world pulses of lesser economic importance. Exp Agric 21(2):1–18

Sodani SN, Paliwal RV, Jain LK (2004) Phenotypic stability for seed yield in rainfed horse gram (*Macrotyloma uniflorum* (Lam.) Verdc.). In: Proceedings of the national symposium on arid legumes for sustainable agriculture and trade, Central Arid Zone Research Institute, Jodhpur, Rajasthan, India, 5–7 Nov

Staples IB (1966) A promising new legume. Qld Agric J 92:388–392

Staples IB (1982) Intra and interspecific crosses in *Macrotyloma* axillare. In: Proceedings of the second Australian Agronomy Conference, Wagga Wagga, p 340

Sudha N, Mushtari Begum J, Shambulingappa KG, Babu CK (1995) Nutrients and some anti-nutrients in horsegram (*Macrotyloma uniflorum* (Lam.) Verdc.). Food Nutr Bull 16:81–83

Sudhakar C, Bai LS, Veeranjaneyulu K (1992) Lead tolerance of certain legume crops grown on lead ore tailings. Agric Ecosyst Environ 41:253–261

Sudhakar C, Ramanjulu S, Reddy PS, Veeranjaneyulu K (1997) Response of some Calvin cycle enzymes subjected to salinity shock *in vitro*. Indian J Exp Biol 35:665–667

Varma VS, Durga KK, Ankaiah R (2013) Assessment of genetic diversity in horsegram (*Dolichos uniflorus*). EbSCO online library

Vavilov NI (1951) Phytogeographic basis of plant breeding. Chron Bot 13:13–54

Verdcourt B (1970) Studies in the *Leguminosae-papilionoideae* for the 'Flora of Tropical East Africa'. Kew Bull 24(3):379–447

Zeven AC, de Wet JMJ (1982) Dictionary of cultivated plants and their regions of diversity. Centre for Agricultural Publishing and Documentation, Wageningen

Zohary D (1970) Centers of diversity and centers of origin. In: Frankel OH, Bennett F (eds) Genetic resources of plants. Blackwell Scientific, Oxford, pp 33–42